International Symposium Advances in Dark Matter and Particle Physics 2016

EPJ Web of Conferences Volume 142 (2017)

Messina, Italy
24 - 27 October 2016

Editors:

M. Battaglieri
F. Curciarellop
G. Mandaglio
M. Raggi
P. Valente

ISBN: 978-1-5108-3941-0

Printed from e-media with permission by:

Curran Associates, Inc.
57 Morehouse Lane
Red Hook, NY 12571

Some format issues inherent in the e-media version may also appear in this print version.

Printed by Curran Associates, Inc. (2017)

For additional information, please contact EDP Sciences – Web of Conferences at the address below.

EDP Sciences – Web of Conferences
17, Avenue du Hoggar
Parc d'Activité de Courtabœuf
BP 112
F-91944 Les Ulis Cedex A
France

Phone: +33 (0) 1 69 18 75 75
Fax: +33 (0) 1 69 28 84 91

contact-edps@webofconferences.org

Additional copies of this publication are available from:

Curran Associates, Inc.
57 Morehouse Lane
Red Hook, NY 12571 USA
Phone: 845-758-0400
Fax: 845-758-2633
Email: curran@proceedings.com
Web: www.proceedings.com

TABLE OF CONTENTS

INTERNATIONAL SYMPOSIUM

Advances in Dark Matter and Particle Physics

Messina (Italy) , October 24-27, 2016

Scientific Editors:

Marco Battaglieri, Francesca Curciarello, Giuseppe Mandaglio, Mauro Raggi, Paolo Valente

Symposium presentation:

The Symposium will focus on the latest results of running experiments and new ideas in dark matter and dark force searches at accelerators. This new and rapidly growing field is based on the idea of the existence of light dark matter with particle-like nature but with a very weak coupling to the Standard-Model world. Light dark matter may be produced in the interaction of moderate energy (~GeV) but high intensity (~10^{20} EOT/year) electron and proton beam with ordinary matter and detected via its decay to e+e- pairs (visible decay) or via the effect of the interaction with a detection volume (invisible decay). In a consistent way, a new boson (the A' or dark photon) is expected to mediate the interaction between the light dark matter and the SM particles. An A', with mass in the MeV-GeV range, is predicted in various extensions of the Standard Model of elementary particles by adding a new gauge symmetry interacting with SM particles, the A' provides a "portal" between the ordinary matter and the secluded sector. Over the past few years, an intense activity has produced a significant number of experimental programs focused on dark photon search with different techniques. The A' search is being pursued in the major high-energy and nuclear physics facilities world-wide, with a notable concentration in USA and Europe.

The latest results in dark photon searches both in the invisible and visible will be reviewed in the broader context of the current searches for dark forces and light dark matter using beam dump experiments, meson decays, and other techniques. Special emphasis will be placed to the complementarity of the different approaches and the perspectives of joint efforts between major players on the two sides of the Atlantic Ocean.

The Symposium will also review and discuss, from experimental and theoretical point of view, other topics relevant in astrophysics, high precision test of Standard Model, meson resonances formation in hadronic and non-hadronic channels, hadron spectroscopy, leptons and electromagnetic radiation.

Organized by:

University of Messina

Ministero degli Affari Esteri e della Cooperazione Internazionale

Fondazione Bonino-Pulejo (Messina)

Symposium Organization and Committees

Honorary Chairman : Giorgio Giardina (Messina)

Chairman: Giuseppe Mandaglio (Messina - Italy)

Chairman: Simon Eidelman (Novosibirsk – Russia)

Chairman: Marco Battaglieri (Genova – Italy)

Chairman: Paolo Valente (Roma – Italy)

Scientific Secretary: Antonio Anastasi (Messina)

International Advisory Committee

· J. Alexander (Cornell - USA)

· M. Battaglieri (INFN–Genova, Italy)

· H. Czyz (Katowice University, Poland)

· F. Curciarello (INFN- Frascati, Italy)

· A. D'Angelo (University of Rome "Tor Vergata", Italy)

· A. Denig (University of Mainz- Germany)

· R. De Vita (INFN–Genova, Italy)

· A. Di Domenico (University of Rome "La Sapienza", Italy)

· V. Druzhinin (BINP and University of Novosibirsk – Russia)

· S. Eidelman (BINP and University of Novosibirsk – Russia)

· G. Giardina (University of Messina, Italy)

· A. Kupsc (Uppsala University, Sweden)

· I. Logashenko (BINP and University of Novosibirsk – Russia)

· B. Lucini (Swansea University, UK)

· M. Mandaglio (University of Messina, Italy)

· D. Moricciani (INFN Sezione Roma 2, Italy)

· P. Pedroni (INFN – Pavia, Italy)

· M. Perelstein (Cornell – USA)

· S. Paul (Munich, Germany)

· M. Raggi (INFN – Frascati, Italy)

· F. Riggi (University of Catania, Italy)

· A. Tricomi (University of Catania, Italy)

· P. Valente (INFN – Sezione Roma 1, Italy)

· G. Venanzoni (INFN – Frascati, Italy)

· B. Wojtsekhowski (JLab, USA)

Local Organizing Committee: A. Anastasi, M. Battaglieri, F. Curciarello, G. Giardina, G. Mandaglio, M. Raggi, P. Valente

Topics:

· Dark matter and dark forces

· Hadron Spectroscopy

· R measurements

· Searches of new particles and interaction forces

· Precision Tests of Standard Model

Website: http://newcleo.unime.it/Events/ADMPP2016

edp sciences

EPJ Web of Conferences

Statement of Peer review

In submitting conference proceedings to EPJ Web of Conferences, I certify to the Publisher that I adhere to the **Policy on Publishing Integrity** of the journal in order to safeguard good scientific practice in publishing.

1. All articles have been subjected to peer review administered by the proceedings editors.
2. Reviews have been conducted by expert referees, who have been requested to provide unbiased and constructive comments aimed, whenever possible, at improving the work.
3. Proceedings editors have taken all reasonable steps to ensure the quality of the materials they publish and their decision to accept or reject a paper for publication has been based only on the merits of the work and the relevance to the journal.

Title, date and place of the conference

International Symposium
Advances in Dark Matter and Particle Physics, October 24-27 2016,
Messina (Italy)

Proceedings editor(s):

M. Battaglieri, F. Curciarello, G. Mandaglio, M. Raggi, P. Valente

Date and editor's signature

30/03/2017

EPJ Web of Conferences **142**, 00001 (2017) DOI: 10.1051/epjconf/201714200001
Advances in Dark Matter and Particle Physics

PREFACE

The International Symposium "Advances in Dark Matter and Particle Physics" was held from October 24th to 27th 2016 at the University of Messina in Sicily (Italy). The Symposium took place in the suggestive library room of the "Accademia Peoloritana dei Pericolanti" where sixty researchers, specialist in particle physics and accelerator machines, coming from various Universities and Laboratories of the world, have presented and discussed their projects sharing the own experience in order to overcome common technical problems. The latest results of running experiments and new ideas in dark matter and dark force searches at accelerators, precision measurements in particle physics testing the Standard Model, the most advanced simulation tools, event generator codes, and theoretical predictions, have been presented by 39 talks followed by intense discussions and round tables.

Particularly, the latest limits on a visibly and invisibly-decaying dark photon have been shown in the broader context of dark forces and light dark matter searches performed by using complementary approaches as beam dump experiments (APEX, BDX and HPS at JLAB, LDMX at SLAC, MMAPS at Cornell, PADME at LNF), radiative meson decays and the study of ISR processes at colliders (BABAR at SLAC, KLOE2 at LNF). Moreover, new promising studies such as the ^{8}Be anomaly and other search proposals able to probe the hypothesis of the anti-gravitation of antimatter, have been reported. The opportunities offered by different facilities providing different intense beams have been also presented. The synergy with the most important simulation tools (Geant4) and event generators (Phokhara and Babayaga) have been presented and discussed in a dedicated session. In the symposium, it has been reviewed and discussed, from both experimental and theoretical point of view, other important topics relevant in astrophysics, high precision tests of Standard Model (anomalous magnetic moment of muon and hadronic contribution measurements), meson resonances formation in hadronic and non-hadronic channels, hadron spectroscopy, leptons and electromagnetic radiation. In particular, the latest results of the A2, BaBar, BESII, CMD-3, SND, KLOE2 Collaborations have been presented and discussed.

The Symposium also hosted a satellite meeting of participants to dark photon initiatives in Italy (PADME at LNF) and USA (MMAPS at Cornell and JLAB), together with colleagues from other European labs (VEPP-3), in order to establish a stronger coordination and collaboration among the different experimental groups on both sides of the Atlantic. The PADME and MMAPS initiative, in particular, started the process of joining forces on the realization of the invisible dark photon experiments with the missing mass technique. The Symposium and the satellite Italy-USA meeting has been sponsored by the Italian Ministry for Foreign Affairs and International Cooperation, CUP I86D16000060005.

M. Battaglieri, F. Curciarello, G. Mandaglio, M. Raggi, P. Valente
Editors of the International Symposium ADMPP2016

EPJ Web of Conferences **142**, 01001 (2017)   DOI: 10.1051/epjconf/201714201001
Advances in Dark Matter and Particle Physics

MMAPS: Missing-Mass A-Prime Search

Jim Alexander

[1] *Cornell University*

Abstract. An experiment is proposed to search for dark photons via electron-positron annihilation into photon plus dark photon. The dark photon would be reconstructed in a missing-mass technique, thus ensuring complete model independent. The Cornell University synchrotron can provide a positron beam suitable to probe the interesting mass range for dark photons. A detector based on CsI calorimetry is proposed, and the potential experimental reach in coupling constant and dark photon mass is shown.

1 Introduction and Motivation

Although dark matter is generally thought to be noninteracting, a feeble analog of the standard model electromagnetic interaction could be active in the dark sector, and could be produced in e^+e^- annihilations. Many experiments are currently in progress or under construction to search for the "dark photon". The experiment discussed here, based on the 5.3 GeV positron beam at Cornell University's Wilson Laboratory (soon to be upgraded to 6.0 GeV), would extend the current limits in dark photon parameter space by almost two orders of magnitude. The experiment employs a missing-mass strategy [2] to search for $e^+e^- \to \gamma A'$, reconstructing the dark photon mass with a missing mass technique that depends only on the e^+, e^-, and γ kinematics. No assumptions about the A' decay mode need be made. A detector has been proposed, constructed primarily from components recovered from the retired CLEO and Babar experiments; similar detectors proposed at LNF (PADME) [3] and the Budker Institute (VEPP3) [4] could also serve. The initially proposed experiment will be referred to here as the "Missing-Mass A-Prime Search", or MMAPS.

2 Experimental Strategy

In the MMAPS experiment proposed here, we search for the process $e^+e^- \to \gamma A'$ in events produced by a positron beam incident on a fixed target. The ordinary photon in the final state is observed and its four-momentum measured; with a sufficiently large dataset the unseen A' will appear as a bump in the missing-mass spectrum, $M^2_{\rm miss} \equiv (p_{e^+} + p_{e^-} - p_\gamma)^2$.

The figure of merit for a bump-hunting experiment is the sensitivity $S/\sqrt{B}$ at any bump mass $M_{\rm miss}$. For luminosity $\mathcal{L}$, runtime T, and mass resolution σ_M, the figure of merit for an apparatus of acceptance $\mathcal{A}$ can be defined as F.O.M. $\equiv \sqrt{\mathcal{L}T\mathcal{A}}/\sigma_{M\rm rec}$. The experimental goals are therefore, in broad-brush terms, to obtain the highest possible beam intensity, highest practical acceptance, and the lowest possible mass resolution.

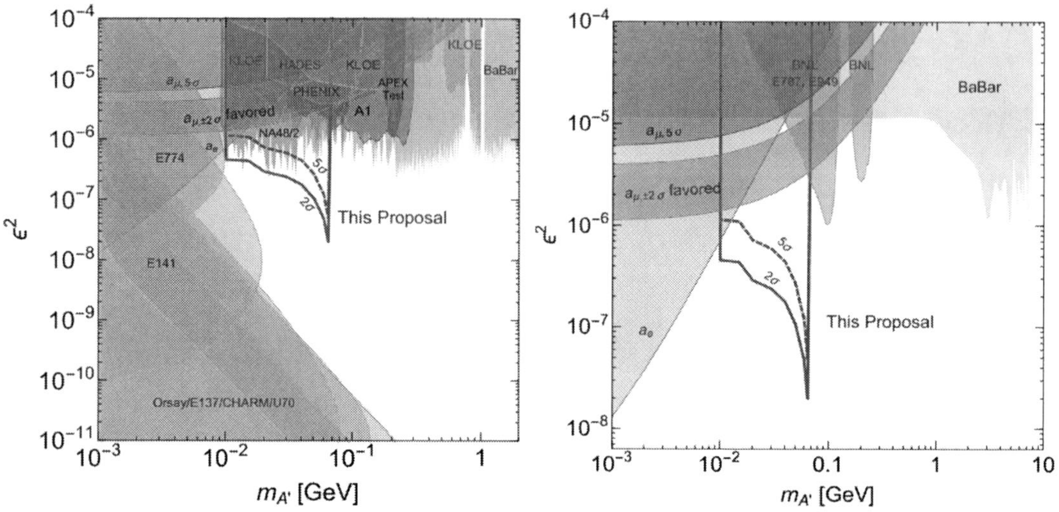

Figure 1. Current experimental bounds[5] on the dark photon mass $M_{A'}$ and the square of the coupling constant to electrons and positrons, ε^2. Left panel: experiments sensitive to the *visible* $A' \to e^+e^-$ final state. Right panel: experiments sensitive to *invisible* A' decays. Estimated 95% c.l. exclusion reach of the experiment proposed here is shown by the teal curve labelled "This Proposal".

MMAPS can be realized at Cornell's Wilson Lab where the existing synchrotron provides a positron source with properties well-matched to the goals of this experiment. A comparable positron beam is not available elsewhere in the world. The positrons will be extracted slowly from the synchrotron over an energy range around 6 GeV, and directed to a fixed target; the main detector element, a calorimeter of CsI crystals, is positioned 10 m further downstream. Figures 2 and 5 illustrate the planned location and configuration of the detector. A small number of magnets need to be designed and fabricated to extract the positron beam from the synchrotron, and some vacuum work is needed when we reposition the existing synchrotron magnets to accomodate new elements.

Most of the detector components will be adapted from existing equipment inherited from previous experiments. The CsI crystals are taken from the now-retired CLEO experiment, and many of the accelerator components already exist and require only reconfiguration or other limited modifications. For improved performance, the photodiodes used to readout the CLEO calorimeter will be replaced by phototubes.

The synchrotron today fills the CESR storage ring with e^+ and e^- for x-ray production, and the experiment proposed here will operate entirely in the shadow of the x-ray program, providing positrons for dark photons only in the 3-minute intervals between positron top-ups carried out for the x-ray program. A modest duty factor of 75% is expected to cover the edge-effects associated with switching between two kinds of synchrotron operation.

The anticipated performance of MMAPS is based on GEANT4 simulation [6]. The sensitivity is shown in Fig. 1, which indicates (in brown) the region of the parameter space of coupling constant versus mass that this experiment can exclude at 2σ or 5σ with a run of 10^7 seconds, together with the existing set of exclusion limits from other experiments.

3 Beam Extraction

To provide the positron beam needed for the MMAPS project we use the Cornell synchrotron [7]. For MMAPS we will extract positrons from the synchrotron and deliver them to a fixed target; the acceleration and extraction take place in the shadow of xray operations, ensuring efficient use of the machine without interference in xray production.

Positrons are created and brought to 150 MeV in a short linac prior to injection into the synchrotron, where they are accelerated to 6 GeV. The average positron current in the synchrotron is $I^{+}_{avg} \sim 7.7$ nA. The synchrotron acceleration cycle repeats 60 times per second; the high energy beam is normally extracted at the peak of the ramp. The orbital period is 2.52 μs, with 180 bunches at 14 ns intervals. Maximum positron current is realized in 15 bunches spaced 168 ns apart. A bunch completes ~3000 turns from injection to extraction at the peak of cycle.

For routine injection into the CESR storage ring, the entire beam in the synchrotron is extracted on a single turn by a fast kicker. For MMAPS operation a slow spill is prefered to minimize pileup; we plan to use a resonant extraction scheme, whereby the synchrotron beam is slowly extracted over about 1000 turns (~ 2.5ms). To accomplish this, quadrupoles shift the betatron tune of the synchrotron beam onto a third-integer ($Q = 10\frac{2}{3}$), and a localized sextupole field drives a nonlinear resonance that shrinks the stable region of phase space in a controlled manner. The amplitude of the particles outside of the shrinking stable region increases over the course of several turns, until the horizontal displacement of particles outside the stable region is sufficient to clear a thin septum. The displacement is increased by the transverse kick from the first thin septum.Two stronger septa then steer the particles into the extraction beam line. A resonantly extracted electron beam was used in fixed target experiments at Cornell in the period 1968-1979 [8], and is also in routine use today at the 3.5 GeV ELSA ring in Bonn [9].

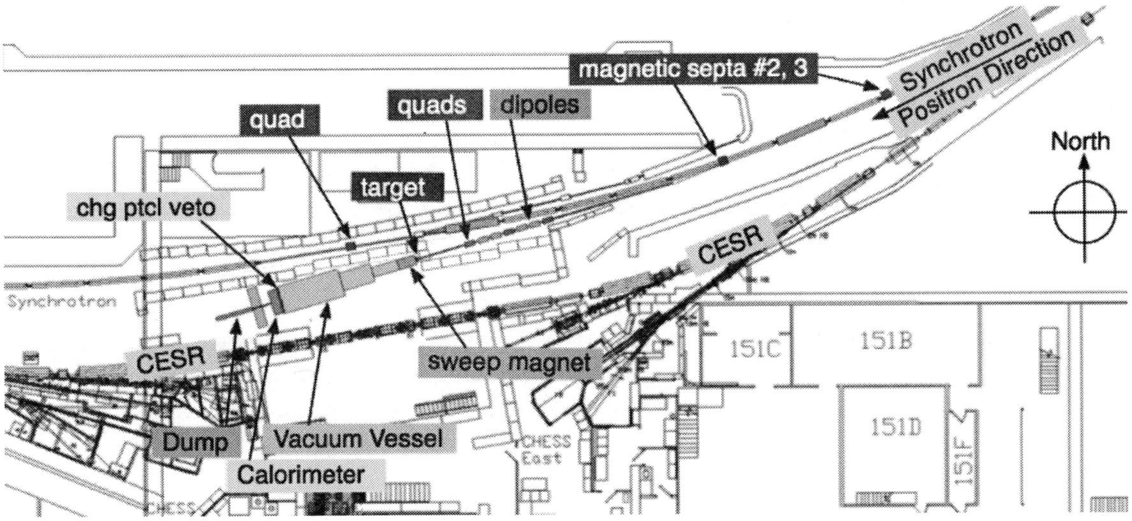

Figure 2. Approximate configuration of dark photon apparatus. Plan view shows south arc of synchrotron and CESR storage ring with MMAPS detector and beamline superimposed.

Extraction efficiency depends on our ability to manipulate the betatron tune of the synchrotron with quadrupoles, the strength of the resonance with the sextupole, and the performance of the extraction septa. Simulation studies indicate that the extraction efficiency will be around 80% with an

0.8 mm thick septum, which is consistent with performance achieved under a similar extraction strategy in the pre-1979 era. For a fixed extraction efficiency, the emittance of the extracted beam scales with septum thickness; the simulations show that the extracted beam emittance is well below the experimental requirement, even with a 1 mm septum. To minimize the emittance of the extracted beam, the high energy beam is moved to within ~ 10mm of the septum with a 3-component pulsed bump that turns on at the start of the extraction cycle. Evolution of the phase space of the positron beam through acceleration cycle is shown in Fig . 3.

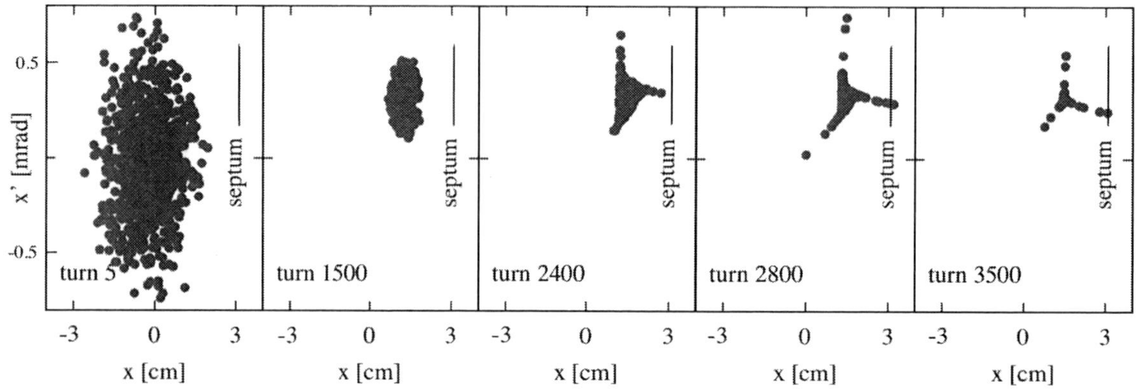

Figure 3. Beam phase space in the synchrotron at five points during the acceleration and extraction cycle. At turn 5 (leftmost panel) injection from the linac into the sychrotron the phase space fills the aperture. By turn 1500 the emittance has shrunk as the particles gained energy, and the bump has shifted it towards the extraction septum. Around turn 2400, the pulsed quadrupole and sextupole shrink the stable phase space. During turns 2500-3500, (right two panels) the particles outside the stable region step beyond the septum and into the extraction line.

The capability to resonantly extract beam from the synchrotron requires installation of a new sextupole and two quadrupole magnets, together with associated current sources. A single sextupole will drive the resonance. One trim quadrupole will be needed to shift the horizontal betatron tune onto the resonance and another to fix the vertical tune. To move the beam close to the first septum for extraction, a closed-orbit bump is generated with the existing trim windings on three of the synchrotron magnets. The uniformity with which positrons are extracted through the ~ 1000 turn extraction cycle is sensitive to the tune shift (quadrupole), sextupole strength, and bump amplitude from one turn to the next. Programmable power supplies for the extraction magnets will enable the requisite fine tuning.

The first septum is relatively thin. It's kick is enough to separate the outgoing particles from the circulating beam so that they clear a second, thicker septum, which then provides sufficient kick to direct the beam into the extraction line. The septum magnets will be inserted between synchrotron magnets as shown in Figure 2.

The sextupole, quadrupoles, and trim windings for the closed-orbit bump will all require programmable current sources. Uniform extraction over a thousand turns is sensitive to the proximity to the third integer resonance and the strength of the sextupole. The septa on the other hand, can be powered in series with the synchrotron dipoles, in the same way that the septa are currently powered to extract beam for injection into CESR. The beam energy varies over the course of the extraction cycle, but at any moment the synchrotron dipole magnetic field is precisely known, given the phase in the cycle, and the beam energy is fixed by the synchrotron ~ 1% energy acceptance. A measurement of the synchrotron beam position, with the turn by turn capable beam position monitor, gives the beam energy to $< 0.1\%$. The magnets in the extraction beam line will likewise be powered in series with the synchrotron magnets in order to maintain energy independent focusing and steering of the extracted beam.

The beam that exits the synchrotron will see a pair of dipole correctors and quadrupoles in the extraction beam line to steer and focus it onto the target. The quadrupoles will be assembled from synchrotron laminations. Combined function magnets and quadrupoles will be powered in series with the synchrotron magnets so that the energy of the beam line is consistent with the beam energy. The dipole correctors will be air core (low impedance) magnets that permit near turn by turn correction of the varying launch angle of the extracted particles.

A model of the synchrotron guide field [10], including the magnets required for resonant extraction, was used to study resonant extraction. The time-dependent fields of the sextupole, quadrupole, and displacment bump magnets were tuned in the simulation to produce a nearly uniform spill, thus defining the required time dependence of the magnets.

To benchmark the simulation results with measurements, we have recently installed a high bandwidth stripline detector in the synchrotron to measure turn-by-turn transverse position of the accelerating beam. The measured displacement corresponds to the turn dependent energy offset. We are preparing to install two additional strip line detectors near the extraction sextupole and septa respectively to measure phase advance.

The synchrotron will continue to serve as the injector for the storage ring for its primary role as an xray source. Every three minutes the delivery of positrons to the MMAPS target will be interrupted for top-off of the positron beam in the storage ring; this will result in a duty cycle for MMAPS operations of 80%. We also anticipate that with a conservative design of the components for the resonant extraction scheme, that we will extract $\sim 30\%$ of the positrons produced in the linac, corresponding to an average current $\langle I_+ \rangle \sim 2.3$nA.

3.1 The External Beam

A measurement of the missing mass, $M^2_{\text{miss}} \equiv (p_{e^+} + p_{e^-} - p_\gamma)^2$, requires knowledge of both the final and initial state kinematics. The extraction system described above will provide an incoming positron beam whose energy, direction, and position lie well within the specifications, $\sigma_{E_+} < 15$ MeV, $\theta_+ \ll 0.4$ mrad, and $\sigma_x, \sigma_y \ll 4$ mm. For beam parameters within these limits, the spread in beam phase space has no discernible impact on missing mass resolution.

The beam extracted from the synchrotron needs to be steered to the target. The fast BPMs will be used for tuning the beam extraction and monitoring the pass through the exit beam line with a 160 ns response time. We plan to use four beam monitors in the vacuum channel between the accelerator and production target and an additional one in front of the beam dump. Each monitor will be of the "SciFi" type [11] will have 64 scintillator fibers with multianode PMT readout.[12]

The 2.5 msec beam spill repeats every $\frac{1}{60}$ sec = 16.7 msec, and each spill is further substructured into 15,000 bunches, uniformly spaced by 168ns; each bunch delivers about 17,000 positrons to the target. Given a Beryllium target of 12.7 mm length, the luminosity can be calculated. We distinguish between the *average* luminosity, which determines final statistical sensitivity for the experiment, from the *peak* luminosity, which drives the occupancy in the calorimeter during the spill:

$$\mathcal{L}_{\text{av}} = 0.9 \times 10^{34}\,\text{cm}^{-2}\text{s}^{-1} \qquad \mathcal{L}_{\text{pk}} = 6.1 \times 10^{34}\,\text{cm}^{-2}\text{s}^{-1}. \tag{1}$$

4 Detector

The extracted beam line directs the positrons onto a beryllium target, and a calorimeter sitting 10 m further downstream measures the energy and direction of photons produced by the e^+e^- collisions.

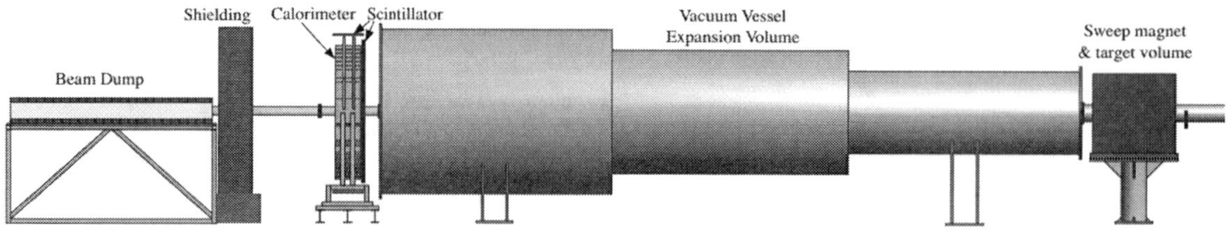

Figure 4. Elevation view of MMAPS.

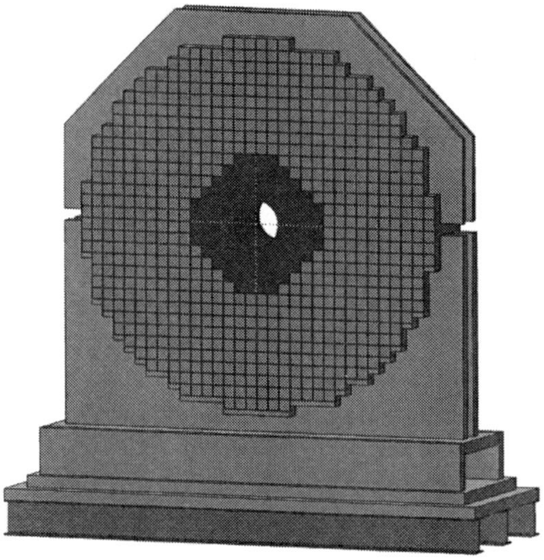

Figure 5. Front view of crystal calorimeter.

The experimental setup, illustrated in Fig. 5, includes a large vacuum vessel between the target and calorimeter, a scintillator veto wall to exclude charged particles, and a beam dump. The active area of the calorimeter and veto wall covers a solid angle of $2° - 5°$ in the lab frame (roughly $136° - 161°$ in the CM frame).

4.1 Calorimeter

The calorimeter is a stack of CsI crystals located 10 m from the target. The crystals, recovered from the CLEO endcap calorimeters, are $5 \times 5 \times 30\,\text{cm}^3$ rectangular solids, well suited to stacking in simple arrays. The inner and outer radii of the annular configuration correspond to lab angles for the photon of $2° \lesssim \theta_\gamma \lesssim 5°$.

The Thallium-doped CLEO CsI crystals have a high scintillation yield of 54,000 photons per MeV. The emission spectrum peaks around 560 nm, with a rise time of $\sim$ 30 ns and fall time of $\tau \approx 1\,\mu$s [13].

The particle identification system in the Babar [14] experiment depended on fast, high-gain, low-noise PMTs [15]. For the emission spectrum of CsI(Tl), the Babar tubes have a quantum efficiency of 8% which, together with the photon collection efficiency of the crystal+tube system, leads to a photoelectron yield of 600 p.e. per MeV. A typical 10 MeV deposition in one crystal will therefore be measured with 1% precision (*i.e.* limited by photoelectron statistics), yielding an order of magnitude

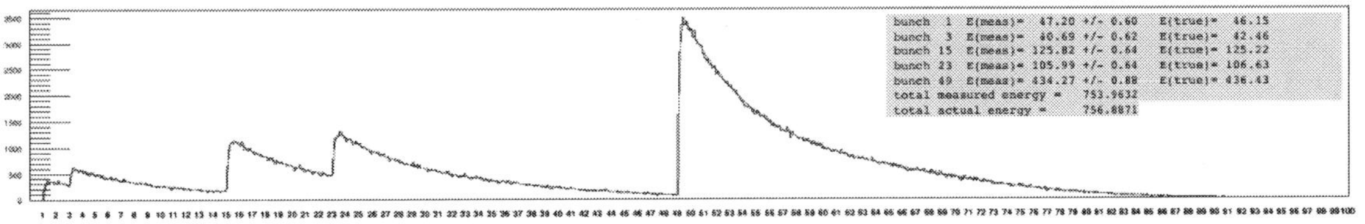

Figure 6. Pulse train after digitization, for 100 beam bunches (16 μs). Vertical axis is in photoelectrons per 20 ns bin; successive beam bunches are separated by 8 bins.

improvement over photodiode performance. Furthermore, the fast response of the PMT gaurantees that it tracks the CsI(Tl) signal rise time without distortion, a feature that plays a crucial role in handling signal pileup.

Photon energy resolution is the dominant contributor to the missing-mass resolution over most of the angular acceptance. The low intrinsic noise of PMTs ensures that electronic noise (a significant consideration for low energy photons in CLEO) is a minor contribution in MMAPS. In the 5-700MeV range of interest, shower leakage out the back of the crystal is also not important. Thus side-leakage, due the lateral spread of the shower and the consequent fluctuation in energy recovered by clustering, dominates photon energy resolution.

An additional contribution comes from signal pileup, which can affect the measurement of energy deposited in a single crystal when two or more photons happen to deposit energy in the same crystal in nearby beam bunches. The PMT outputs will be continuously digitized by FADCs, and the resulting waveforms analyzed to extract individual photon energies and arrival times. Fig. 6 shows a sample waveform from 16 microseconds of simulated data, showing 5 pulses in one crystal. (The waveform shown is chosen to emphasize extreme pileup conditions: for nominal operating conditions, $\lambda = 2.2$, a random sampling of 16 μs of data has less than 1% chance of yielding 5 or more pulses.) Decomposing such a pulse train into individual hits is done by an algorithm that takes advantage of the fixed pulse shape and discrete arrival times (coincident with beam bunches).

The average amount of photon pileup in the calorimeter is determined by the peak luminosity discussed in Sec. 3.1, the 1μs characteristic time profile of CsI(Tl) emission noted above, and the various background cross sections, acceptances, and spatial distributions. For the intended running conditions, *i.e.* the peak luminosity as given in Eq. 1, we find the average number of photons in the calorimeter for any beam bunch is $\lambda = 2.2$.

4.2 Calibration

Calibration constants that convert ADC counts to MeV must be measured precisely and frequently. Several independent methods allow for plenty of cross-checking:

Two-gamma calibration. In $e^+e^- \to \gamma\gamma$ events, only one photon is in the calorimeter acceptance, and its angle θ_γ is correlated with its energy: $E_\gamma = m_e/(1 - \beta_{CM} \cos\theta_\gamma)$. The calorimeter geometry will be known to arbitrary precision, allowing us to use the measured angle to determine energy with a resolution $\sigma_E = |dE/d\theta| \sigma_\theta$. In practice $\sigma_\theta \approx 1$ mrad, and $|dE/d\theta|$ varies from 3 MeV/mrad at the outer edge to 36 MeV/mrad at the inner radius of the calorimeter. This process calibrates the entire calorimeter continuously, to a precision well below the intrinsic resolution single-photon resolution: a few seconds of data is sufficient to calibrate every crystal to a fraction of a percent accuracy at the energy corresponding to the crystal's position.

Cosmic calibration. Each crystal in the calorimeter sees a ~ 2 Hz rate of cosmic rays. The most probable value of the energy deposition for a 5 cm path length in CsI is 28 MeV. The actual path length will vary but can be accurately determined event-by-event by using the full crystal array and the two horizontal scintillator layers for tracking. With one day of cosmic tracks, the peak of the Landau distribution can be determined in every crystal to better than 0.5%. This provides a calibration at a fixed low energy point.

Beam calibration. The sweeping magnet plays two roles, the main one being to sweep Bhabhas out of the calorimeter acceptance. The second is to steer low-current, low-energy beam, provided during dedicated calibration runs, at selected crystals. Independently controlled horizontal and vertical coils allow one to choose the (ϕ, θ) direction to strike a selected crystal. This provides absolute calibration across the middle range of energy; but the process is slow and will be done infrequently.

5 Conclusions and Outlook

The thinking about dark matter has grown beyond the traditional WIMP paradigm in recent years, opening new vistas and new opportunities for experimental work and, hopefully, discovery. Dark photons in the Mev-to-GeV mass range are a target of particular interest, and many experiments underway or in construction are designed to search for evidence of dark photons. MMAPS is one possible realization of such an experiment, exploiting the missing-mass technique, and depending on the presence of a high-energy, high-intensity positron beam. A beam of this description could be delivered at Cornell University, by equipping the existing synchrotron with additional magnets to create a resonant extraction scheme. This beam could support either MMAPS or one of the similar experiments, such as PADME, that are already in design and construction.

References

[1] S. Ritz *et al.*, "Building for Discovery: Strategic Plan for U.S. Particle Physics in the Global Context", Report of the Particle Physics Project Prioritization Panel (P5), available online (June 2014).
[2] B. Wojtsekhowski, "Searching for a U-boson with a positron beam", arXiv:0906.5265
[3] Mauro Raggi, Venelin Kozhuharov, P. Valente, "The PADME experiment at LNF", arXiv:1501.01867
[4] B. Wojtsekhowski, D. Nikolenko, I. Rachek, "Searching for a new force at VEPP-3", arXiv:1207.5089
[5] Plots courtesy of Rouven Essig; private communication.
[6] S. Agostinelli *et al.* [GEANT4 Collaboration], "GEANT4: A Simulation toolkit," Nucl. Instrum. Meth. A **506**, 250 (2003).
[7] R. R. Wilson, "The 10-GeV TO 20-GeV Cornell Electron Synchrotron," CS-33, NP-17179.
[8] J. DeWire, H. Edwards, R. Littauer, G. Rouse and R. Sundelin, "An External Beam from the Cornell 10 GeV Synchrotron," Proceedings 1971 Particle Accelerator Conference, p. 964-965.
[9] W. Hillert, "The Bonn electron stretcher accelerator ELSA: Past and future, Eur. Phys. J. A **28S1**, 139 (2006).
[10] D. Sagan, Nucl. Instrum. Methods Phys. Res. A558, 356 (2006). BMAD
[11] "A 1mm Scintillating Fibre Tracker Readout by a Multi-anode Photomultiplier" arXiv:1106.5649v1 [physics.ins-det] 28 Jun 2011
[12] http://www.hamamatsu.com/resources/pdf/etd/H8804_TPMH1333E.pdf

DOI: 10.1051/epjconf/201714201001

[13] J.D. Valentine, W.W. Moses, S.E. Derenzo, D. Wehe, G. F. Knoll, "Temperature dependence of CsI(Tl) gamma-ray scintillation decay time constants and emission spectrum" NIM A325 (1993) 147 ScienceDirect link

[14] I. Adam *et al.* [BaBar DIRC Collaboration], "The DIRC particle identification system for the BaBar experiment," Nucl. Instrum. Meth. A **538**, 281 (2005).

[15] EMI 9125FLB17, made by Electron Tube Ltd.

The new g-2 experiment at Fermilab

A. Anastasi[1-2,a] on behalf of the g-2 Collaboration

[1] *Dipartimento MIFT, Universitá degli Studi di Messina, Via Salita Sperone 31, 98166 Messina ME, Italy*
[2] *INFN - Laboratori Nazionali di Frascati, Via Enrico Fermi 40, 00044 Frascati RM, Italy*

Abstract. There is a long standing discrepancy between the Standard Model prediction for the muon g-2 and the value measured by the Brookhaven E821 Experiment. At present the discrepancy stands at about three standard deviations, with an uncertainty dominated by the theoretical error. Two new proposals – at Fermilab and J-PARC – plan to improve the experimental uncertainty by a factor of 4, and it is expected that there will be a significant reduction in the uncertainty of the Standard Model prediction. I will review the status of the planned experiment at Fermilab, E989, which will analyse 21 times more muons than the BNL experiment and discuss how the systematic uncertainty will be reduced by a factor of 3 such that a precision of 0.14 ppm can be achieved.

1 Introduction

The muon anomaly $a_\mu = (g-2)/2$ is a low-energy observable, which can be both measured and computed to high precision [1, 2]. Therefore it provides an important test of the Standard Model (SM) and it is a sensitive search for new physics [3]. Since the first precision measurement of a_μ from the E821 experiment at BNL in 2001 [4], there has been a discrepancy between its experimental value and the SM prediction. The significance of this discrepancy has been slowly growing due to reductions in the theory uncertainty. Figure 1 (taken from [5]) shows a recent comparison of the SM predictions of different groups and the BNL measurement for a_μ. The a_μ determinations of the different groups are in very good agreement and show a consistent $\approx 3\,\sigma$ discrepancy [5–7], despite many recent iterations in the SM calculation. It should be noted that with the final E821 measurement and advances in the theoretical SM calculation that both the theory and experiment uncertainties have been reduced by more than a factor two in the last ten years [8]. The accuracy of the theoretical prediction ($\delta a_\mu^{\mathrm{TH}}$, 5 $\times 10^{-10}$) is limited by the strong interaction effects which cannot be computed perturbatively at low energies. The leading-order hadronic vacuum polarization contribution, a_μ^{HLO}, gives the main uncertainty ($\sim$3 $\times 10^{-10}$) [9]. It can be related by a dispersion integral to the measured hadronic cross sections, and it is known with a fractional accuracy of 0.6%, i.e. to about 0.4 ppm. The $O(\alpha^3)$ hadronic light-by-light contribution, a_μ^{HLbL}, is the second dominant contribution ti error in the theoretical evaluation. It cannot at present be determined from data, and relies on using specific models. Although its value is almost two orders of magnitude smaller than a_μ^{HLO}, it is much worse known (with a fractional error of the order of 30%) and therefore it still give a significant contribution to $\delta a_\mu^{\mathrm{TH}}$ (between 2.5 and 4 $\times 10^{-10}$).

From the experimental side, the error achieved by the BNL E821 experiment is $\delta a_\mu^{\mathrm{EXP}} = 6.3 \times 10^{-10}$ (0.54 ppm) [10]. This impressive result is still limited by the statistical errors, and a new experiment, E989 [11], to measure the muon anomaly to a precision of 1.6×10^{-10} (0.14 ppm) is under construction at Fermilab. If the central value remains unchanged, then the statistical significance of the discrepancy with respect to the SM prediction would be over 5σ, see Ref. [2], and would be larger than this with the expected improvements in the theoretical calculation.

2 Measuring a_μ

The measurement of a_μ uses the spin precession resulting from the torque experienced by the magnetic moment when placed in a magnetic field. An ensemble of polarized muons is introduced into a magnetic field, where they are stored for the measurement period. With the assumption that the muon velocity is transverse to the magnetic field ($\vec{\beta} \cdot \vec{B} = 0$), the rate at which the spin turns relative to the momentum vector is given by the difference frequency between the spin precession and cyclotron frequencies. Because electric quadrupoles are used to provide vertical focusing in the storage ring, their electric field is seen in the muon rest frame as a motional magnetic field that can affect the spin precession frequency. In the presence of both $\vec{E}$ and $\vec{B}$ fields, and in the case that $\vec{\beta}$ is perpendicular to both, the anomalous precession frequency (*i.e.* the frequency at which the muon's spin advances relative to its

[a] e-mail: antonioanastasi89@gmail.com

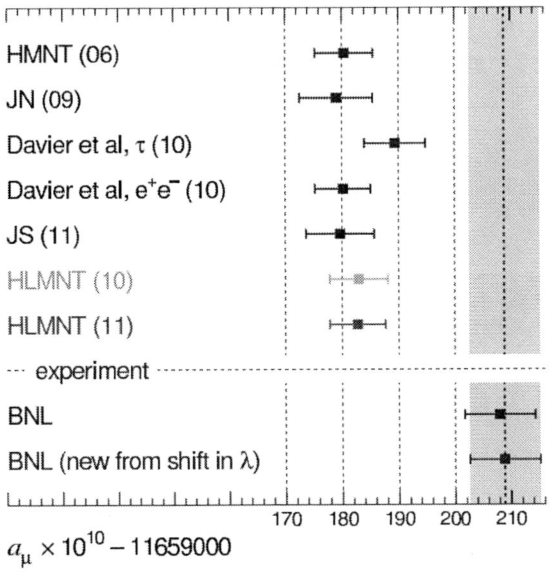

Figure 1. Standard model predictions of a_μ by several groups compared to the measurement from BNL (taken from [5]).

momentum) is

$$\vec{\omega_a} = \vec{\omega_S} - \vec{\omega_C} = -\frac{q}{m}\left[a_\mu \vec{B} - \left(a_\mu - \frac{1}{\gamma^2 - 1}\right)\frac{\vec{\beta} \times \vec{E}}{c}\right] \quad (1)$$

The experimentally measured numbers are the muon spin frequency ω_a and the magnetic field, which is measured with proton NMR, calibrated to the Larmor precession frequency, ω_p, of a free proton. The anomaly is related to these two frequencies by

$$a_\mu = \frac{\tilde{\omega}_a/\tilde{\omega}_p}{\lambda - \tilde{\omega}_a/\tilde{\omega}_p} = \frac{R}{\lambda - R}, \quad (2)$$

where $\lambda = \mu_\mu/\mu_p = 3.183345137(85)$ (determined experimentally from the hyperfine structure of muonium), and $R = \tilde{\omega}_a/\omega_p$. The tilde over ω_a means it has been corrected for the spread in the beam momentum (the so-called electric-field correction) and for the vertical betatron oscillations which mean that $\vec{\beta} \cdot \vec{B} \neq$ (the so-called pitch corrections): these are the only corrections made to the measurement. The magnetic field in Eq. (1) is an average that can be expressed as an integral of the product of the muon distribution times the magnetic field distribution over the storage region. Since the moments of the muon distribution couple to the respective multipoles of the magnetic field, either one needs an exceedingly uniform magnetic field, or exceptionally good information on the muon orbits in the storage ring, to determine $< B >_{\mu-dist}$ to sub-ppm precision. This was possible in E821 where the uncertainty on the magnetic field averaged over the muon distribution was 170 ppb (parts per billion). The coefficient of the $\vec{\beta} \times \vec{E}$ term in Eq. (1) vanishes at the "magic" momentum of 3.094 GeV/c where $\gamma = 29.3$. Thus a_μ can be determined by a precision measurement of ω_a and B. At this magic momentum, the electric field is used only for muon storage and the magnetic field alone determines the precession frequency. The finite spread in beam momentum and vertical betatron oscillations introduce small (sub ppm) corrections to the precession frequency. These are the only corrections made to the measurement.

The experiment consists of repeated fills of the storage ring, each one introducing an ensemble of muons into a magnetic storage ring, and then measuring the two frequencies ω_a and ω_p. The muon lifetime is 64.4 μs, and the data collection period is typically 700 μs. The g-2 precession period is 4.37 μs, and the cyclotron period ω_C is 149 ns.

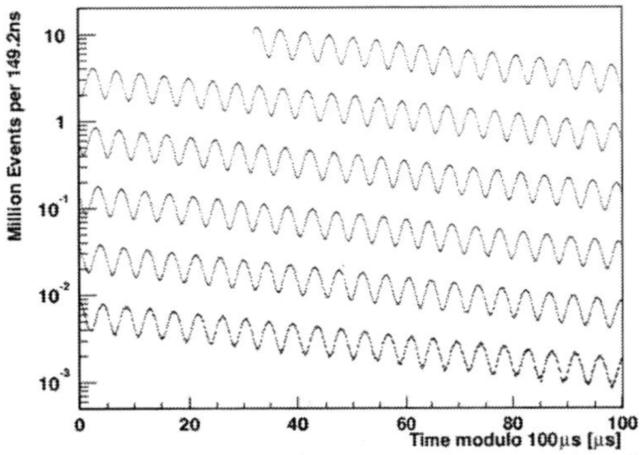

Figure 2. Distribution of electron counts versus time for 3.6 billion muon decays from the E821 experiment. The data are wrapped around modulo 100 μs [10].

Because of parity violation in the weak decay of the muon, a correlation exists between the muon spin and the direction of the high-energy decay electrons. Thus as the spin turns relative to the momentum, the number of high-energy decay electrons is modulated by the frequency ω_a, as shown in Fig. 2. The E821 storage ring was constructed as a "super-ferric" magnet, meaning that the iron determined the shape of the magnetic field. Thus B_0 needed to be well below saturation and was chosen to be 1.45 T. The resulting ring had a central orbit radius of 7.112 m, and 24 detector stations were placed symmetrically around the inner radius of the storage ring. The detectors were made of Pb/SciFi electromagnetic calorimeters which measured the decay electron energy and time of arrival. The detector geometry and number were optimized to detect the high energy decay electrons, which carry the largest asymmetry, and thus information on the muon spin direction at the time of decay. In this design many of the lower-energy electrons miss the detectors, reducing background and pileup.

3 The Fermilab experiment

The E989 experiment at Fermilab plans to measure a_μ to an uncertainty of 16×10^{11} (0.14 ppm), derived from a 0.10 ppm statistical error and roughly equal 0.07 ppm systematic uncertainties on ω_a and ω_p.

The proposal efficiently uses the unique properties of the Fermilab beam complex to produce the necessary flux of muons, which will be injected and stored in the (relocated) muon storage ring. To achieve a statistical uncertainty of 0.1 ppm, the total data set must contain more than 1.8×10^{11} detected positrons with energy greater than 1.8 GeV, and arrival time greater than 30 μs after injection into the storage ring. Four out of 20 of the 8-GeV Booster proton batches in 15 Hz operational mode, each subdivided into four bunches of intensity 10^{12} p/bunch, will be used to provide muons. The proton bunches fill the muon storage ring at a repetition rate of 12 Hz, to be compared to the 4.4 Hz at BNL. The proton bunch hits a target in the antiproton area, producing a 3.1 GeV/c pion beam that is directed along a nearly 2000 m decay line, including several revolutions around the Delivery Ring, which are used to further eliminate pions and to displace secondary protons from muons using time of flight and a kicker to sweep out the protons. The resulting pure muon beam is injected into the storage ring. The muons enter the ring through a superconducting inflector magnet. At present it is envisaged that the BNL inflector will be used but there is a vigorous R&D programme underway investigating the possible use of a new large aperture inflector that would increase the number of stored muons and reduce the multiple scattering. A better optimized pulse-forming network will energize the storage ring kicker to place the beam on a stable orbit. The pion flash (caused by pions entering the ring at injection) will be eliminated owing to the long beamline, and the muon flux will be significantly increased because of the ability to take zero-degree muons.

Figure 3. The new MC-1 building at Fermilab, where the muon g-2 storage ring is being reassembled in the larger part to the left. The part to the right houses the counting room, electronics, etc, with cryogenics services further right. (Image credit: Fermilab.)

In the summer of 2013 the E821 muon storage has been moved from Brookhaven to Fermilab and it has been already relocated in the newly completed MC-1 building at Fermilab (see Figs. 3 and 4) with a stable floor and good temperature control, neither of which were available at Brookhaven.

The new experiment will require upgrades of detectors, electronic systems and data acquisition equipment to handle the much higher data volumes and slightly higher instantaneous rate and also to reduce systematic errors on the measurement of ω_a and ω_p. 24 Electromagnetic calorimeters stations with large area SiPM (1.2×1.2 cm^2) readout will be used. Each calorimeter is composed by 9 wide $\times$ 6 height Cerenkov lead fluoride (PbF_2) crystals wrapped in a black absorptive material (tedlar). The

Figure 4. Re-assembly of the g-2 storage-ring magnet at Fermilab, after the three superconducting coils were positioned gently on top of the newly assembled bottom ring of steel yoke segments. The coils and their complex interconnect system (top right in photo) were transported as a single unit from Brookhaven to Fermilab by land, sea and river, in 2013. (Image credit: Fermilab.)

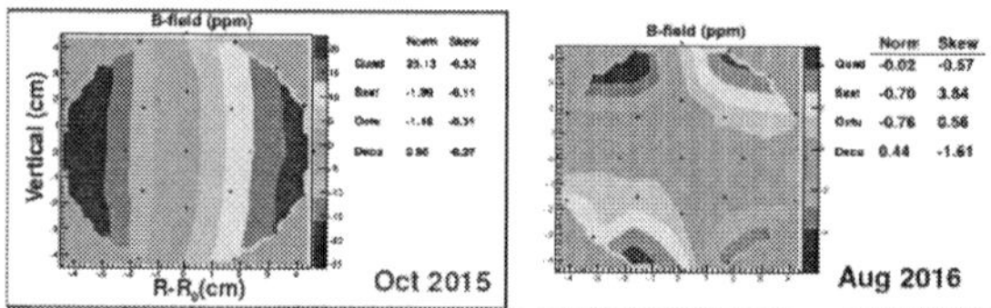

Figure 5. Comparison of the shimming from 2015 and 2016.

high level of segmentation improves pileup recognition. Three 1500 in vacuum straw tracker tubes will be used to precise monitor properties of the muon beam via tracking of Michel decay positrons. Also the provide data for an improvement of the electric dipole moment mesurement, which can be obtained in parallel [12]. In addition an high precision laser calibration system will be used to untangling gain from other systematics and reduce gain fluctuation contribution to the systematic error at the level of 0.02 ppm. A modern data acquisition system will be used to read out waveform digitizer data at 800 MHz and to store it in order to be able to use in parallel different analysis technique. Another important contribution to reduce the error on the determination of a_μ is related to the measurement of ω_p. The storage ring magnetic field will be shimmed to an impressive uniformity. The magnet acheived full power in September 21 of 2015 with a peak to peak variation of 1400 ppm; in August 2016 the peak variation was reduced to 50 ppm (5). The goal of shimming is a peak to peak variation of 50 ppm with a muon weighted systematic uncertainty of 70 ppb, in order to improve by a factor 2 with respect to the 170 ppb of BNL.

In less than two years of running, the statistical goal of 4×10^{20} protons on target can be achieved for positive muons. A follow-up measurement using negative muons is possible. Two additional physics results will be obtained from the same data: a new limit on the muon's electric dipole moment; and, a more stringent limit on possible CPT or Lorentz violation in muon spin precession. The first physics data-taking is expected in 2017. A BNL-level result in terms of precision is expected in 2018, while the final measurement with an uncertainty of 140 ppb (an improvement by a factor of 4 over E821) is expected in 2020.

4 Conclusion

The measurements of the muon g-2 have been an important benchmark for the development of QED and the Standard Model. In the recent years, following the impressed accuracy (0.54 ppm) reached by E821 experiment at BNL, a worldwide effort from different theoretical and experimental groups have significant improved its SM prediction. At present there appears to be a 3σ difference between the theoretical (SM) and the experimental value. This discrepancy, which would fit well with SUSY expectations and other beyond the Standard Model theories, is a valuable constraint in restricting physics beyond the standard model and guiding the interpretation of LHC results. In order to clarify the nature of the observed discrepancy between theory and experiment and eventually firmly establish (or constrain) new physics effects, new direct measurements of the muon g-2 with a fourfold improvement in accuracy have been proposed at Fermilab by E989 experiment, and J-PARC. Data taking for the E989 experiment is expected to start in 2017.

References

[1] F. Jegerlehner, "The anomalous magnetic moment of the muon," Springer Tracts Mod. Phys. **226** (2008) 1

[2] T. Blum, A. Denig, I. Logashenko, E. de Rafael, B. Lee Roberts, T. Teubner and G. Venanzoni, arXiv:1311.2198 [hep-ph]

[3] D. Stöckinger, "Muon (g-2) and physics beyond the standard model," In Roberts, Lee B., Marciano, William J. (eds.): Lepton dipole moments 393-438 (Advanced series on directions in high energy physics. 20)

[4] H. N. Brown *et al.* [Muon g-2 Collaboration], Phys. Rev. Lett. **86** (2001) 2227

[5] K. Hagiwara, R. Liao, A. D. Martin, D. Nomura and T. Teubner, J. Phys. G **38** (2011) 085003

[6] F. Jegerlehner and A. Nyffeler, Phys. Rept. **477** (2009) 1

[7] M. Davier, A. Hoecker, B. Malaescu and Z. Zhang, Eur. Phys. J. C **71** (2011) 1515 [Erratum-ibid. C **72** (2012) 1874]

[8] J. Prades, Acta Phys. Polon. Supp. **3** (2010) 75

[9] M. Davier, arXiv:1612.02743

[10] G. W. Bennett *et al.* [Muon G-2 Collaboration], Phys. Rev. D **73** (2006) 072003

[11] New Muon (g-2) Collaboration, R.M. Carey *et. al.*, see http://lss.fnal.gov/archive/testproposal/0000/fermilab-proposal0989.shtml

[12] G. W. Bennett *et al.* [Muon (g-2) Collaboration], Phys. Rev. D **80** (2009) 052008

EPJ Web of Conferences **142**, 01003 (2017) DOI: 10.1051/epjconf/201714201003
Advances in Dark Matter and Particle Physics

General features of spontaneous baryogenesis

Elena Arbuzova*

[1] *Department of Higher Mathematics, Dubna State University, 141980 Dubna, Russia,*
[2] *Novosibirsk State University, 630090 Novosibirsk, Russia*

Abstract. The classical version of spontaneous baryogenesis is studied in details. It is shown that the relation between the time derivative of the (pseudo)goldstone field and the baryonic chemical potential essentially depends upon the representation chosen for the fermionic fields with non-zero baryonic number (quarks). The kinetic equation, used for the calculations of the cosmological baryon asymmetry, is generalized to the case of non-stationary background. The effects of the finite interval of the integration over time are also included into consideration.

1 Introduction

The Standard Model of cosmology (also called Standard Cosmological Model) is currently the best theory for the description of the Universe. It is based on two fundamental ingredients: the Standard Model of particle physics, which is used for the matter content, and General Relativity, which describes gravitational interactions. It also requires the inflationary paradigm, which is an elegant mechanism to fix a few problems of the scenario. The Standard Model of cosmology very successfully explains a huge amount of observational data, including as the most remarkable ones Hubble's law, the primordial abundances of light elements, and the cosmic microwave background.

However, there are several puzzles that strongly suggest to look for new physics. It is striking, that only 5% of the Universe is made of the known matter (mainly protons and neutrons). About 25% of the Universe is likely made of some weakly interactive particles not belonging to the Minimal Standard Model of particle physics. For the time being this component is called *dark matter*. The other 70% of the Universe is really a mystery: it looks like a uniformly distributed perfect fluid with an unusual equation of state $P = -\rho$, where P is the pressure and ρ is the energy density, and it is responsible for the present accelerated expansion rate of the Universe. It is usually indicated with the name *dark energy*, but its origin is not clear at all and a breakdown of General Relativity at large scales is also a possibility. The inflation mechanism cannot be cast in the framework of the Standard Model of particle physics and there are also some possible alternative scenarios that fix the Standard Model of cosmology in a different way. Lastly, we do not understand the origin of the matter-antimatter asymmetry around us. The local Universe is clearly matter dominated, but such an asymmetry cannot be created within the Minimal Standard Model of particle physics.

The term "baryogenesis" is used to indicate the generation of the asymmetry between baryons and antibaryons. In 1967, Andrey Sakharov pointed out three ingredients, today known as Sakharov

*e-mail: arbuzova@uni-dubna.ru

principles [1], to produce a matter-antimatter asymmetry from an initially symmetric Universe. These conditions include: a) non-conservation of baryonic number; b) breaking of symmetry between particles and antiparticles; c) deviation from thermal equilibrium. (For the details see e.g. the reviews [2, 3]). However, none of these conditions is obligatory. Of particular interest is the scenario of spontaneous baryogenesis (SBG) [4], which can proceed in thermal equilibrium, moreover, it is usually most efficient in thermal equilibrium. The term "spontaneous" is related to spontaneous breaking of underlying symmetry of the theory.

It is assumed that in the unbroken phase the theory is invariant with respect to the global $U(1)$-symmetry, which ensures conservation of baryonic number. This symmetry is spontaneously broken and in the broken phase the Lagrangian density acquires the term

$$\mathcal{L}_{SB} = (\partial_\mu \theta) J^\mu_B , \tag{1}$$

where θ is the Goldstone field and J^μ_B is the baryonic current. Due to the spontaneous symmetry breaking (SSB) this current is not conserved. The next step is the statement that the Hamiltonian density corresponding to $\mathcal{L}_{SB}$ is simply the Lagrangian density taken with the opposite sign:

$$\mathcal{H}_{SB} = -\mathcal{L}_{SB} = -(\partial_\mu \theta) J^\mu_B . \tag{2}$$

For the spatially homogeneous field $\theta = \theta(t)$ this Hamiltonian is reduced to $\mathcal{H}_{SB} = -\dot\theta\, n_B$, where $n_B \equiv J^4_B$ is the baryonic number density, so it is tempting to identify $\dot\theta$ with the chemical potential, μ, of the corresponding system. If this is the case, then in thermal equilibrium the baryon asymmetry would evolve to:

$$n_B = \frac{g_S B_Q}{6}\left(\mu T^2 + \frac{\mu^3}{\pi^2}\right) \to \frac{g_S B_Q}{6}\left(\dot\theta T^2 + \frac{\dot\theta^3}{\pi^2}\right), \tag{3}$$

where T is the cosmological plasma temperature, g_S and B_Q are respectively the number of the spin states and the baryonic number of quarks, which are supposed to be the bearers of the baryonic number.

In presented work the classical version of spontaneous baryogenesis is studied [5]. The talk is organized as follows. In Section 2 the general features of the spontaneous breaking of baryonic $U(1)$-symmetry are described, the equations of motion for the quarks fields and for the (pseudo)Goldstone mode are presented. In Sec. 3 the baryonic chemical potential is introduced, and the baryon number density is calculated in equilibrium with respect to baryo-nonconserving interactions for different representations of the quark fields. Next, in Sec. 4 the standard kinetic equation in stationary background is presented. In Sec. 5 we derive kinetic equation in time dependent external field and/or for the case when energy is not conserved because of finite limits of integration over time. Several examples, when such kinetic equation is relevant, are presented in Sec. 6. Lastly in Sec. 7 we conclude.

2 Spontaneous symmetry breaking and goldstone mode

Let us consider the theory of complex scalar field Φ interacting with "quarks", Q, and "leptons", L, with the Lagrangian:

$$\mathcal{L}(\Phi) = g^{\mu\nu}\partial_\mu \Phi^* \partial_\nu \Phi - V(\Phi^*\Phi) + \bar{Q}(i\gamma^\mu\partial_\mu - m_Q)\,Q + \bar{L}(i\gamma^\mu\partial_\mu - m_L)L + \mathcal{L}_{int}(\Phi, Q, L)\,, \tag{4}$$

where $\mathcal{L}_{int}$ describes the interaction between Φ and fermionic fields. In the toy model studied below we take it in the form:

$$\mathcal{L}_{int} = \frac{\sqrt{2}}{m_X^2}\frac{\Phi}{f}\,(\bar{L}\gamma_\mu Q)(\bar{Q}^c\gamma_\mu Q) + h.c.\,, \tag{5}$$

where Q^c is charged conjugated quark spinor, m_X is a parameter with dimension of mass, and f is related to the vacuum expectation value of Φ defined below in Eq. (6). Such an interaction can appear e.g. in $SU(5)$ Grand Unified Theory. For simplicity, in our toy model we do not take into account the quark colors.

Q and L can be any fermions, not necessarily quarks and leptons of the Standard Model. They can be e.g. new heavy fermions possessing similar or the same quantum numbers as the quarks and leptons of the Standard Model and may couple to the ordinary quarks and leptons.

The global $U(1)$-symmetry is assumed to be spontaneously broken at the energy scale f via the potential of the form:

$$V(\Phi^*\Phi) = \lambda\left(\Phi^*\Phi - f^2/2\right)^2 . \tag{6}$$

This potential reaches minimum at the vacuum expectation value of Φ equal to $\langle\Phi\rangle = fe^{i\phi_0/f}/\sqrt{2}$ with an arbitrary constant phase ϕ_0.

Below scale f we can neglect the heavy radial mode of Φ with the mass $m_{radial} = \lambda^{1/2} f$, since being very massive it is frozen out, but this simplification is not necessary and is not essential for the baryogenesis. The remaining light degree of freedom is the variable field ϕ, which is the Goldstone boson of the spontaneously broken $U(1)$. Up to a constant factor the field ϕ is the angle around the bottom of the Mexican hat potential given by eq. (6). Correspondingly we introduce the dimensionless angular field $\theta \equiv \phi/f$ and thus $\Phi = \langle\Phi\rangle \exp(i\theta)$.

The low energy limit of the Lagrangian (4) in the broken phase, which effectively describes the dynamics of θ-field, has the form:

$$\mathcal{L}_1(\theta) = \frac{f^2}{2}\partial_\mu\theta\,\partial^\mu\theta + \bar{Q}_1(i\gamma^\mu\partial_\mu - m_Q)Q_1 + \bar{L}(i\gamma^\mu\partial_\mu - m_L)L + \left(\frac{e^{i\theta}}{m_X^2}(\bar{L}\gamma_\mu Q_1)(\bar{Q}_1^c\gamma_\mu Q_1) + h.c.\right) - U(\theta) . \tag{7}$$

Here we added the potential $U(\theta)$, which may be induced by an explicit symmetry breaking and can lead, in particular, to a nonzero mass of θ. We use the notation Q_1 for the quark field to distinguish it from the phase rotated field Q_2 introduced below in Eq. (9). In a realistic model the quark fields should be (anti)symmetrized with respect to color indices, omitted here for simplicity.

If $U(\theta) = 0$, the theory still remains invariant under the global transformations (i.e. with $\alpha = const$):

$$Q \to e^{-i\alpha/3}Q, \quad L \to L, \quad \theta \to \theta + \alpha . \tag{8}$$

If we only rotate the quark field as above but with coordinate dependent $\alpha = \theta(t, \mathbf{x})$, introducing the new field $Q_1 = e^{-i\theta/3}Q_2$, then the Lagrangian (7) is transformed into:

$$\mathcal{L}_2(\theta) = \frac{f^2}{2}\partial_\mu\theta\partial^\mu\theta + \bar{Q}_2(i\gamma^\mu\partial_\mu - m_Q)Q_2 + \bar{L}(i\gamma^\mu\partial_\mu - m_L)L + \left(\frac{1}{m_X^2}(\bar{Q}_2\gamma_\mu L)(\bar{Q}_2\gamma_\mu Q_2^c) + h.c.\right) + (\partial_\mu\theta)J^\mu - U(\theta) , \tag{9}$$

where the quark baryonic current is $J_\mu = (1/3)\bar{Q}\gamma_\mu Q$. Note that the current has the same form in terms of Q_1 and Q_2.

The equation of motion for the quark field Q_1 obtained from Lagrangian (7) has the form:

$$(i\gamma^\mu\partial_\mu - m_Q)Q_1 + \frac{e^{-i\theta}}{m_X^2}\left[\gamma_\mu L(\bar{Q}_1\gamma_\mu Q_1^c) + 2\gamma_\mu Q_1^c(\bar{Q}_1\gamma_\mu L)\right] = 0 . \tag{10}$$

Analogously the equation of motion for the phase rotated field Q_2 derived from Lagrangian (9) is:

$$\left(i\gamma^\mu\partial_\mu - m_Q + \frac{1}{3}\gamma^\mu\partial_\mu\theta\right)Q_2 + \frac{1}{m_X^2}\left[\gamma_\mu L(\bar{Q}_2\gamma_\mu Q_2^c) + 2\gamma_\mu Q_2^c(\bar{Q}_2\gamma_\mu L)\right] = 0\,. \tag{11}$$

Equations for θ-field derived from these two Lagrangians in flat space-time have respectively the forms:

$$f^2(\partial_t^2 - \Delta)\theta + U'(\theta) + \left[\frac{i\,e^{-i\theta}}{m_X^2}(\bar{Q}_1\gamma_\mu L)(\bar{Q}_1\gamma_\mu Q_1^c) + h.c.\right] = 0 \tag{12}$$

and

$$f^2(\partial_t^2 - \Delta)\theta + U'(\theta) + \partial_\mu J_B^\mu = 0\,, \tag{13}$$

where $U'(\theta) = dU/d\theta$.

Using either the equation of motion (10) or (11) we can check that the baryonic current is not conserved. Indeed, its divergence is:

$$\partial_\mu J_B^\mu = \frac{i\,e^{-i\theta}}{m_X^2}(\bar{Q}_1\gamma_\mu Q_1^c)(\bar{Q}_1\gamma^\mu L) + h.c. \tag{14}$$

(and similarly for Q_2 but without the factor $\exp(-i\theta)$). So the equations of motion for θ in both cases (12) and (13) coincide, as expected.

Eq. (13) expresses the law of the total baryonic current conservation in the unbroken phase. When the symmetry is broken, the non-conservation of the physical baryons (in our case of "quarks") becomes essential and may lead to the observed cosmological baryon asymmetry. Such B-non conserving interaction may have many different forms. The one presented above describes transition of three quark-type fermions into (anti)lepton. There may be transformation of two or three quarks into equal number of antiquarks. Such interaction describes neutron-antineutron oscillations, now actively looked for [6]. There even can be a "quark" transition into three "leptons". Depending on the interaction type the relation between $\dot\theta$ and the effective chemical potential would have different forms (see below).

In the spatially homogeneous case, when $\partial_\mu J_B^\mu = \dot n_B$ and $\theta = \theta(t)$, and if $U(\theta) = 0$, equation (13) can be easily integrated giving:

$$f^2\left[\dot\theta(t) - \dot\theta(t_{in})\right] = -n_B(t) + n_B(t_{in})\,. \tag{15}$$

It is usually assumed that the initial baryon asymmetry vanishes, $n_B(t_{in}) = 0$.

The evolution of $n_B(t)$ is governed by the kinetic equation discussed in Sec. 4, which allows to express n_B through $\theta(t)$ and thus to obtain the closed systems of, generally speaking, integro-differential equations. In thermal equilibrium the relation between $\dot\theta$ and n_B may become an algebraic one, but this is true only in the case when the integration over time is sufficiently long and if $\dot\theta$ is constant or slowly varying function of time.

In the cosmological Friedmann-Robertson-Walker (FRW) background the equation of motion of θ (13) becomes:

$$f^2(\partial_t + 3H)\dot\theta - a^{-2}(t)\,\Delta\theta + U'(\theta) = -(\partial_t + 3H)n_B, \tag{16}$$

where $a(t)$ is the cosmological scale factor and $H = \dot a/a$ is the Hubble parameter. For the homogeneous theta-field, $\theta = \theta(t)$, this equation turns into:

$$f^2(\partial_t + 3H)\dot\theta + U'(\theta) = -(\partial_t + 3H)n_B. \tag{17}$$

We do not include the curvature effects into the Dirac equations because this is not necessary for what follows. Still we are using expression for the current divergence in the form $\mathcal{D}_\mu J^\mu = \dot n_B + 3Hn_B$, but not just $\dot n_B$.

 DOI: 10.1051/epjconf/201714201003

3 Kinetic equilibrium

The study of kinetics of fermions in the cosmological background is grossly simplified if the particles are in equilibrium with respect to elastic scattering, to their possible annihilation e.g. into photons, and to other baryo-conserving interactions. The equilibrium with respect to elastic scattering implies the following form of the phase space distribution functions:

$$f_{eq} = \left[1 + \exp(E/T - \xi)\right]^{-1}, \tag{18}$$

where the dimensionless chemical potential $\xi = \mu/T$ has equal magnitude but opposite signs for particles and antiparticles. The baryonic number density for small ξ is usually given by the expression

$$n_B = g_S B_Q \xi_B T^3/6 \tag{19}$$

(compare to eq. (3)). Here ξ_B is the baryonic chemical potential. This equation which expresses baryonic number density through chemical potential is true only for the normal relation between the energy and three-momentum, $E = \sqrt{p^2 + m^2}$, with equal masses of particles and antiparticles.

Vanishing baryon asymmetry implies $\xi_B = 0$, as is usually the case. If the baryonic number of quarks is conserved, n_B remains constant in the comoving volume and it means in turn that $\xi = const$ for massless particles. If $n_B = 0$ initially, then ξ_B remains identically zero. If baryonic number is not conserved, then as we see below from the kinetic equation, equilibrium with respect to B-nonconserving processes leads to $\xi_B = c\dot{\theta}/T$, as is envisaged by SBG. The constant c depends upon the concrete type of reaction. Complete thermal equilibrium in the standard theory demands $n_B \to 0$, but a deviation from thermal equilibrium of B-nonconserving interaction leads to generation of non-zero ξ_B and correspondingly to non-zero n_B.

The situation changes, if quarks and antiquarks satisfy the equation of motion (11), for which the following dispersion relation is valid

$$E = \sqrt{p^2 + m^2} \mp \dot{\theta}/3, \tag{20}$$

where the signs $\mp$ refer to particles or antiparticles respectively. So the energies of quarks and anti-quarks with the same three-momentum are different. This is similar to mass difference which may be induced by CPT violation. It is noteworthy that the above dispersion relation is derived under assumption of constant or slow varying $\dot{\theta}$. Otherwise the Fourier transformed Dirac equation cannot be reduced to the algebraic one and the particle energy is not well defined.

The baryon number density corresponding to the dispersion relation (20) is given by the expression

$$n_B \equiv g_S B_Q \int \frac{d^3p}{(2\pi)^3}\left[f(p) - \bar{f}(p)\right] = \frac{g_S B_Q}{6}\left(\xi_B + \frac{\dot{\theta}}{3T}\right)T^3. \tag{21}$$

where $\bar{f}$ is the distribution function of antiparticles. If the baryon number is conserved and is zero initially, the condition $\xi_B + \dot{\theta}/(3T) = 0$ would be always fulfilled. If B is not conserved, then the equilibrium with respect to B-nonconserving processes demands $\xi_B = 0$, as it follows from kinetic equation presented below. So evidently $\xi_B \neq \dot{\theta}$ but nevertheless the baryon asymmetry is proportional to $\dot{\theta}$ as follows from eq. (21).

4 Kinetic equation in (quasi)stationary background

The temporal evolution of the distribution function of i-th type particle, $f_i(t, p)$, in an arbitrary process $i + Y \leftrightarrow Z$ in the FRW background, is governed by the equation:

$$\frac{df_i}{dt} = (\partial_t - H\, p_i \partial_{p_i}) f_i = I_i^{coll}, \tag{22}$$

DOI: 10.1051/epjconf/201714201003

with the collision integral equal to:

$$I_i^{coll} = \frac{(2\pi)^4}{2E_i} \sum_{Z,Y} \int d\nu_Z \, d\nu_Y \delta^4(p_i + p_Y - p_Z)$$

$$\left[|A(Z \to i + Y)|^2 \prod_Z f \prod_{i+Y} (1 \pm f) - |A(i + Y \to Z)|^2 f_i \prod_Y f \prod_Z (1 \pm f) \right], \tag{23}$$

where $A(a \to b)$ is the amplitude of the transition from state a to state b, Y and Z are arbitrary, generally multi-particle states, $(\prod_Y f)$ is the product of the phase space densities of particles forming the state Y, and

$$d\nu_Y = \prod_Y \overline{dp} \equiv \prod_Y \frac{d^3 p}{(2\pi)^3 2E}. \tag{24}$$

The signs '+' or '−' in $\prod(1 \pm f)$ are chosen for bosons and fermions respectively. We neglect the effects of space-time curvature in the collision integral, which is generally a good approximation.

We are interested in the evolution of the baryon number density, which is the time component of the baryonic current J^μ: $n_B \equiv J^4$. Due to the quark-lepton transitions the current is non-conserved and its divergence is given by eq. (14). The similar expression is evidently true in terms of Q_2 but without the factor $\exp(-i\theta)$. Let us first consider the latter case, when the interaction described by the Lagrangian (9), which contains the product of three "quark" and one "lepton" operators, and take as an example the process $q_1 + q_2 \leftrightarrow \bar{q} + l$.

Since the interaction in this representation does not depend on time, the energy is conserved and the collision integral has the usual form with conserved four-momentum. Quarks are supposed to be in kinetic equilibrium but probably not in equilibrium with respect to B-nonconserving interactions, so their distribution functions have the form:

$$f_q = \exp\left(-\frac{E}{T} + \xi_B\right) \text{ and } f_{\bar{q}} = \exp\left(-\frac{E}{T} - \xi_B\right). \tag{25}$$

Here and in what follows the Boltzmann statistics is used. According to ref. [7], Fermi corrections are typically at the 10% level. Since the dispersion relation for quarks and antiquarks (20) depends upon $\dot{\theta}$, the baryon asymmetry in this case is given by eq. (21) and the kinetic equation takes the form:

$$\frac{g_S B_Q}{6} \frac{d}{dt} \left(\xi_B + \frac{\dot{\theta}}{3T} \right) = -c_1 \Gamma \xi_B, \tag{26}$$

where c_1 is a numerical factor of order unity and Γ is the rate of baryo-nonconserving reactions. If the amplitude of this reaction has the form determined by the Lagrangian (9), then $\Gamma \sim T^5/m_X^4$.

For constant or slow varying temperature the equilibrium solution to this equation is $\xi_B = 0$ and the baryon number density (21) is proportional to $\dot{\theta}$, $n_B = (g_S B_Q/18)\dot{\theta} T^2$, with $\dot{\theta}$ evolving as:

$$\dot{\theta} = \frac{f^2}{f^2 + g_S B_Q T^2/18} \left(\frac{T}{T_{in}} \right)^3 \dot{\theta}(t_{in}). \tag{27}$$

We see that the equilibrium value of n_B drops down with decreasing temperature as T^5. However at small temperatures baryon non-conserving processes switch-off and n_B tends to a constant value in comoving volume.

Let us check now what happens if the dependence on θ is moved from the quark dispersion relation to the B-nonconserving interaction term (12). The collision integral (23) contains delta-functions imposing conservation of energy and momentum if there is no external field which depends upon coordinates. In our case, when quarks "live" in the $\theta(t)$-field, the collision integral should be modified. Since the interaction proceeds in time dependent background, the energy conservation delta-function does not emerge, and we have now an additional factor, namely, $\exp[\pm i\theta(t)]$, under the integral by calculation of the amplitude of transition from the initial state $|in\rangle$ to a final state $|fin\rangle$. In general case this integral cannot be taken analytically, but if we can approximate $\theta(t)$ as $\theta(t) \approx \dot{\theta}t$ with a constant or slowly varying $\dot{\theta}$, the integral is simply taken giving e.g. for the process of two quark transformation into antiquark and lepton, $q_1 + q_2 \leftrightarrow \bar{q} + l$, the energy balance condition imposed by $\delta(E_{q_1} + E_{q_2} - E_{\bar{q}} - E_l - \dot{\theta})$. In other words the energy is non-conserved due to the action of the external field $\theta(t)$. The approximation of linear evolution of θ with time can be valid if the reactions are fast in comparison with the rate of the θ-evolution.

Returning to our case we can see that the collision integral taken over the three-momentum of the particle under scrutiny (i.e. particle i in eq. (23)) e.g. for process the $q_1 + q_2 \to l + \bar{q}$ turns into:

$$\dot{n}_B + 3Hn_B \sim \int d\tau_{l\bar{q}} d\tau_{q_1 q_2} |A|^2 \delta(E_{q_1} + E_{q_2} - E_l - E_{\bar{q}} - \dot{\theta}) \delta(\mathbf{P}_{in} - \mathbf{P}_{fin}) e^{-E_{in}/T} \left(e^{\xi_L - \xi_B + \dot{\theta}/T} - e^{2\xi_B} \right), \tag{28}$$

where $d\tau_{l,\bar{q}} = d^3 p_l d^3 p_{\bar{q}} / [4 E_l E_{\bar{q}} (2\pi)^6]$. We assumed here that all participating particles are in kinetic equilibrium, i.e. their distribution functions have the form (25). In expression (28) ξ_B and ξ_L denote baryonic and leptonic chemical potentials respectively and the effects of quantum statistics are neglected but only for brevity of notations. The assumption of kinetic equilibrium is well justified because it is enforced by the very efficient elastic scattering. Another implicit assumption is the usual equilibrium relation between chemical potentials of particles and antiparticles, $\bar{\mu} = -\mu$, imposed e.g. by the fast annihilation of quark-antiquark or lepton-antilepton pairs into two and three photons. Anyhow the assumption of kinetic equilibrium is one of the cornerstones of the spontaneous baryogenesis.

The conservation of $(B + L)$ implies the following relation: $\xi_L = -\xi_B/3$. Keeping this in mind, we find

$$\dot{n}_B + 3Hn_B \approx -\left(1 - e^{\dot{\theta}/T - 3\xi_B + \xi_L}\right) I \approx \left(\frac{\dot{\theta}}{T} - \frac{10}{3}\xi_B\right) I, \tag{29}$$

where we assumed that ξ_B and $\dot{\theta}/T$ are small. In relativistic plasma with temperature T the factor I, coming from the collision integral, can be estimated as $I = T^8/m^4$, where m is a numerical constant with dimension of mass. It differs from m_X, introduced in eq. (7), by a numerical coefficient.

For a large factor I we expect the equilibrium solution

$$\xi_B = \frac{3}{10}\frac{\dot{\theta}}{T}, \tag{30}$$

so $\dot{\theta}$ up to the numerical factor seems to be the baryonic chemical potential, as expected in the usually assumed SBG scenario. The value of the coefficient $c = 3/10$ in Eq. (30) may be different for other types of B-nonconcerving reactions, e.g. for the reaction $3q \leftrightarrow 3\bar{q}$ one can find that $c = 1/6$. Let us remind that for the dispersion relation (20) the baryonic chemical potential is not proportional to $\dot{\theta}(t)$, but is equal to zero, see Eq. (26) and comments below.

5 Kinetic equation for time-varying amplitude

The canonical kinetic equation (22) is usually presented for scattering or decay processes in time independent or slowly varying background with the collision integral giving by eq. (23).

DOI: 10.1051/epjconf/201714201003

As it was mentioned above, if the interaction proceeds in time dependent background and/or the time duration of the process is finite, then the energy conservation delta-function does not emerge, so one has to make the time integration with an account of time-varying background and integrate over the phase space without energy conservation.

In what follows we consider two-body inelastic process with baryonic number non-conservation with the amplitude obtained from the last term in Lagrangian (7). At the moment we will not specify the concrete form of the reaction but only will say that it is the two-body reaction

$$a + b \leftrightarrow c + d, \tag{31}$$

where a, b, c, and d are some quarks and leptons or their antiparticles. The expression for the evolution of the baryonic number density, n_B, follows from eq. (22) after integration of its both sides over $d^3 p_i/(2\pi)^3$. Thus we obtain:

$$\dot{n}_B + 3Hn_B = -\frac{(2\pi)^3}{t_{max}} \int d\nu_{in} d\nu_{fin}\, \delta(\mathbf{P}_{in} - \mathbf{P}_{fin})\, |A|^2 \left(f_a f_b - f_c f_d\right) \tag{32}$$

where e.g. $d\nu_{in} = d^3 p_a d^3 p_b/[4E_a E_b (2\pi)^6]$ and the amplitude of the process is defined as

$$A = \left(\int_0^{t_{max}} dt\, e^{i[(E_c+E_d-E_a-E_b)t+\theta(t)]} \right) F(p_a, p_b, p_c, p_d), \tag{33}$$

and F is a function of 4-momenta of the participating particles, determined by the concrete form of the interaction Lagrangian. In what follows we consider two possibilities: $F = const$ and $F = \psi^4\, m_X^{-2}$, where in the last case ψ^4 symbolically denotes the product of the Dirac spinors of particles a, b, c, d.

In the case of equilibrium with respect to baryon conserving reactions the distribution functions have the canonical form $f_a = \exp(-E_a/T + \xi_a)$, where $\xi_a \equiv \mu_a/T$ is the dimensionless chemical potential. So for constant F the product $|A|^2(f_a f_b - f_c f_d)$ depends upon the particle 4-momenta only through E_{in} and E_{fin}, where

$$E_{in} = E_a + E_b, \quad \text{and} \quad E_{fin} = E_c + E_d. \tag{34}$$

Note that the amplitude A (33) depends only on $E_- = E_{in} - E_{fin}$ but not on $E_+ = E_{in} + E_{fin}$, while the products of the particle densities in the phase space are

$$f_a f_b = \exp\left(-\frac{E_+ + E_-}{2T} + \xi_a + \xi_b\right) \quad \text{and} \quad f_c f_d = \exp\left(-\frac{E_+ - E_-}{2T} + \xi_c + \xi_d\right). \tag{35}$$

Now we can perform almost all (but one) integrations over the phase space in eq. (32). Collecting all the factors (35), we finally obtain:

$$\dot{n}_B + 3Hn_B = -\frac{T^5}{2^5 \pi^6\, t_{max}} \int_0^\infty dy \left[e^{\xi_a+\xi_b} \left(|A_+|^2 + |A_-|^2 e^{-y} \right) - e^{\xi_c+\xi_d} \left(|A_-|^2 + |A_+|^2 e^{-y} \right) \right], \tag{36}$$

where A_+ is the amplitude taken at positive E_-, while A_- is taken at negative E_-. With the substitution $E_- \to |E_-|$ the only difference between A_+ and A_- is that $A_-(\theta) = A_+(-\theta)$.

The equilibrium is achieved when the integral in eq. (36) vanishes. This point determines the equilibrium values of the chemical potentials in external $\dot{\theta}$ field. Clearly it takes place at:

$$\xi_a + \xi_b - \xi_c - \xi_d = \frac{\langle |A_+|^2 e^{-y} + |A_-|^2 \rangle}{\langle |A_+|^2 + |A_-|^2 e^{-y} \rangle} - 1, \tag{37}$$

where the angular brackets mean integration over dy as indicated in eq. (36).

DOI: 10.1051/epjconf/201714201003

6 Examples of time-varying θ

6.1 Constant $\dot{\theta}$

This is the case usually considered in the literature and the simplest one. The integral (33) is taken analytically resulting in:

$$|A|^2 \sim \frac{2 - 2\cos[(\dot{\theta} - E_-)t_{max}]}{(\dot{\theta} - E_-)^2}, \tag{38}$$

where E_- is running over the positive semi-axis.

For large t_{max} this expression tends to $\delta(E_- - \dot{\theta})$, so $|A_+|^2 = 2\pi\delta(E_- - \dot{\theta})t_{max}$ and $|A_-|^2 = 2\pi\delta(E_- + \dot{\theta})t_{max} = 0$, if $\dot{\theta} > 0$ and vice versa otherwise. Hence the equilibrium solution is

$$\xi_a + \xi_b - \xi_c - \xi_d - \dot{\theta}/T = 0, \tag{39}$$

coinciding with the standard result.

The limit of $\dot{\theta} = const$ corresponds to the energy non-conservation by the rise (or drop) of the energy of the final state in reaction (31) exactly by $\dot{\theta}$. However if t_{max} is not sufficiently large, the non-conservation of energy is not equal to $\dot{\theta}$ but somewhat spread out and the equilibrium solution would be different. There is no simple analytical expression in this case, so we have to take the integrals over y in eq. (37) numerically.

The results of the calculations are presented in Fig 1. In the left panel the values of the r.h.s. of eq. (37) are presented as a function of $\dot{\theta}/T$ for the cut-off of the time integration in eq. (38) equal to: $\tau \equiv t_{max}T = 30; 10; 3$. The larger is the integration time the closer are the lines to $\dot{\theta}/T$, which is also depicted.

In the right panel the relative differences between the r.h.s. of eq. (37) and $\dot{\theta}/T$, normalized to $\dot{\theta}/T$, as a function of $\dot{\theta}/T$ for different maximum time of the integration are presented. We see that for $\tau = 30$ the deviations are less than 10%, while for $\tau = 3$ the deviations are about 30%. If we take τ close to unity, the deviations are about 100%. The value of $\dot{\theta}/T$ is bounded from above by approximately 0.3 because at large $\dot{\theta}/T$ the linear expansion, used in our estimates, is invalid.

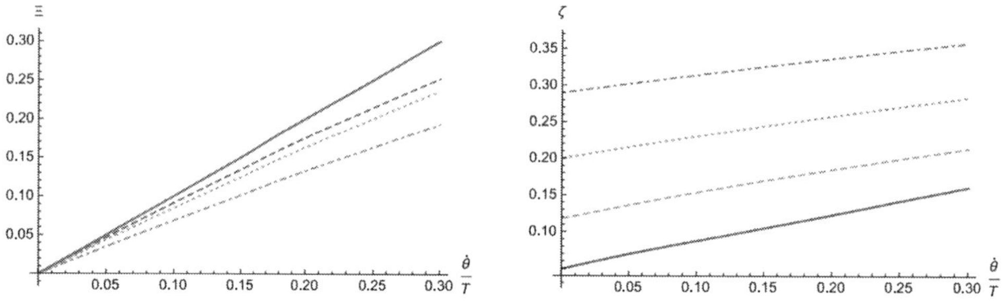

Figure 1. *Left*: The r.h.s. of eq. (37), denoted by Ξ, as a function of $\dot{\theta}/T$ for the cut-off of the time integration in eq. (38): $\tau \equiv t_{max}T = 30$ (dashed); 10 (dotted); 3 (dotdashed) and $\dot{\theta}/T$ (thick). *Right*: The relative difference: $\zeta = \Xi/(\dot{\theta}/T) - 1$, as a function of $\dot{\theta}/T$ for $\tau = 30$ (thick), 10 (dashed), 5 (dotted), 3 (dotdashed).

The realistic values of τ depend upon the model parameters. There is one evident limit related to the cosmological expansion, which implies $\tau < t_{cosm}T \sim T/H \sim m_{Pl}/T$. Here m_{Pl} is the Planck mass, H is the Hubble parameter, and $T_{cosm} \sim 1/H$, so the effects of the expansion may be significant only near the Planck temperature. Another upper bound on τ is presented by the kinetic equations which demands the characteristic time variation to be close (at least initially) to the inverse reaction rate $\gamma \sim T^5/m_X^4$. The discussed effects would have an essential impact on the approach to equilibrium for $T \sim m_X$ which might be realistic.

6.2 Second order Taylor expansion of $\theta(t)$

As we have seen in the previous subsection the approximation $\dot\theta = const$ is noticeably violated. Here we assume that $\theta(t)$ can be approximated as

$$\theta(t) = \dot\theta\, t + \ddot\theta\, t^2/2, \tag{40}$$

where $\dot\theta$ and $\ddot\theta$ are supposed to be constant or slowly varying. In this case the integral over time (33) can also be taken analytically but the result is rather complicated. We need to take the integral

$$\int_0^{t_{max}} dt \exp[i\theta(t)]. \tag{41}$$

Its real and imaginary parts are easily expressed though the Fresnel functions. So the amplitude squared is given by the functions tabulated in Mathematica and the position of the equilibrium point can be calculated, as in the previous case, by numerical calculation of one dimensional integral.

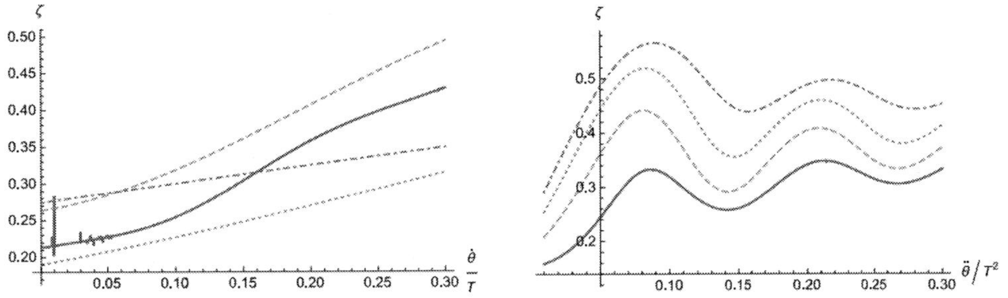

Figure 2. *Left*: The relative difference: $\zeta = \Xi/(\dot\theta/T) - 1$, as a function of $\dot\theta/T$ for the cut-off of the time integration $\tau = 30$ (thick), 10 (dashed), 5 (dotted), 3 (dotdashed) for fixed $\ddot\theta/T^2 = 0.1$. *Right*: The same difference as a function of $\ddot\theta/T^2$ for fixed time of integration $\tau = 10$ and different $\dot\theta/T = 0.1$ (thick), 0.2 (dashed), 0.3 (dotted), 0.4 (dotdashed).

The r.h.s. of eq. (37) as a function of $\dot\theta/T$ for different values of τ is presented in Fig. 2, at the left panel. It is interesting that the dependence on τ is non-monotonic. This may be understood by diminishing of the impact of $\ddot\theta t^2$ at smaller time interval.

To check the dependence on $\ddot\theta$ we calculated again the r.h.s. of eq. (37) but now as a function of $\ddot\theta/T^2$ presented at the right panel of Fig. 2 for fixed time of integration and different values of $\dot\theta/T$. We see that the equilibrium point oscillates as a function $\ddot\theta$.

7 Conclusion

We argue that in the standard description $\dot\theta$ is not formally the chemical potential, though in thermal equilibrium $\dot\theta$ may tend to the chemical potential with the numerical coefficient which depends upon the model. However, this result is not always true but depends upon the chosen representation of the "quark" fields. In the theory described by the Lagrangian (7) which appears "immediately" after the spontaneous symmetry breaking, $\theta(t)$ directly enters the interaction term and in equilibrium $\mu_B \sim \dot\theta$ indeed. On the other hand, if we transform the quark field, so that the dependence on θ is shifted to the bilinear product of the quark fields (9), then chemical potential in equilibrium does not tend to $\dot\theta$, but to zero. Still, the magnitude of the baryon asymmetry in equilibrium is always proportional to $\dot\theta$.

The assumption of a constant or slowly varying $\dot\theta$, which is usually done in the SBG scenario, may be not fulfilled and to include the effects of an arbitrary variation of $\theta(t)$, as well as the effects of the finite time integration, we transformed the kinetic equation in such a way that it becomes operative in non-stationary background. A shift of the equilibrium value of the baryonic chemical potential due to this effect is numerically calculated.

In spite of these corrections to the standard SBG scenario, it remains a viable mechanism for creation of the observed cosmological excess of matter over antimatter. However, this mechanism is not particularly efficient in the case of pure spontaneous symmetry breaking, when the potential of the θ-field is absent. Non-zero potential $U(\theta)$, which can appear as a result of an explicit breaking of the baryonic $U(1)$-symmetry in addition to the spontaneous breaking may grossly enhance the efficiency of the spontaneous baryogenesis. The evaluation of the efficiency demands numerical solution of the ordinary differential equation of motion for the θ-field together with the integral kinetic equation. In the case of thermal equilibrium the kinetic equation is reduced to an algebraic one and the system is trivially investigated. The out-of-equilibrium situation is much more complicated technically and will be studied elsewhere.

We assumed that the symmetry breaking phase transition in the early universe occurred instantly. It may be a reasonable approximation, but still the corrections can be significant. This can be also a subject of future work.

Acknowledgement. This work was supported by the Grant of President of Russian Federation for the leading scientific schools of the Russian Federation, NSh-9022.2016.2.

References

[1] A.D. Sakharov, Pis'ma ZhETF, 5 (1967) 32.

[2] A.D.Dolgov, Phys. Repts **222** (1992) No. 6;
A.D. Dolgov, Surveys in High Energy Physics, **13** (1998) 83, hep-ph/9707419.

[3] V.A. Rubakov, M.E. Shaposhnikov, Usp. Fiz. Nauk, **166** (1996) 493, hep-ph/9603208;
A. Riotto, M. Trodden, Ann. Rev. Nucl. Part. Sci. **49** (1999), 35, hep-ph/9901362;
M. Dine, A. Kusenko, Rev. Mod. Phys. **76** (2004) 1.

[4] A. Cohen, D. Kaplan, Phys. Lett. B **199**, 251 (1987);
A. Cohen, D. Kaplan, Nucl.Phys. B **308** (1988) 913;
A. G. Cohen, D.B. Kaplan, A.E. Nelson, Phys.Lett. B **263** (1991) 86;
A. G. Cohen, D.B. Kaplan, A.E. Nelson, Phys.Lett. B **336** (1994) 41, hep-ph/940634.

[5] E.V. Arbuzova, A.D. Dolgov, V.A. Novikov, Phys. Rev. D **94** (2016) 123501, arXiv:1607.01247.

[6] D.G. Phillips, II, *et al*, Phys.Rept. **612** (2016) 1-45.

[7] A.D. Dolgov, K. Kainulainen, Nucl.Phys. B**402** (1993) 349-359.

Overview of ALICE results on hadronic resonance production

Angela Badalà[1,a] for the ALICE Collaboration

[1] *INFN - Sezione di Catania , Via S. Sofia 64, 95123, Catania (Italy)*

Abstract. The measurement of hadronic resonance production in heavy-ion collisions is a valuable tool to study the properties of the hadronic phase. In addition, these measurements contribute to the study of particle production mechanisms, such as recombination and statistical hadronization, and can give information on the parton energy loss in the hot QCD medium. Measurement of a wide set of resonances with different lifetimes is useful to better characterize the hadronic phase and the time span between chemical and thermal freeze-out. Proton-proton (pp) collisions have been used extensively as a reference for the study of larger colliding systems, but recent measurements performed in high-multiplicity pp and proton-lead (p–Pb) collisions at the LHC have shown features that are reminiscent of those observed in lead-lead (Pb–Pb) collisions. Resonance measurements in small systems serve as a reference for heavy-ion collisions and contribute to searches for collective effects. An overview of recent results on hadronic resonance production measured in ALICE will be presented. Transverse momentum (p_T) spectra, ratios of yield to that of long-lived hadrons of the $K^*(892)^0$ and $\phi(1020)$ mesons in pp, p–Pb, and Pb–Pb collisions at LHC energies will be discussed. The most recent results include the measurement of resonance production in pp collisions at 7 TeV as a function of the charged-particle multiplicity, that will be compared to the results for other light hadrons as pions, kaons, protons, K^0_S, Λ, Ξ^-, Ω^-. In addition, the p_T spectra and yields of the $\rho(770)^0$ meson in pp and Pb–Pb collisions at $\sqrt{s_{NN}}$ = 2.76 TeV and of $\Sigma(1385)^{\pm}$ and $\Xi(1530)^0$ baryons in p–Pb collisions at $\sqrt{s_{NN}}$ = 5.02 TeV will be presented.

1 Introduction

Hadronic resonances, along with stable hadrons, allow the study of properties of heavy-ion collisions, both in the early (quark-gluon plasma) and late (hadronic) stage of their evolution. Modification of their yield after chemical freeze-out is expected due to the regeneration and re-scattering effects in the hadronic phase, which are expected to have their greatest strength at low transverse momentum. Final resonance yields depend on the scattering cross sections of their decay products and the timescale between the chemical and the kinetic freeze-out compared to the resonance lifetime, which determine the fraction of 'undetected' particles. However resonances may be regenerated by pseudo-elastic interactions in the hadronic medium. Therefore, resonances with different lifetimes are good candidates to probe the interplay of particle re-scattering and regeneration in the hadronic phase. Comparison of such measurements to theoretical models [1, 2] allows for an estimation of the timescale between chemical and kinetic freeeze-out.

Measurements of resonance and stable-hadron yields may be used to estimate properties of the hadronic phase of heavy-ion collisions, such as temperature and lifetime. Information on the particle production mechanisms can be derived from the comparison of particles with similar mass but different baryon number and/or strangeness content. Then comparison of resonance production with that of long lived hadrons can be interesting in this respect. In particular, for strangeness production it is worth studying the ϕ meson with its hidden strangeness content.

Both meson and baryon resonances have been measured by the ALICE experiment [3] in different collisions systems (pp, p–Pb, Pb–Pb) at LHC energies [4–8]. Resonance measurements in pp and p–Pb systems are useful as references and may help in disentangling initial-state effects from genuine in-medium effects, which may occur in Pb–Pb collisions. At LHC in p–Pb collisions final state dense matter effects could be present considering that the pseudo-rapidity density of final state particles reaches value similar to central collisions at top RHIC energy [9]. Moreover parton shadowing and/or novel phenomena like saturation as implemented in the Color Glass Condensate model [10] may become apparent due to the low fractional parton momentum x and the high gluon density reached in the initial state. Deconfinement and collectivity effects are not expected in pp collisions and for this reason they have been often used as reference for larger systems. Recently, in high-multiplicity pp and p–Pb collisions LHC experiments have observed features tradionally associated to the formation of a strongly-interacting quark-gluon medium and hinting at the presence of collective phenomena [11–13]. This has triggered the investigation of collectivity-driven features in small systems and the study of strangeness production as a function of the multiplicity in p–Pb and pp collisions [14–17].

[a] e-mail: Angela.Badala@ct.infn.it

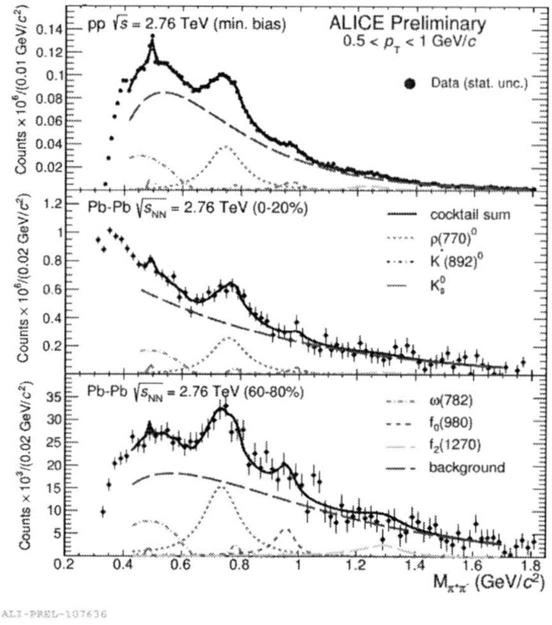

Figure 1. (color online) Invariant mass distributions for π^++π^- pairs with p_T in the range 0.5-1.0 GeV/c in minimum bias pp collisions at $\sqrt{s}$ = 2.76 TeV (upper panel) and in Pb–Pb collisions at $\sqrt{s_{NN}}$ = 2.76 TeV in the centrality range 0-20% (middle panel) and 60-80% (lower panel).

2 Resonance identification in ALICE

In pp and in Pb–Pb collisions resonances are measured in one unit of rapidity $|y| < 0.5$ in the centre-of-mass reference frame, while in p–Pb the rapidity range is restricted to $-0.5 < y < 0$, in order to ensure the best detector acceptance with the shifted centre-of-mass of the system. The position of the primary vertex is estimated using the tracks reconstructed in the Inner Tracking System (ITS) and in the Time Projection Chamber (TPC) and its component along the beam axis is required to be within 10 cm from the centre of the ALICE detector. A detailed review of the ALICE detector and its particle identification capabilities can be found in [3, 18]. The V0A and V0C detectors, two forward-rapidity scintillator hodoscopes, were used for event triggering and beam-gas rejection. To avoid auto-correlation bias the event class selection is based on the total charge deposited in the V0A and V0C detectors and for each event class the value of $\langle dN_{ch}/d\eta \rangle$ is estimated as the average number of primary charged tracks at mid-rapidity.

The $\rho(770)^0$, $K^*(892)^0$ and $\phi(1020)$ mesons (hereafter referred to as ρ^0, K^{*0}, ϕ) and the $\Sigma(1385)^\pm$ and $\Xi(1530)^0$ baryons (hereafter referred to as $\Sigma^{*\pm}$ and Ξ^{*0}) and their anti-particles are reconstructed by the invariant mass spectrum of their hadronic decays products ($\rho^0 \longrightarrow \pi^+ \pi^-$, $K^{*0} \longrightarrow \pi^\pm K^\pm$, $\phi \longrightarrow K^+K^-$, $\Sigma^{*\pm} \longrightarrow \Lambda\pi^\pm$, and $\Xi^{*0} \longrightarrow \Xi^-\pi^+$). Identification of pions and kaons is carried out using the measurement of the specific energy loss (dE/dx) in the TPC. The TPC dE/dx measurement allows pions to be separated from kaons for momenta up to $p \sim 0.7$ GeV/c, while the proton/antiproton band starts to overlap with the pion/kaon band at $p \sim 1$ GeV/c. An improvement in the significance of the signal has been achieved using the information from the Time-Of-Flight (TOF) detector for tracks for which it is available. The TOF allows pions and kaons to be unambiguously identified up to $p \sim$ 1.5-2.0 GeV/c. The two mesons can be distinguished from (anti)protons up to $p \sim 2.5$ GeV/c. For the baryonic resonances, the intermediate decay daughters Λ and Ξ^- are identified through selection based on their decay topologies.

Combinatorial backgrounds are estimated using either like-charge pairs or event mixing. They are subtracted from the distribution of the unlike-charge pairs for the different p_T and centrality/multiplicity intervals. The resulting invariant mass distributions are then fitted with a relativistic Breit-Wigner (K^{*0}, $\Sigma^{*\pm}$) or a Voigtian (ϕ, Ξ^{*0}) function added to a first or second order polynomial to describe the residual background (see [4–8] for further details). The background-subtracted invariant mass distributions obtained for π^++π^- pairs with p_T in the range 0.5-1.0 GeV/c in minimum bias pp collisions at $\sqrt{s}$ =2.76 TeV and in Pb–Pb collisions at $\sqrt{s_{NN}}$ = 2.76 TeV in the centrality range 0-20% and 60-80% are shown in the three panels of Fig. 1. The black curves in this figure represent the fit result. The fit function is given by a cocktail including a smooth function to describe a continuum residual background (dashed red curves in Fig. 1) plus peaks accounting for the contribution of the K^0_S, K^{*0}, 2- and 3-body decay of $\omega(782)$, $f_0(980)$ and $f_2(1270)$. The shape of the ρ^0 peak is described by the product of a relativistic p-wave Breit-Wigner function, a phase-space factor, a mass-dependent reconstruction efficiency, and a Söding interference term [19].

3 Resonance results in pp, p–Pb and Pb–Pb collisions

The procedure used to estimate the resonanance p_T spectrum has been extensively explained in [4–8]. To extract the particle yields and the $\langle p_T \rangle$, the spectra are fitted using a Lévy-Tsallis parameterization [20] (for pp and p–Pb collisions) or Boltzmann-Gibbs blast-wave functions [21] (for Pb–Pb collisions). To extract the dN/dy and the $\langle p_T \rangle$ the measured p_T distributions are integrated, while the fits are used to estimate the resonance yield at low and high p_T, where no signal could be measured. It may be noted that the extrapolated fraction of the total yield for the K^{*0} is lower than 0.1%.

The p_T spectra and the yields of the K^{*0} and ϕ have been measured in pp, p–Pb and Pb–Pb collisions at the various energies studied at LHC both in inclusive and in different multiplicity or centrality intervals. Similar information has been obtained for the ρ^0 in pp and Pb–Pb collisions at $\sqrt{s_{NN}}$ = 2.76 TeV and for the $\Sigma^{*\pm}$ and Ξ^{*0} in pp collisions at $\sqrt{s}$ = 7 TeV and in p–Pb collisions at $\sqrt{s_{NN}}$ = 5.02 TeV.

Using Run 2 data the K^{*0} and ϕ p_T spectra have been measured in inelastic pp collisions at $\sqrt{s}$ = 13 TeV. The

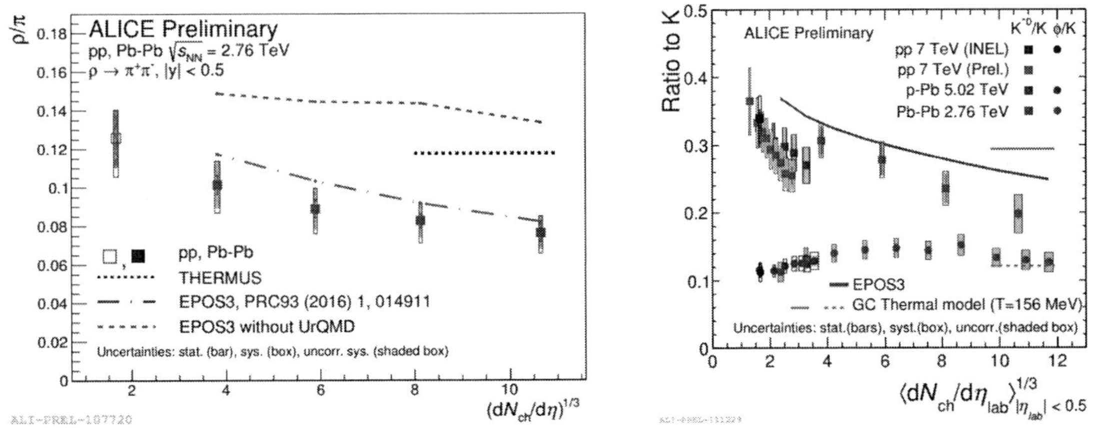

Figure 2. Ratios $\rho^0/\langle\pi^\pm\rangle$ (left panel), K^{*0}/K and ϕ/K (right panel) as a function of the cubic root of the charged particle multiplicity density $dN_{ch}/d\eta$ for various collision systems. For central collisions the values predicted by a grand-canonical thermal model are also shown [23]. The behaviour estimated by EPOS model [2] for K^{*0}/K ratio (continuous blue line) and $\rho^0/\langle\pi^\pm\rangle$ ratio (dashed blue line) are shown. The green dashed line in left panel represents the values of the $\rho^0/\langle\pi^\pm\rangle$ ratio estimated by EPOS if no final state interaction are taken into account.

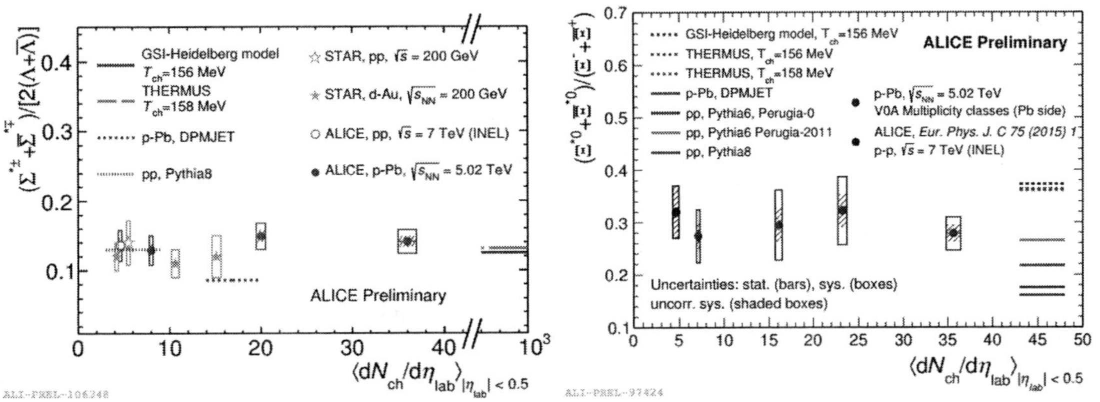

Figure 3. Ratios $(\Sigma^{*\pm}+\overline{\Sigma}^{*\mp})/2(\Lambda+\overline{\Lambda})$ (left panel) and $(\Xi^{*0}+\overline{\Xi}^{*0})/(\Xi^-+\overline{\Xi}^+)$ (right panel) measured in pp [8, 37], d–Au [38] and p–Pb collisions as a function of average charged particle density measured at mid-rapidity. Statistical uncertainties (bars) are shown together with total systematic uncertainties (hollow boxes) and systematic uncertainties uncorrelated across multiplicity (shaded boxes). Some model predictions are also shown [24–27] as lines at their appropriate abscissa, respectively.

comparison of the p_T-integrated K^{*0}/K and ϕ/K ratios obtained at this energy with the existing data show no energy dependence through 2-3 orders of magnitude in the collision energy.

In order to check the presence of a suppression in the production of the resonances and to study whether the strength of the suppression is related to the system size, the ratios of the p_T-integrated particle yields $\rho^0/\langle\pi^\pm\rangle$, K^{*0}/K and ϕ/K have been reported as a function of the cubic root of the charged particle multiplicity density $\langle dN_{ch}/d\eta\rangle^{1/3}$, for pp, p–Pb and Pb–Pb collisions, respectively, at $\sqrt{s}$ = 7 TeV and $\sqrt{s_{NN}}$ = 5.02 and 2.76 TeV (Fig. 2). A centrality-dependent suppression is clearly observed for $\rho^0/\langle\pi^\pm\rangle$ and K^{*0}/K in Pb–Pb collisions. In particular in central collisions the measured ratios are about 60% of the estimate of a grand-canonical thermal model [23], which does not include re-scattering effects. The observed suppression may be related to the pion rescattering mechanism $\sigma(\pi,\pi)$, which destroys the pion-pion and pion-kaon correlations. It is interesting to note that the behaviour of the ratios is at least qualitatively reproduced by calculations using the EPOS model [2], which takes the regeneration and rescattering effects of the resonance decay particles in the hadronic phase explicitly into account by UrQMD [22] (see blue curves in left and right panels of Fig. 2). On the contrary the ϕ, which lives 10 times longer than the K^{*0} and 35 times longer than the ρ^0, decays predominantly after the end of the hadronic phase and its yield should not be affected by regeneration and re-scattering effects. This seems to be

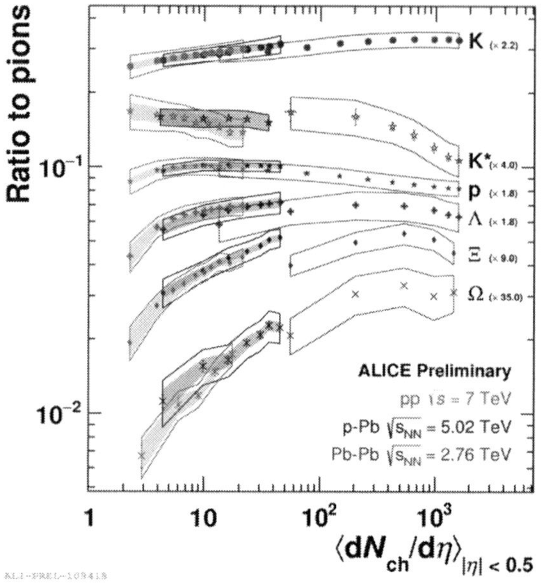

Figure 4. (color online) p_T-integrated yield ratios of K, K^{*0}, p, Λ, Ξ^-, Ω^- to pion as a function of average charged particle multiplicity at mid-rapidity in different collision systems: pp collisions at $\sqrt{s}$ = 7 TeV [17] (green symbols), p–Pb collisions [14, 15] at $\sqrt{s_{NN}}$ = 5.02 TeV (blue symbols) and Pb–Pb at $\sqrt{s_{NN}}$ = 2.76 TeV [33, 34] (red symbols). The hollow bands indicate the total systematic uncertainties, while the shaded ones represent the systematic uncertainties uncorrelated across multiplicity. These are shown only for pp and p–Pb collisions.

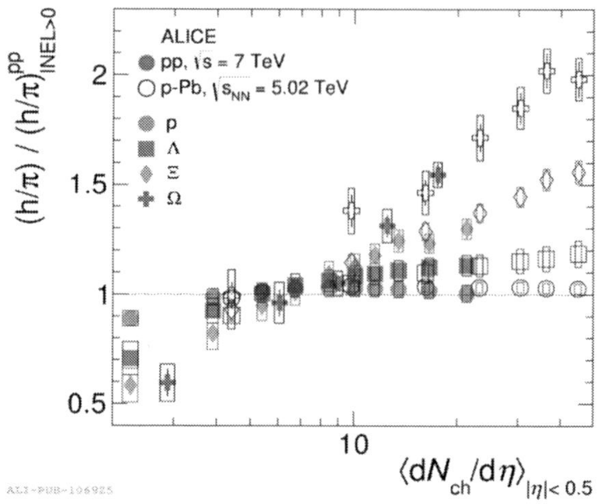

Figure 5. (color online) Particle yield ratios to pions of strange and multi-strange baryons and protons normalised to the values measured in the inclusive INEL>0 pp sample as a function of the average charged particle multiplicity, for pp at $\sqrt{s}$ = 7 TeV [17] and p–Pb at $\sqrt{s_{NN}}$ = 5.02 TeV [14, 15].

confirmed by the ϕ/K behavior. In fact it is rather independent of the event multiplicity class and in Pb–Pb collisions the distribution is almost flat and it is consistent with the estimate of a grand-canonical thermal model [23]. Furthermore, an apparent multiplicity-dependent suppression of the K^{*0}/K ratio has been observed in pp and p–Pb collisions, which may be an indication of the presence of a hadron-gas phase in high-multiplicity pp and p–Pb collisions.

In Fig. 3 the integrated particle ratios of excited hyperon to stable one as $(\Sigma^{*\pm}+\overline{\Sigma}^{*\mp})/2(\Lambda+\overline{\Lambda})$ (left panel) and $(\Xi^{*0}+\overline{\Xi}^{*0})/(\Xi^-+\overline{\Xi}^+)$ (right panel) measured in pp at $\sqrt{s}$ = 7 TeV [8] and p–Pb at $\sqrt{s_{NN}}$ = 5.02 TeV collisions are reported as a function of the average charged particle density and compared to results obtained from STAR in pp and d–Au collisions at $\sqrt{s_{NN}}$ = 200 GeV [38]. No multiplicity dependence is observed for these ratios in p–Pb collisions and they are consistent with the values measured in pp and d–Au collisions. The observed constant behaviour of the yield ratios of excited to ground-state hyperons with same strangeness content may indicate that neither regeneration nor re-scattering dominates with increasing collision system size, even for $\Sigma^{*\pm}$, which has a shorter lifetime than Ξ^{*0} by a factor of 4. Furthermore the constant behaviour of $\Sigma^{*\pm}/\Lambda$ ratio is in contrast with the apparent decrease observed for the K^{*0}/K ratio in a similar multiplicity range, considering the similarly short lifetimes of $\Sigma^{*\pm}$ and K^{*0}.

While the $\Sigma^{*\pm}/\Lambda$ ratios are consistent with PYTHIA8 [24] and thermal model predictions [26, 27] in pp and in p–Pb collisions, DPMJET [25] underpredicts the ratios obtained in p–Pb collisions. The same models are not able to estimate the Ξ^{*0}/Ξ^-values both in pp and in p–Pb collisions. In particular, PYTHIA8 [24] and DPMJET [25] estimates are lower while thermal model predictions [26, 27] are higher. For small systems a canonical treatment is a priori required to take into account exact strangeness conservation. However, for the chosen ratios the canonical corrections are identical for numerator and denominator, due to the same strangeness quantum number. Therefore, the grand canonical values can be used.

4 Strangeness production in pp, p–Pb and Pb–Pb

Strangeness enhancement was one of the first proposed QGP signatures [28]. At LHC (as SPS [29–31] and RHIC [32] energies) a clear increase of strangeness production from pp to Pb-Pb was observed [33, 34]. This strangeness enhancement is expected to be more pronounced for multi-strange baryons, and this was indeed observed in collisions of heavy nuclei. The abundances of strange particles in heavy-ion collisions are compatible with those of a hadron gas in thermal and chemical equilibrium and can be described using a grand canonical statistical model [23, 26]. The p–Pb results [14, 15] as a function of the average charged particle multiplicity density at mid-rapidity are consistent with pp results at low multiplicity and with central values of Pb–Pb collisions at high multiplicity.

 DOI: 10.1051/epjconf/201714201004

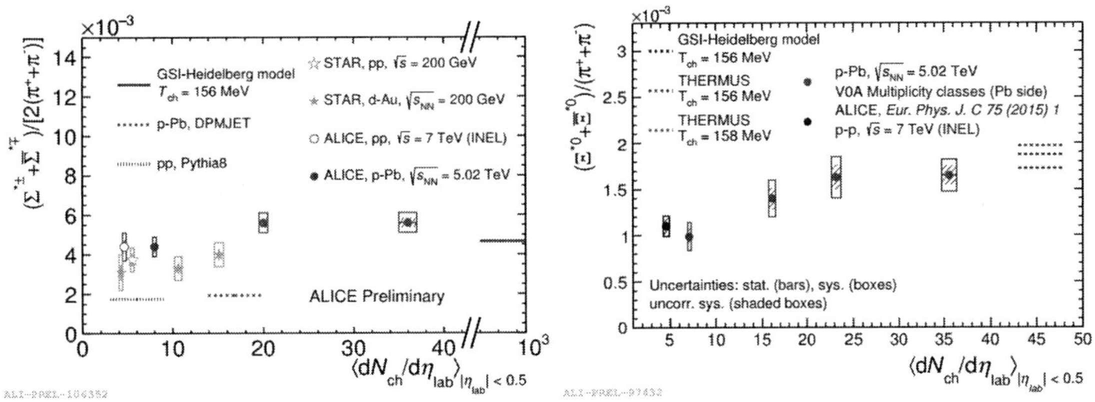

Figure 6. Ratios $(\Sigma^{*\pm}+\overline{\Sigma}^{*\mp})/2(\pi^+ + \pi^-)$ (left panel) and $(\Xi^{*0}+\overline{\Xi}^{*0})/(\pi^+ + \pi^-)$ (right panel) measured in pp [8, 37], d–Au [38] and p–Pb as a function of average charged particle density measured at mid-rapidity. Yields of pions for p–Pb collisions are from [14]. Statistical uncertainties (bars) are shown together with total systematic uncertainties (hollow boxes) and systematic uncertainties uncorrelated across multiplicity (shaded boxes). Some model predictions are also shown [24–27] as lines at their appropriate abscissa, respectively.

In Fig. 4 the yields of K, K^{*0}, Λ, Ξ^-, Ω^- relative to the pion yield obtained in pp, p–Pb and Pb–Pb collisions are reported as a function of $\langle dN_{ch}/d\eta \rangle$ measured at mid-rapidity. In pp collisions [17] strange to non-strange integrated particle ratios increase with multiplicity, showing significant enhancement of strange and multi-strange particle production and following the same trend observed in p–Pb collisions despite differences in the initial state and reaching values similar to those observed in Pb–Pb collisions. It is worth noting that the K^{*0}/π ratio in pp collisions at $\sqrt{s}$ = 7 TeV shows hint of the presence of a decreasing trend with multiplicity although not significant given the present uncertainties (Fig. 4). In Fig. 5 the pp and p–Pb hadron-to-pion ratios normalised by the multiplicity-integrated ratio in pp, corresponding to the INEL>0 event class, are reported as a function of multiplicity. While the p/π ratio is practically constant, the increase of strange and multi-strange particles follows a clear hierarchy with strangeness content. Monte Carlo models commonly used for pp collisions at LHC are not able to describe satisfactorily the behaviour of strange and multi-strange particles (see for example Fig. 2 in [17]). While PYTHIA8 [24] doesn't reproduce the observed enhancement with multiplicity in pp collisions, DIPSY [35] and EPOS LHC [36] predict a rising trend with multiplicity, but fail to reproduce the rise correctly. Moreover the predicted proton-to-pion ratio is multiplicity-dependent contrary to the measured one (see Fig. 4)

In Fig. 6 the integrated particle ratios of excited hyperon to pion as $(\Sigma^{*\pm}+\overline{\Sigma}^{*\mp})/2(\pi^+ + \pi^-)$ (left panel) and $(\Xi^{*0}+\overline{\Xi}^{*0})/(\pi^+ + \pi^-)$ (right panel) measured in pp at $\sqrt{s}$ = 7 TeV and p–Pb at $\sqrt{s_{NN}}$ = 5.02 TeV collisions are reported as a function of the average charged particle density and compared to results obtained from STAR in pp and d–Au collisions at $\sqrt{s_{NN}}$ = 200 GeV. An increase of the relative strangeness production with the multiplicity similar to the one observed for the ratio of hyperon to pion (as Λ/π and Ξ^-/π) is observed. The ratios increase gradually and approach thermal model values [26, 27] in the highest multiplicity event class. The constant behaviour of the $\Sigma^{*\pm}/\Lambda$ and Ξ^{*0}/Ξ^- ratios indicates that the observed strangeness enhancement depends predominantly on the hyperon strangeness content rather than on their mass. QCD-inspired predictions clearly underestimate the observed $\Sigma^{*\pm}/\Lambda$ ratios for both pp [24] and p–Pb [25] collisions.

5 Conclusions

In this paper the latest results obtained from ALICE on hadronic resonances have been presented. In particular results on $K^*(892)^0$ and $\phi(1020)$ in pp, p–Pb and Pb–Pb collisions at the various energies studied at LHC both in inclusive and in different multiplicity or centrality intervals, on the $\rho(770)^0$ measured in pp and Pb–Pb collisions at $\sqrt{s_{NN}}$ = 2.76 TeV and on the $\Sigma(1385)^\pm$ and $\Xi(1530)^0$ measured in pp collisions at $\sqrt{s}$ = 7 TeV and in p–Pb collisions at $\sqrt{s_{NN}}$ = 5.02 TeV have been shown.

The ALICE Collaboration has measured a centrality-dependent suppression of ρ^0/π and K^{*0}/K ratios in Pb–Pb collisions, which can be described by EPOS calculations with UrQMD to describe hadronic interactions in the hadronic phase [2]. A multiplicity-dependent suppression of the K^{*0}/K ratio is also observed in pp and p–Pb collisions. In contrast, the ϕ/K ratio does not exhibit a centrality- or multiplicity-dependent suppression. This behaviour may be due to the loss of the K^{*0} and ρ^0 signal caused by re-scattering of decay products in the hadronic phase, while the ϕ yield is not affected because of its long lifetime as it decays mostly after the hadronic phase.

From the K^{*0}/K ratio in central Pb–Pb collisions it is possible to estimate a lower limit of hadronic phase lifetime (τ >2 fm/c) [6] using a thermodynamical model [1] which takes into account only the loss of resonances due to re-scattering effects in the hadronic medium. An hadronic lifetime of about 9 fm/c for most central Pb–Pb collisions is estimated by EPOS followed by a hadronic afterburner phase modeled via UrQMD [2].

The $\Sigma^{*\pm}/\Lambda$ and Ξ^{*0}/Ξ^{-} ratios are not observed to depend on system size or activity in pp and p–Pb collisions, which could mean that the effects of re-scattering and regeneration are not so important for the charged particle densities reached in these colliding systems.

The results in high multiplicity pp collisions have revealed interesting features that are similar to those observed in p–Pb and Pb–Pb collisions. While these features are usually interpreted in Pb–Pb collisions as due to collectivity, their origin in smaller systems is still to be fully understood. An enhanced production of strange and multi-strange particles in high-multiplicity pp collisions with respect to inelastic events has been observed. The magnitude of this strangeness enhancement increases with the event activity, quantified by $\langle dN_{ch}/d\eta \rangle$, and with hadron strangeness. No enhancement is observed for particles with no strange quark content, demonstrating that the observed effect is strangeness rather than mass related. The multiplicity dependence of strangeness production is strikingly similar in pp and p–Pb, and approaches values similar to those obtained in Pb–Pb collisions. None of the current MC models are successful at fully describing these observations, suggesting that further developments are needed for a complete microscopic understanding of strangeness production and indicating the presence of a phenomenon novel in high-multiplicity pp collisions.

References

[1] G. Torrieri and J. Rafelski, J. Phys. G **28**, 1911 (2002); C. Markert *et al.*, aXiv:hep:ph/0206260v2 (2002)

[2] A. G. Knospe *et al.*, Phys. Rev. C **93**, 014911 (2016)

[3] K. Aamodt *et al.* (ALICE Coll.), J. Instrum. **3**, S08002 (2008)

[4] B. Abelev *et al.* (ALICE Coll.), Eur. Phys. J. C **71**, 1594 (2011)

[5] B. Abelev *et al.* (ALICE Coll.), Eur. Phys, J. C **72**, 2183 (2012)

[6] B. Abelev *et al.* (ALICE Coll.), Phys. Rev. C **91**,024609 (2015)

[7] J. Adam *et al.* (ALICE Coll.), Eur. Phys. J. C **76**, 245 (2016)

[8] B. Abelev *et al.* (ALICE Coll.), Eur. Phys. J. C **75**, 1 (2015)

[9] B. Abelev *et al.* (ALICE Coll.), Phys. Rev. Lett. **110**, 032301 (2013)

[10] F. Gelis *et al.*, Annu. Rev. Nucl. Part. Sci. **60**, 463 (2010)

[11] B. Abelev *et al.* (ALICE Coll.), Phys. Lett. B **719**, 29 (2013); B. Abelev *et al.* (ALICE Coll.), Phys. Lett. B **726**, 164 (2013)

[12] G. Aad *et al.* (ATLAS Coll.), Phys. Rev. Lett. **110**, 182302 (2013); G. Aad *et al.* (ATLAS Coll.), Phys. Lett. B **725**, 60 (2013)

[13] V. Khachatryan *et al.* (CMS Coll.), JHEP **09**, 091 (2010); S. Chatrchyan *et al.* (CMS Coll.), Phys. Lett. B **718**, 795 (2013); V. Khachatryan *et al.* (CMS Coll.), Phys. Lett. B **765**, 193 (2017)

[14] B. Abelev *et al.* (ALICE Coll.), Phys. Lett. B **728**, 25 (2014)

[15] J. Adam *et al.* (ALICE Coll.), Phys. Lett. B **758** (2016) 389

[16] V. Khachatryan *et al.* (CMS Coll.), arXiv:1605.06699

[17] J. Adam *et al.* (ALICE Coll.), arXiv:1606.07424 (2016)

[18] B. Abelev *et al.* (ALICE Coll.), Int. J. Mod. Phys. A **29**, 1430044 (2014)

[19] A. Beddall *et al.*, Acta Phys. Polon. B **39**, 173 (2008)

[20] C. Tsallis, J. Stat. Phys. **52**, 479 (1988)

[21] E. Schnedermann *et al.*, Phys. Rev. C **48**, 2462 (1993)

[22] S. A. Bass *et al.*, Prog. Part. Nucl. Phys. **41**, 255 (1998); M. Bleicher *et al.*, J. Phys. G **25**, 1859 (1999)

[23] J. Stachel *et al.*, J. Phys. Conf. Ser. **509**, 012019 (2014)

[24] T. Sjostrand *et al.*, Comput. Phys. Commun. **178**, 852 (2008)

[25] S. Roesler *et al.*, *Conference Proceedings MC2000* (Lisbon, Portugal, October 23-26, 2000) 1033–1038, hep-ph/0012252

[26] A. Andronic *et al.*, Phys. Lett. B **673**, 1420145 (2009)

[27] S. Wheaton *et al.*, Comp. Phys. Commun. **180**, 84 (2009)

[28] J. Rafelski *et al.*, Phys. Rev. Lett. **48**, 1066 (1982); P. Koch *et al.*, Phys. Rep. **142**, 167 (1986)

[29] E. Andersen *et al.*, Phys. Lett. B **433**, 209 (1998); E. Andersen *et al.* (WA97 Coll.), Phys. Lett. B **449**, 401 (1999)

[30] S. V. Afanasiev *et al.* (NA49 Coll.), Phys. Lett. B **538**, 275 (2002); T. Anticic *et al.* (NA49 Coll.), Phys. Rev. Lett. **93**, 022302 (2004)

[31] F. Antinori *et al.* (NA57 Coll.), Phys. Lett. B **595**, 68 (2004); F. Antinori *et al.* (NA57 Coll.), J. Phys. G **37**, 045105 (2010)

[32] J. Adams *et al.* (STAR Coll.), Phys. Rev. Lett. **92**, 182301 (2004); J. Adams *et al.* (STAR Coll.), Phys. Rev. Lett. **98**, 062301 (2007); B. I. Abelev *et al.* (STAR Coll.), Phys. Rev. C **77**, 044908 (2008)

[33] B. Abelev *et al.* (ALICE Coll.), Phys. Lett. B **728**, 216 (2014) [Erratum: Phys. Lett. B **734**, 409 (2014)]

[34] B. Abelev *et al.* (ALICE Coll.), Phys. Rev. Lett. **111**, 222301 (2013)

[35] C. Bierlich *et al.*, JHEP **03**, 148 (2015)

[36] T. Pierog *et al.*, Phys. Rev. C **92**, 034906 (2015)

[37] J. Adam *et al.* (ALICE Coll.), Eur. Phys. J. C **75**, 226 (2015)

[38] B. I. Abelev *et al.* (STAR Coll.), Phys. Rev. C **78**, 044906 (2008)

EPJ Web of Conferences **142**, 01005 (2017) DOI: 10.1051/epjconf/201714201005
Advances in Dark Matter and Particle Physics

Light Dark Matter search in a beam-dump experiment: BDX at Jefferson Lab

Mariangela Bondí[1,a] for the BDX Collaboration
[1] INFN - Catania, Via S. Sofia 64, 95125 Catania (Italy)

Abstract. The Beam Dump eXperiment (BDX) aims to investigate the existence of light Dark Matter (DM) particles in the MeV-GeV mass range, theoretically well motivated but still barely explored. In this mass range BDX will be sensitive to large regions of DM parameter space, exceeding the discovery potential of existing and planned experiments by up to two orders of magnitude.
BDX is an e^--beam thick-target experiment, expected to run at Jefferson Lab, downstream of the Hall A beam dump, receiving up to 10^{22} electron-on-target (EOT). The detector is made of two main components: an electromagnetic calorimeter (Ecal), and a veto detector used to suppress the background. BDX will be sensitive to elastic and inelastic DM-electron scattering at the level of 10 counts per year, reaching the limit of the irreducible background due to beam-produced neutrinos. The expected signature of the DM interaction in the Ecal is a ~ GeV electromagnetic shower paired with a null activity in the surrounding active veto counters.
In order to evaluate the background contributions two approaches were used: the cosmogenic background was extrapolated from results obtained with a prototype running at INFN-LNS (Italy), while the beam- related background was evaluated by GEANT4 Monte Carlo simulations.
This paper describes the experimental setup and the results of the background studies along with the status and prospectives of the experiment.

1 Introduction

The search for Dark Matter (DM) is one of the hottest topics of modern physics. Despite the copious number of astrophysical and cosmological observations proving its existence, its elementary properties remain largely elusive [1]. Although the Standard Model (SM) contains several neutral particles, none serves as a viable dark matter candidate. Thus the existence of DM is an overwhelming evidence of physics beyond the SM and uncovering its particles identity is top priority in fundamental physics.
Various extensions of the SM propose different candidates for the role of DM particles, like the Weakly Interacting Massive Particles (WIMPs) (~10 GeV - 10 TeV mass range) expected to weakly interact with SM particles [2]. Due to the lack of evidence for WIMPs so for from LHC and direct DM searches, other well motivated models of DM gained recently the interest of the physics community. These models take into account DM particles in MeV-GeV mass range (Light Dark Matter (LDM)), whose existence is motivated by both empirical and theoretical considerations. On the theory side, simple SM extensions can accommodate DM-SM interactions that yield the observed DM cosmological abundance. On the empirical side, such models feature particles that explain the currently discrepant value of the muon's anomalous magnetic moment and resolve anomalies in astrophysical observations, while simultaneously evading cosmological and direct-production constraints.
By and large the most viable LDM model to probe is that where LDM interacts with the SM through a "dark photon" A'. Dark photon can mix with the ordinary photon through kinetic mixing, which induces their weak coupling to electric charge, ϵe, where $\epsilon<10^{-3}$ [3][4]. The minimal parameter space of vector-mediated LDM is characterized by ϵ, the coupling α_D of the A' to the LDM particle χ and two masses $m_{A'}$ and m_χ.
In this scenario, LDM searches at collider are limited by the huge SM backgrounds producing missing energy.
Instead, high intensity ~ GeV electron-beam fixed-target experiments offer large sensitivity to a broad class of dark sector scenarios that feature particles in the elusive MeV-GeV mass range [5]. Electron beam-dump experiments are complementary to dedicated efforts at proton beam facilities, have comparable DM scattering yield, can run on a smaller scale than proton-beam counterparts, and benefit of reduced beam-related backgrounds.
A generic sketch of a beam dump experiment is shown in Fig.1. The interaction between the energetic electrons and the atoms of the dump leads to the production of Dark photons through a Bremsstrahlung-like radiative process. Subsequently, the A' decays to LDM particle-antiparticle pairs ($\chi\bar{\chi}$) (see Fig.2a), which unaltered pass through the dump and the shielding between the dump and the detector. A fraction of them will then scatter off nucleons, nuclei, or electrons in the detector volume, giving rise to a detectable signal. For e^--χ interaction (see Fig.2b), since

[a] e-mail: mariangela.bondi@ct.infn.it

 DOI: 10.1051/epjconf/201714201005

$m_e << m_\chi$, the typically scattered electron carries GeV-scale energy producing an electromagnetic shower in the detector. For χ-N interactions the much higher mass of nucleons result in a typically recoil energy of the order of ~ MeV.

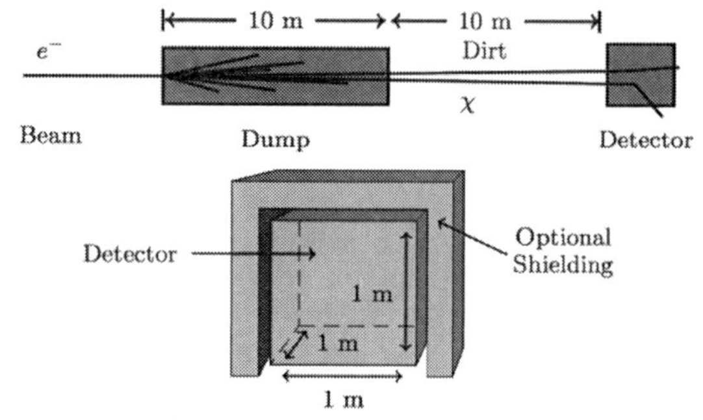

Figure 1. Schematic of the experimental setup. A high-intensity multi-GeV electron beam impinging on a beam-dump produces a secondary beam of dark sector states. In the basic setup, a small detector is placed downstream the beam-dump so that muons and energetic neutrons are entirely ranged out.

The considerable sensitivity of beam-dump experiments to LDM is underscored by the reach of running neutrino experiments [6–9] . For example, LSND measurement of meson production and decay [10] can be used to derive the most stringent constraints to date on the parameter space for invisibly-decaying dark mediators that couple to both baryons and leptons [9].

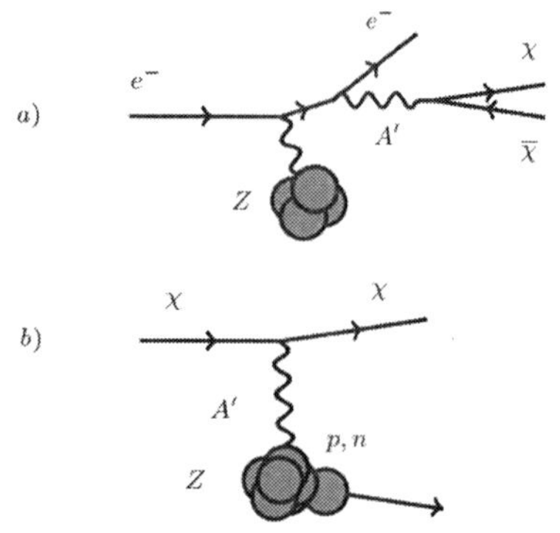

Figure 2. a) $\chi\bar{\chi}$ pair production in electron-nucleus collisions via the Cabibbo-Parisi radiative process (with A' on - off-shell) and b) χ scattering off an electron in the detectors

The Beam Dump experiment (BDX) at JeffersonLab [11] aims to produce and detect LDM, assuming valid the above cited theoretical paradigm. Tacking advantage of the high intensity ~ GeV electron beam available at JLab, BDX has the unique capability of dramatically improve sensitivity to MeV - GeV DM, extending well beyond the reach of existing and proposed experiments, reaching the limit of the un-reducible background due to beam-produced neutrinos.

2 BDX at Jefferson Lab

2.1 Experimental setup

BDX at Jefferson Lab will make use of the CEBAF (Continuous Electron Beam Accelerator Facility) high-intensity electron beam, impinging on the JLab Hall-A beam dump which is enclosed in a concrete tunnel at the end of the beam transport line. The Hall-A at JLab can receive from CEBAF a 11 GeV electron beam with a maximum current of about 65μA. The possibility of running to the maximum energy available at Jlab together with a sizeable current, that allows to collect 10^{22} electron-on-target (EOT) in 285 days, in full parasitic runs, makes Hall-A the optimal choice.

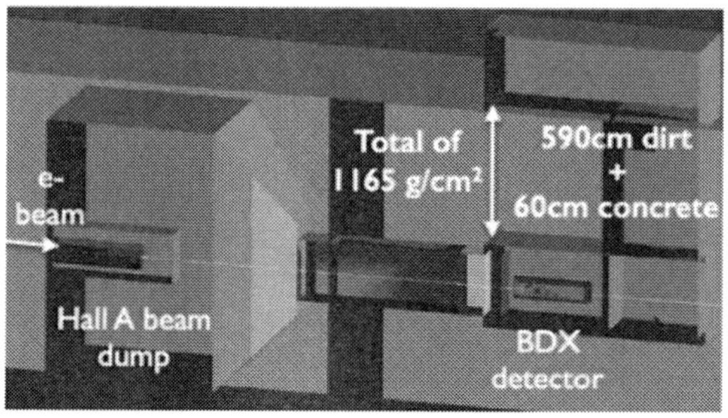

Figure 3. The BDX experimental setup.

The BDX detector will be placed along the LDM beam trajectory, in a new underground facility located 20 meters downstream of the Hall-A dump (Fig. 3). The extremely low signal rate expected, due to the two-step processes involving weak mixing between the SM photon and A', makes background rejection the critical issue of BDX. To range out all the SM beam products, except neutrinos, the space between the dump and the detector will be filled with dirt plus ~ 7 m of concrete and ~ 7 m of iron; to reduce the beam-unrelated background, mainly due to cosmic neutrons and muons, the detector will be placed in a bunker 8 m underground with 10 meters of water equivalent (mwe) overburden. Moreover, the detector will include an active veto and passive shielding.
The detection of the low-energy scattered protons is a secondary goal of the experiment because it provides an alternative probe of LDM production. However, background at ~ MeV energy scale expected to limit the sensitivity in this channel.

2.2 Detector design

The BDX detector is made of two main components: an electromagnetic calorimeter used to detect signals pro-

DOI: 10.1051/epjconf/201714201005

duced by the interacting DM particles, and a veto detector used to reduce the cosmogenic background (see Fig.4 for a sketch of the detector). The calorimeter consists of 8 modules of 10 x 10 CsI(Tl) crystals each, arranged with the long size along the beam direction. The average size of each crystal is (4.7 x 5.4 x 32.5) cm^3. This arrangement has a cross section of 50x55 cm^2 for a total length of 295 cm. SiPM are used as light-readout sensors.
The BDX EM calorimeter is operated inside two hermetic layers of active veto counters, made of plastic scintillators: the outermost called Outer Veto (OV) and the innermost Inner Veto (IV). The OV consists of 2cm-thick plastic scintillators coupled to PMT through plexiglass light guides or wavelength shifting plastic. Due to the large size requested to cover the whole calorimeter and to preserve the possibility of changing the geometry, the OV is segmented in many different paddles. In particular, the top and the bottom layers are divided in two parts while the lateral sides are made by 11 paddles per side.

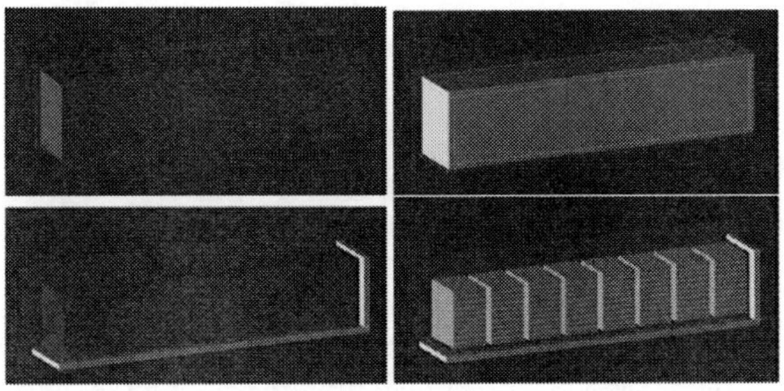

Figure 4. The BDX detector. The Outer Veto is shown in green, the Inner Veto is blue and the lead vault in gray. Crystals arranged in 8 blocks of 10x10, are shown in light blue. Each crystal is read by a SiPM directly attached to the front face (not shown).

The IV consists of a hermetic box comprised of six 1-cm thick plastic scintillator paddles. The lateral paddles are read on one side by 4 SiPMs, directly glued to the scintillator surface. Grooves on the surface host two 1 mm WLS fibers to convey the scintillation light to the SiPM. The upstream and downstream covers have a spiral-like groove. The WLS fibers are read by one SiPM per paddle located on the scintillator surface.
A 5 cm thick layer of lead is placed between the two active vetoes. The "sandwich" configuration, with the lead between the IV and the OV, has been chosen to detect with the OV the cosmic muons that may produce spallation in the lead. It also acts as a radiator for high energy photons, allowing the detection of charged secondaries in the IV.
This detector concept has been optimized and validated in various campaign of measurements at INFN - Sezione di Catania and Laboratori Nazionali del Sud (LNS) with two prototypes described in Section 3 and 4.

2.3 Expected Background

2.3.1 Beam-Related Background

The beam-related background for the electron recoil measurements was studied through MC simulations, considering a detection threshold of 300 MeV. The interaction of the primary 11 GeV electron beam in the dump was simulated with a GEANT4-based code and the flux of secondaries was studied as a function of the distance from the dump. For this purpose, the flux of particles was sampled over planes perpendicular to the beam direction. Results are shown in Fig. 5: Z = 0 corresponds to the upstream end of the aluminum dump, while the detector is located at Z = 20 m. The only particles that are found to reach the detector area are ν and $\bar{\nu}$, predominantly from pion and muon decays at rest. In fact, the number of neutrons and muons decreases rapidly as the distance from the dump increases and, performing an exponential extrapolation of the projected counts, we estimated that no neutrons and muons would reach the detector with energy $\geq$ 300 MeV.

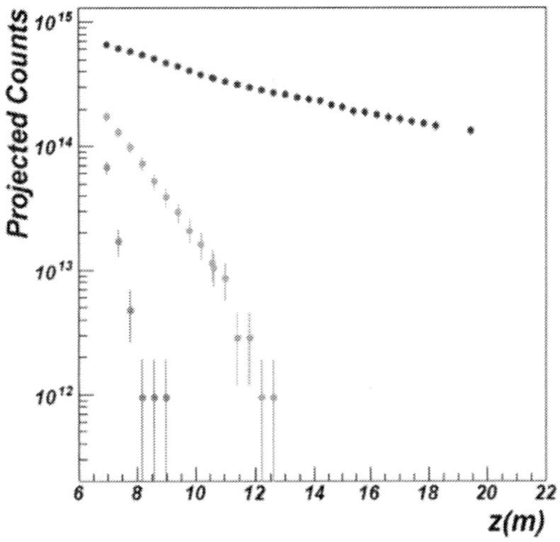

Figure 5. Number of secondary particles produced by the interaction of the 11 GeV electrons as a function of the distance from the upstream end of the dump. The reported counts are normalized to 10^2 2 EOT. The threshold on the particles energy is 500 MeV. The different colors correspond to different particle types: neutrinos in red, muons in cyan, neutrons in green and photons in blue.

The neutrino reaching the detector are mostly ν_μ and $\bar{\nu}_\mu$, with only about 3-4% ν_e and $\bar{\nu}_e$. In this energy range, neutrinos interact mostly via charged current, with the dominant processes being quasi-elastic scattering and resonance production on the nucleon, whose cross section is of the order of $\sigma_{\nu N} \sim 10^{-38}$ cm^2 [12]. These neutrinos can scatter in the BDX detector producing a muon or electron that can be detected. Muons with energy of order of hundreds of MeV loose energy via ionization and can be distinguished from the electron recoil signal. The scattered electrons instead, induce electromagnetic showers, resulting in the same signature of the χ-e$^-$ interaction. The result of the simulation indicates that, for an energy threshold of 300 MeV, the expected neutrino background is of order of ~ 10 events for 10^{22} EOT.

2.3.2 Cosmogenic Background

Beam-unrelated background is mainly due to cosmic neutrons, cosmic muons and their decay products, including rare decays of muons producing gamma's between the passive shielding and the active veto. Both direct cosmic flow and secondary particles contribute to the beam-unrelated background rate in the detector.
This rate has been evaluated by extrapolating the results obtained with the BDX prototype in Catania. The similar experimental set-up (including overburden) as well as a full prototype that incorporates all the elements of the BDX detector (active veto's, lead shielding, CsI(Tl) crystal) provide a solid base for a realistic, although conservative, estimate of the expected rates. Details of the experimental conditions and data analysis are reported in the next sections. The extrapolation has been performed by scaling the experimental rates of a single crystal to the 800 crystals comprising the full detector. This is certainly an upper limit on the expected rates since this assumes crystal-to-crystal fully uncorrelated counts, which overestimates the case for χ-e$^-$ scattering. The results show that for energy thresholds higher enough, between 300-350 MeV, the number of cosmogenic background counts, in 285 days, reduces to zero. Thus, by choosing the appropriate energy threshold we could expect to have zero background. However, in order to be conservative, we considered a value of cosmic events $N_{cosmic} = 3$ in the evaluation of the BDX reach and sensitivity.

2.4 BDX Expected Reach

In this Section we present an example of the expected reach of BDX for the χ - e$^-$ channel. Results are reported as upper limit on exclusion plot. More excluding plots for different parameter spaces and/or DM scenarios, together with the detail of the analysis, are reported in [11]. The regions above the lines are excluded since the model, with the chosen parameters, predicts a number of counts greater than two times the sigma of the background. This is true in case of null results. If any excess is observed, a thorough statistical analysis will be necessary to claim a positive result.
In case of no positive observation, the accumulated data would provide very stringent limits on the DM parameter space. In Fig.6 we show the BDX reach in the $\epsilon^2 - m_{A'}$ parameter space (for $m_\chi << m_{A'}$). The gray region is excluded by the re-analysis of old data from non-optimized experiments. The red curves show 3, 10 and 20 event BDX yield projections, including the detection efficiency estimated with MC simulations, for electron scattering with 10^{22} EOT and a 300 MeV recoil energy threshold.
As shown in Fig. 6, BDX at JLAB extends considerably the possible reach with respect to previous experiments, probing wide unexplored DM parameter space regions.

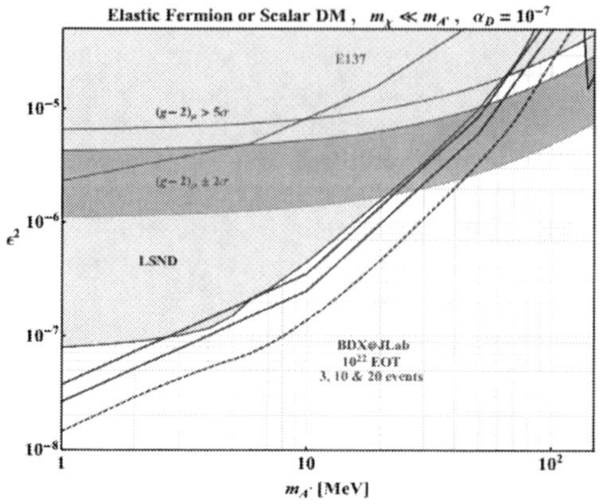

Figure 6. Red curves show 3, 10, 20 event BDX yield projections for χ-e$^-$ scattering for 10^{22} EOT and a 300 MeV recoil energy threshold. Grey regions have already been ruled out by previous experiments.

3 Cosmic measurements campaign in Catania

In this section, we report some details and the results of the dedicated campaign of measurements performed in Catania, Italy (INFN - CT) and Laboratori Nazionali del Sud (INFN - LNS) with first prototype of the BDX detector. These measurements provided information useful for the full detector design and its expected performance in a realistic configuration. We tested the proposed technologies: CsI(Tl) crystals read by SIPMs, plastic scintillators read by PMTs for the OV and plastic scintillator coupled to SIPMs by WLS fibers for the IV; we checked the effect of the lead shielding and the overburden, we measured the single crystal rates as a function of the energy threshold in combination (anti-coincidence) with the veto systems. In order to test and commission whole setup the data were initially taken exposing the prototype to cosmic rays with minimal shielding (15 cm of concrete roof at INFN-CT). Then the prototype was moved into a bunker at LNS with a concrete overburden of 470 cm corresponding to an effective thickness of 1080 g/cm^2 similar to 1165 g/cm^2 expected at JLab.

3.1 The BDX prototype

The first BDX prototype was made by a single CsI(Tl) crystal surrounded on all sides by a layer of veto detectors (IV), a vault of lead bricks, and a second layer of veto detectors forming the OV. The combined use of two charged-particle veto-counter systems allows to compensate for their inefficiencies and better reject background events. Between the inner and the outer veto, the 5cm thick lead vault shields the crystal from radiogenic low energy gammas.

DOI: 10.1051/epjconf/201714201005

Figure 7. The CsI(Tl) crystal inside the first prototype. The two charge-preamplifiers coupled to the SiPMs are visible.

The CsI(Tl) crystal, shown in Fig. 7, is 31cm long and has a trapezoidal shape with a 4.7x4.7cm^2 front-face and a 6x6 cm^2 back-face. Two 3x3mm^2 SiPMs (Hamamatsu S13360-3025CS and S13360-3050CS), with pixel size of 25 and 50 μm, are glued to the crystal front-face and connected each to a custom trans-impedance amplifier.

Figure 8. A picture of the Inner Veto mounted inside the mechanical structure of the prototype (top) and a picture of the top paddle (bottom).

The IV, shown in Fig. 8 is made by plastic scintillators, 1 cm thick, forming a nearly hermetic parallelepiped. Two 35x42 cm^2 EJ200 scintillators are used for the downstream and upstream caps. On each of them a spiral groove hosts a WLS fiber used to collect and transfer the light to a SiPM. Three 35x140 cm^2 EJ200 scintillators form the top, left and right sides of the veto. In this case, the WLS fibers are inserted into four linear grooves running parallel to the long side of the plastic. This solution results in an high detection efficiency (> 99.5%), almost independent on the hit point. Finally, in order to test another possible technology for the IV, the bottom side was made by four bars of extruded plastic scintillators, 8x140 cm^2 each, individually readout by a SiPM coupled to WLS fibers inserted in the middle of each bar. The Outer Veto, shown in Fig.9, is made by 2 cm thick NE110 plastic scintillators. The top (bottom) side is made by two 80x40 cm^2 paddles. Three scintillators of the same area are vertically arranged to cover each of the two lateral sides. A "fish tail" shaped PMMA light guide, glued on one side of the scintillator, directs the light to a 2" photomultiplier tube (Thorn EMI 9954A) optically matched with the light guide trough optical grease. A smaller paddle (56x50 cm^2) forms the upstream (downstream) cap. In this case, light is readout by a 1" Photomultiplier tube (R1924A Hamamatsu) placed in the middle of the plastic surface and directly coupled to it through optical grease. A detection efficiency > 99.5%, uniform over the whole scintillator surface, was measured for each OV paddle for cosmic rays selected by triggering on the coincidence of two small paddles placed above and below the scintillator.

Figure 9. The prototype mounted and cabled at the INFN-Sezione di Catania. OV detectors are visible

Data acquisition is based on VME-VXS JLab fa250 digitizers with 12 bit resolution, 250 MHz sampling rate and 2 μs readout window. The main trigger is generated by a signal over threshold in the CsI(Tl) crystal, namely from the logic OR of the two SIPMs. The output signals are split by a 50 Ohm-50 Ohm divider: one copy is sent to the FADC and the other one to a Leading-Edge Discriminator with thresholds set to 20mV and 50mV for the 25 and 50 μm SiPMs, respectively. These thresholds correspond to about 5 photo electrons (phe) for both SIPMs and to an energy threshold for protons of about 2 MeV.

3.2 Cosmic measurement campaign at LNS - INFN

The primary goal of the LNS campaign was to measure the absolute rates of cosmogenic background in a experimental condition similar to that expected at JLAB. The measurement campaign started in April 2016: the results presented here correspond to about 1 month of data-taking.

The following data-reconstruction procedure was adopted. For the CsI crystal, both waveforms of the SiPM were numerically integrated over a gate of 2 μs to obtain the corresponding charge, that was converted in MeV units by using calibration constants deduced from the cosmic-rays Landau distribution most probable value. For the IV paddles with multiple readout we required, after numerical integration of the waveforms, that at least two SiPM of the same counter collect 2.5 phe each, within a coincidence gate of 100 ns. For the IV counter with a single SiPM, an higher threshold ($\sim$ 12.5 phe) was implemented. For the OV each PMT signal was numerically integrated, and after this a 100 keV energy threshold was applied.
We studied events with the following topologies:

- All events;
- Anti-coincidence with IV;
- Anti-coincidence with OV;
- Anti-coincidence with both IV and OV;

where anti-coincidence means that we selected events with no activity in any of the veto counters.
Results corresponding to the low-energy part of the CsI energy spectrum are shown in Fig. 10.

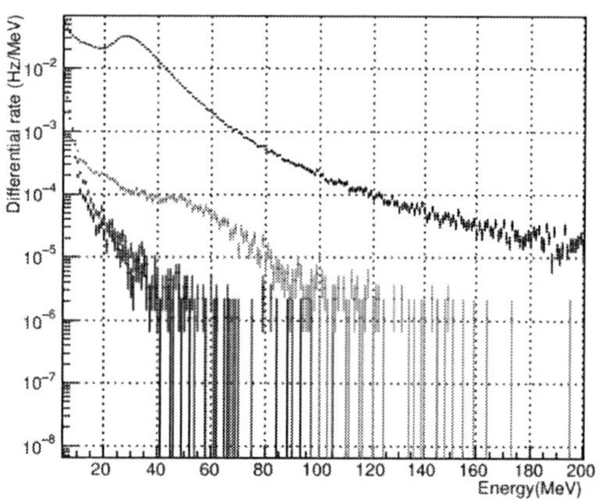

Figure 10. Measured event rate as a function of deposited energy in the crystal. Different colors refer to the different anti-coincidence selections: black -all events, red -anti-coincidence with IV, green - anti-coincidence with OV, blue - anti-coincidence both veto systems.

The integrated rate, requiring the anti-coincidence with both veto systems, for a 10 MeV (20 MeV) threshold, is 1.7 mHz (0.37 mHz).
The analysis of the high-energy part of the spectrum (E>100 MeV) is, instead, more critical, due to the low statistics. To obtain a conservative estimate of the high energy background, we proceeded performing an extrapolation from the lower part of the spectrum (where a non-zero event rate was measured) to the higher regime, trough a Maximum Likelihood fit, using an exponential function. The systematic error has been evaluated by changing the fit range and comparing the obtained results (see Fig. 11 for the differential rate extrapolated for the IV anti-coincidence events).
Results are reported in Tab.1 for the inner-veto anti-coincidence selection and for different energy thresholds.

Table 1. The inner-veto anti-coincidence rate for the BDX-prototype, measured in the LNS configuration, obtained by extrapolating from the low energy part of the spectrum.

Energy threshold	Extrapolated rate
200 MeV	$(3.6 \pm 1.5) \cdot 10^{-8}$ Hz
250 MeV	$(2.9 \pm 1.3) \cdot 10^{-9}$ Hz
300 MeV	$(2.4 \pm 1.1) \cdot 10^{-10}$ Hz
350 MeV	$(1.9 \pm 0.9) \cdot 10^{-12}$ Hz

These results were used to extrapolate the expected background rates of the final BDX experiment, scaling the experimental rates of a single crystal to the 800 crystals comprising the full detector. As already pointed out in section 2.3.2, this is an upper limit on the expected rates since this assumes crystal- to-crystal fully uncorrelated counts, which overestimates the case for χ-e^- scattering. However, these rates allowed to obtain a conservative evaluation the final BDX reach reported in Fig.6.

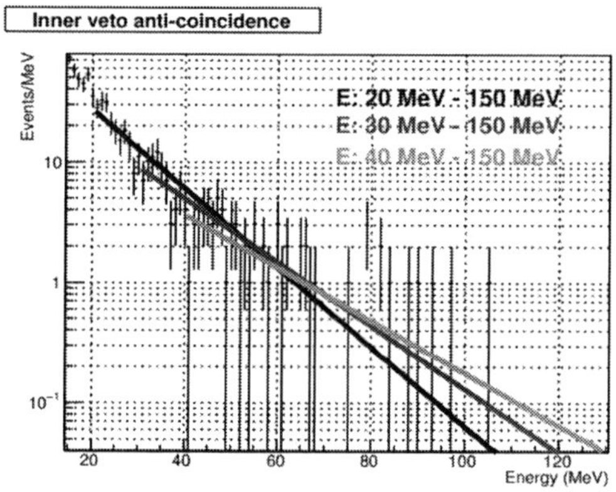

Figure 11. Extrapolation of the measured CsI energy spectrum in anti-coincidence with the IV from low-energy to high-energy region, where no events were detected. The three curves refer to Maximum Likelihood fits performed in different energy ranges.

4 Upgrade of the prototype and the second cosmic measurements campaign in Catania

BDX was proposed to JLab Program Advisory Committee (PAC) on June 2016 [11]. The PAC found the physics case exiting and conditionally approved the proposal, encouraging the collaboration to perform further studies to benchmark beam-related background simulations. In this respect, the collaboration is considering the possibility to propose on site measurements with a much smaller and

simpler detector of the beam-produced muon flux at different distances from the dump. At the same time, the prototype detector was improved in all its components and a new cosmic measurement campaign is currently underway in Catania.
In particular, the single crystal was replaced with a 4x4 matrix of CsI(Tl) crystals (Fig. 12) assembled and tested at the INFN - Sezione di Genova, each read-out by a 6x6 mm^2 SiPM (Hamamatsu S13360-6025PE).

Figure 12. A picture of the 4x4 matrix of CsI(Tl) crystals placed inside the second prototype.

Important changes have also been made to the two veto systems in order to test new solutions and increase their efficiency. In the IV the 4 extruded bars of the bottom side were replaced with a more efficient clear-plastic scintillator 35x140 cm^2, read-out by a system of WLS and SiPM. In particular the WLS fibers are hosted into 4 linear grooves running parallel to the long side of the scintillator, and are used to collect and transfer the light to 4 independent SiPMs.
The upgraded version of the OV is much more hermetic thanks to the following changes: the two 80 x 40 cm^2 paddles on the top side were replaced by a single large area (184 x 54 cm^2) plastic scintillator, read-out on both sides; the upstream and downstream caps were replaced by two new larger plastic scintillators featuring a more efficient light collection technique; one more 80x40 cm^2 paddle was added for each lateral side. For the new detectors (top, upstream and downstream) a different read-out solution is used, with the light guide replaced by slim wavelength-shifting plastic [13]. This solution has the advantage of reducing significantly the space and the mechanics needed by the light read-out system and making the veto more compact and hermetic, preserving, at the same time, a good light collection efficiency. Top detector comprises a large EJ200 scintillator 2.5 cm thick, which is diamond milled on all sides, except the two shortest ones that are 'frosted' and face one strip of EJ280 wavelength shifting plastic (2.5x2.5x62 cm^3). At the end of each EJ280 is a 2.5 cm in diameter, Hamamatsu R1924A photomultiplier tube (PMT) for light collection. Fig.13 shows the OV top detector.

Figure 13. The OV top detector is shown. In the right panel a detail of one of the wavelength shifting plastic is shown.

In the upstream and downstream panels light is collected only on one side. The EJ200 plastic scintillators have dimensions 58x66x2.5 cm^3, while the EJ280 wavelength-shifting plastic are 62 x 2.5x2.5 cm^3, and are also coupled to Hamamatsu R1924A photomultipliers. Fig.14 shows the upgraded BDX prototype.

Figure 14. The prototype mounted and cabled at the INFN-Sezione di Catania for the second cosmic measured campain.

In Fig.15 is shown the differential rate of the event collected by single CsI(Tl) crystal as a function of the energy for following class of events:

- all events;
- anti-coincidence of the OV in the old configuration (first prototype - Proto1);
- anti-coincidence of the OV in the new configuration (second prototype - Proto2);

As shown in Fig.15, thanks to the changes, the efficiency of the OV is increased by a factor ~ 4/5.

 DOI: 10.1051/epjconf/201714201005

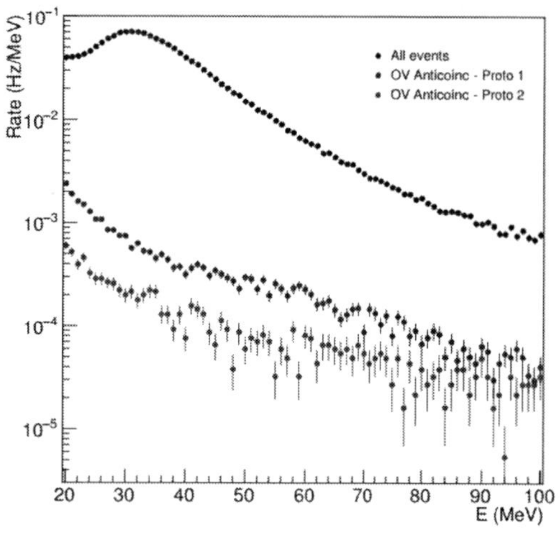

Figure 15. Measured event rate as a function of deposited energy in the crystal. Different colors refer to the anti-coincidence of the OV in the old configuration (blue dots) and in the new configuration (red dots).

5 Conclusion

BDX at Jlab aims to search for dark-matter particles in the MeV - GeV mass range, in an electron beam dump setup. It will look for these particles using an electromagnetic calorimeter surrounded by active vetos and passive shielding in a new underground facility located downstream of the Hall-A beam dump. BDX at JLab will hit the limit of this class of experiment reaching the wall of the irreducible background produced by CC interactions of beam-related ν_e in the detector. The sensitivity of BDX was evaluated by measuring the cosmic background with a prototype under conditions similar to those proposed at Jlab and estimating the beam-related background using GEANT4 MonteCarlo simulations. In the absence of a signal, electromagnetic shower thresholds between 0.3 and 0.5 GeV can be used to set limits on the production of dark matter that exceed the expected sensitivity of previous, and proposed experiments by up to two orders of magnitude.

BDX was conditionally approved by PAC44. The collaboration is considering the possibility of measuring in the near future beam-related background for benchmarking the simulations. Moreover in view of a further optimization of the experimental setup, a second prototype has been assembled and it is currently running cosmic particles in Catania. Further possible optimitations related to the front-end electronics and acquisition system are also under study.

References

[1] K. A. Olive et al., Review of Particle Physics. Chin. Phys., C38:090001, 2014.

[2] Annika H. G. Peter. Dark Matter: A Brief Review. 2012.

[3] B. Holdom, Phys. Lett. B **166B**, 196 (1986).

[4] N. Arkani-Hamed et al., Phys. Rev. D **79**, 015014 (2009)

[5] E. Izaguirre et al., Phys. Rev. D **90**, 014052 (2014)

[6] R. Dharmapalan et al. [MiniBooNE Collaboration], arXiv:1211.2258 [hep-ex].

[7] B. Batell, M. Pospelov and A. Ritz, Phys. Rev. D **80**, 095024 (2009).

[8] P. deNiverville, M. Pospelov and A. Ritz, Phys. Rev.D **84**, 075020 (2011).

[9] P. deNiverville, D. McKeen and A. Ritz, Phys. Rev.D **86**, 035022 (2012).

[10] L. B. Auerbach et al. [LSND Collaboration], Phys.Rev. D **63**, 112001 (2001).

[11] M. Battaglieri et al., *Dark matter search in a Beam-Dump eXperiment (BDX) at Jefferson Lab*, arxiv:1607.01390v1.

[12] J. A. Formaggio and G. P. Zeller, Rev. Mod. Phys., **84**, 1307 (2012).

[13] K.J. Thomas and et al., NIMA, **724**, 47-53 (2013).

EPJ Web of Conferences **142**, 01006 (2017) DOI: 10.1051/epjconf/201714201006
Advances in Dark Matter and Particle Physics

Radiative corrections in e^+e^- collisions with the BabaYaga event generator

Carlo M. Carloni Calame[1,a]

[1] *INFN, Sezione di Pavia, via A. Bassi 6, 27100 Pavia, Italy*

Abstract. The Monte Carlo event generator BabaYaga has been developed in the last decade for high precision simulation of QED processes ($e^+e^- \to e^+e^-$, $e^+e^- \to \mu^+\mu^-$ and $e^+e^- \to \gamma\gamma$) at flavour factories, mainly for luminometry purposes, with an estimated theoretical accuracy at the 0.1% level or better. The relevant QED radiative corrections are included by means of a QED Parton Shower matched with exact next-to-leading order corrections to reach the required accuracy. The main theoretical framework is overviewed and the status of the generator is summarized.

1 Introduction

The knowledge of the luminosity $\mathcal{L}$ is a key ingredient for any measurement at e^+e^- machines. The usual strategy is to calculate it through the relation $\mathcal{L} = N_{obs}/\sigma_{th}$, where σ_{th} is the theoretical cross section of a QED process, namely $e^+e^- \to e^+e^-$ (Bhabha), $e^+e^- \to \mu^+\mu^-$ or $e^+e^- \to \gamma\gamma$, and N_{obs} is the corresponding number of observed events. QED processes are the best choice because of their clean signal, low background and the possibility to push the theoretical accuracy up to the 0.1% level or better. The latter requires the inclusion of the relevant radiative corrections (RCs) in the cross sections calculation and their implementation into Monte Carlo (MC) event generators (EGs) in order to easily account for experimental event selection criteria.

Modern EGs used for luminometry simulate QED processes, including the exact next-to-leading order (NLO) QED corrections and/or a leading-logarithmic (LL) approximation of higher-order (h.o.) effects [1–6]. The consistent inclusion and matching of NLO and h.o. LL contributions is mandatory in view of the required theoretical accuracy. In the following, it is discussed how this is achieved in the BabaYaga EG.

2 The event generator BabaYaga and BabaYaga@NLO

The EG BabaYaga was originally developed for the precise simulation of large-angle Bhabha scattering at low energy e^+e^- colliders, with center of mass energy up to 10 GeV. It was later extended to simulate also $\mu^+\mu^-$ and $\gamma\gamma$ final states in the same energy regime. For the sake of clarity, in this section we focus on Bhabha scattering as reference process to illustrate the theoretical framework of the generator.

In its first version [1], the generator relied upon a QED Parton Shower (PS) to account for the LL photonic corrections, resummed up to all orders in perturbation theory. The PS is a MC algorithm which gives an exact iterative solution of the Dokshitzer-Gribov-Lipatov-Altarelli-Parisi (DGLAP) equation in QED for the non-singlet QED structure function (SF) $D(x, Q^2)$, which reads

$$Q^2 \frac{\partial}{\partial Q^2} D(x, Q^2) = \frac{\alpha}{2\pi} \int_x^1 \frac{dy}{y} P_+(y) D(\frac{x}{y}, Q^2)$$
$$P_+(x) = \frac{1+x^2}{1-x} - \delta(1-x) \int_0^1 dt P(t), \tag{1}$$

where $P_+(x)$ is the regularized Altarelli-Parisi vertex, x is the fraction of energy lost because of radiation and Q^2 is the scale of the process. The QED SFs account for photon radiation emitted by both initial-state and final-state fermions and allow to include the universal virtual and real photonic corrections, resummed up to all orders of perturbation theory. The advantage of the PS solution is that the kinematics of the emitted photons can be recovered (within some approximation) and hence an exclusive event generation can be performed, *i.e.* all the momenta of the final state particles (fermions and an indefinite number of photons) can be reconstructed.

Within the SF approach the corrected cross section can be written as

$$\sigma(s) = \int dx_1 dx_2 dy_1 dy_2 \int d\Omega \times$$
$$D(x_1, Q^2) D(x_2, Q^2) D(y_1, Q^2) D(y_2, Q^2) \times$$
$$\frac{d\sigma_0(x_1 x_2 s)}{d\Omega} \Theta(cuts). \tag{2}$$

Despite its advantages, the PS is essentially accurate at the LL level and a precision better than 0.5-1% can not be expected in the calculation of the cross section (2). In order to improve the accuracy, a matching with the exact NLO RCs is mandatory, in such a way that the good

[a] e-mail: carlo.carloni.calame@pv.infn.it

features of the PS are preserved (namely exclusive event generation and resummation of LL corrections up to all orders) while avoiding the double counting of the $O(\alpha)$ LL corrections, present both in the PS approach and in the NLO calculation.

An original matching algorithm has been implemented in the latest version of BabaYaga [2] (BabaYaga@NLO), which includes exact NLO corrections into a PS framework and achieves an accuracy at the 0.1% level in the calculation of the cross sections.

Without spelling out the details, the fully differential PS cross section implicit in Eq. (2) can be recast in the form

$$d\sigma_{PS}^{\infty} = \Pi(Q^2, \varepsilon) \sum_{n=0}^{\infty} \frac{1}{n!} |\mathcal{M}_{n,PS}|^2 \, d\Phi_n, \tag{3}$$

where $\Pi(Q^2, \varepsilon)$ is the Sudakov form factor accounting for virtual and soft (up to $x = \varepsilon$) radiation, $\mathcal{M}_{n,PS}$ is the amplitude for the emission of n real photons in the PS approximation and $d\Phi_n$ is the exact phase space for the emission of n real photons (with $x \geq \varepsilon$), divided by the incoming flux factor. Eq. (3) can be improved to include the missing NLO contributions by defining the following correction factors

$$\begin{aligned} F_{SV} &= 1 + \frac{d\sigma_{SV}^{NLO} - d\sigma_{SV}^{[\alpha,PS]}}{d\sigma_0} \\ F_{i,H} &= 1 + \frac{d\sigma_{i,H}^{NLO} - d\sigma_{i,H}^{[\alpha,PS]}}{d\sigma_{i,H}^{[\alpha,PS]}}, \end{aligned} \tag{4}$$

where SV stands for soft and virtual photon corrections, H for non-soft real photon corrections, $[\alpha, PS]$ stands for the $O(\alpha)$ expansion of the PS contribution, i runs over the emitted photons and $d\sigma_0$ is the lowest-order differential cross section. With these definitions, the matched differential cross section can be written in the form

$$d\sigma_{matched}^{\infty} = \Pi(Q^2, \varepsilon) \, F_{SV} \sum_{n=0}^{\infty} \frac{1}{n!} |\mathcal{M}_{n,PS}|^2 \, F_{n,H} \, d\Phi_n, \tag{5}$$

which is the master formula according to which event generation and cross section calculation are performed in BabaYaga@NLO.

A few comment are in order with respect to the master formula (5):

1. by construction, the factors in Eq. (4) are infrared and collinear safe quantities and they let the $O(\alpha)$ expansion of $d\sigma_{matched}^{\infty}$ in Eq. (5) coincide with the exact NLO result;
2. the resummation of h.o. LL corrections is preserved;
3. the correction factors (4) tend to be larger in those phase space regions where the PS is more unreliable, typically in phase space regions where the photon is hard and/or not collinear to one of the charged particles;
4. Eq. (5) is cast in a completely differential form, so that events can be generated exclusively as in the pure PS approach;
5. the theoretical error is shifted to corrections of order α^2, *i.e.* at the next-to-next-to-leading-order (NNLO) level.

Eq. (5) is used in BabaYaga@NLO to generate events for Bhabha, $e^+e^- \to \mu^+\mu^-$ and $e^+e^- \to \gamma\gamma$ [3] processes, with an undefinite number of extra photons to account for NLO and h.o. QED radiative corrections. Some numerical results and the estimate of the theoretical accuracy of the approach are sketched in the next section.

3 Results and estimate of the theoretical error

Table (1) shows the impact of different classes of QED RCs in the determination of large-angle Bhabha cross section within typical event selection criteria used for luminometry at flavour factories. Setups (a)-(b) correspond to $\sqrt{s}$ = 1.02 GeV and (c)-(d) to $\sqrt{s}$ = 10 GeV, for large ((a)-(c)) and narrower ((b)-(d)) angular acceptance cuts. In the second paper of Ref. [2] the detailed description of the setups, which tend to select elastic events, is reported.

Table 1. Bhabha cross section (in nb) according to different precision levels. The table is taken from the second paper in Ref. [2], where also the experimental setups are detailed.

set up	(a)	(b)	(c)	(d)
σ_0	6855.74	529.463	71.333	5.502
$\sigma_{0,VP}$	6976.5	542.66	74.763	5.8552
σ_{NLO}	6060.1	451.523	59.90	4.425
σ_{PS}^{α}	6083.6	454.50	60.14	4.456
σ_{PS}^{∞}	6107.6	458.44	60.62	4.530
$\sigma_{matched}^{\infty}$	6086.7	455.86	60.42	4.504

The first row of table (1) is the lowest order cross section, the second includes the effects of vacuum polarization (VP), the third one is the exact NLO result, the fourth is the $O(\alpha)$ expansion of Eq. (3) (*i.e.*, the PS approximation of the NLO result), the fifth is Eq. (3) and the last one corresponds to the most accurate matched formula of Eq. (5). By comparing the cross sections calculated at different theoretical accuracies, it can be inferred that VP affects the cross section at some per-cent level, fixed $O(\alpha)$ QED corrections at the 10-20% level and h.o. QED corrections have an impact of 0.5-1.5%: all these classes of corrections are important to achieve the theoretical precision required by the experiments and are consistently included and simulated in BabaYaga@NLO by means of the master formula (5).

In order to assess the theoretica accuracy of the approach implemented in BabaYaga@NLO, an important step is performing tuned comparisons to independent MC EGs.

As an example, in Figs. (1) and (2), the acollinearity and final-state invariant mass distributions obtained with BabaYaga@NLO and the independent EG Bhwide [4] are compared for setup (a). While the difference on the integrated cross sections is below the 0.01% level, some differences at the 1% level appear (see the insets in the figures)

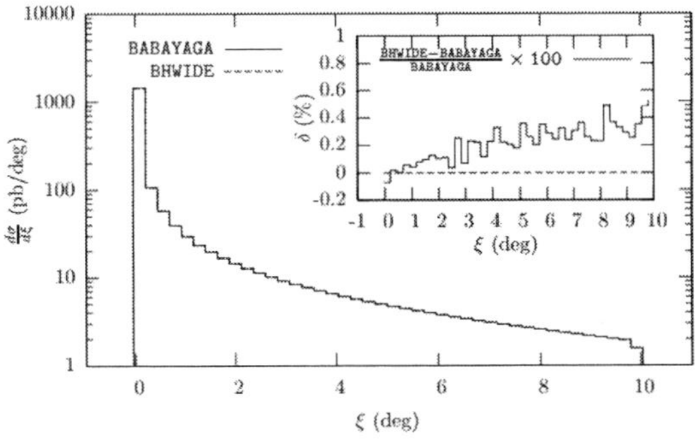

Figure 1. Comparison between `BabaYaga@NLO` and `Bhwide` on the acollinearity distribution.

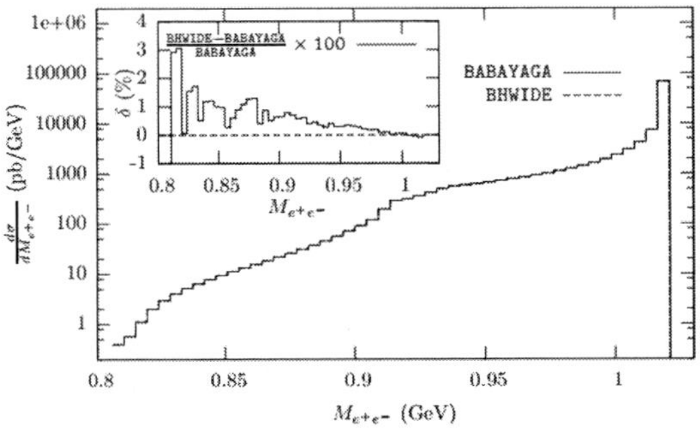

Figure 2. Comparison between `BabaYaga@NLO` and `Bhwide` on the final-state invariant mass distribution.

in the differential distributions and are due to the different theoretical frameworks on which the two EGs are based. However, it is worth noticing that such differences appear where the differential cross section drops by several order of magnitude with respect to the elastic region and where the different treatment of hard radiation beyond NLO can have an impact.

A further step to estimate the theoretical accuracy is to compare with existing calculations of the NNLO corrections to Bhabha scattering, which have been published over the years [7] (see also Ref. [8] and references therein). The dominant part of such corrections are already included in the master formula (5), which can be expanded up to NNLO and whose NNLO terms can be unambigously compared to analytical exact results: any difference should be considered as a theoretical error in the formulation of `BabaYaga@NLO`. A detailed and non-trivial comparison has been carried out in the second paper of Ref. [2], finding that the NNLO corrections not included in Eq. (5) impact the cross section at the level of few units in 10^{-4}, when typical event selection criteria are taken into account.

In the Luminosity section of Ref. [8], also other sources of theoretical uncertainties were considered, such as the uncertainty on the hadronic contribution to VP and extra light-pair emission (for which see also Refs. [9]). The main conclusion of that section is that a sound estimate of the present theoretical error on the luminosity determination via Bhabha scattering at flavour factories lies in the region of 0.1% or slightly below, which is enough for the present experimental requirements. Any further improvement would require the inclusion of the full exact NNLO RCs into MC EGs for Bhabha scattering, matched with LL h.o. contributions. This should be considered a feasible, although not trivial, task.

In a similar way, the matching algorithm of Eq. (5) has been also applied to the $e^+e^- \to \gamma\gamma$ process and a phenomenological study at flavour factories has been presented in Ref. [3].

Table 2. Photon pair production cross sections (in nb) to different accuracy levels (from Ref. [3], see there for details).

$\sqrt{s}$ (GeV)	1	3	10
σ_0	137.53	15.281	1.3753
σ^{α}_{PS}	128.55	14.111	1.2529
σ_{NLO}	129.45	14.211	1.2620
σ^{∞}_{PS}	128.92	14.169	1.2597
$\sigma_{matched}$	129.77	14.263	1.2685

As an example, in table (2) the impact of different classes of radiative corrections for typical event selection criteria at flavour factories is shown, in analogy with table (1) for Bhabha scattering. In the $e^+e^- \to \gamma\gamma$ case, the impact of NLO RC is in the 5-10% range while h.o. corrections change the cross sections by 0.1-0.5%: both effects must be accounted for to reach a theoretical accuracy at the 0.1% level. To the best of our knowledge, `BabaYaga@NLO` is the only MC EG implementing a matching of NLO and h.o. corrections in a PS approach for the $\gamma\gamma$ final state, being able to reach such a theoretical accuracy.

Before concluding, it is worth mentioning that an extension of `BabaYaga` has been developed for dark photon searches at low energy via the radiative return method [10] and has been extensively used by the KLOE-2 collaboration to set limits on the dark photon couplings and mass [11].

4 Conclusions

The theoretical framework and the status of the `BabaYaga` and `BabaYaga@NLO` event generators has been overviewed and summarized. `BabaYaga@NLO` is a Monte Carlo event generator developed for high-precision simulation of QED processes (Bhabha scattering, $e^+e^- \to \mu^+\mu^-$ and $e^+e^- \to \gamma\gamma$) at flavour factories, mainly for luminometry purposes, and it is based on an original algorithm to match exact NLO with higher-order QED radiative corrections in a Parton Shower approach. Within typical event selection criteria for luminometry, the cross sections can be calculated with an overall theoretical error at the level of 0.1% or better. The assessment and the scrutiny of the theoretical error are based on consistent and detailed comparisons with independent generators and existing exact NNLO calculations of the relevant radiative corrections.

Acknowledgements

The author aknowledges the daily and constant collaboration on the long-term BabaYaga project with Guido Montagna, Oreste Nicrosini and Fulvio Piccinini.

The author would like to warmly thank Dr. Mauro Raggi for the kind invitation and Prof. Giorgio Giardina, Dr. Giuseppe Mandaglio and the Local Organizing Committee for the organization of a stimulating and friendly workshop.

References

[1] C. M. Carloni Calame *et al.*, Nucl. Phys. B **584** (2000) 459;
C. M. Carloni Calame, Phys. Lett. B **520** (2001) 16.

[2] C. M. Carloni Calame *et al.*, Nucl. Phys. Proc. Suppl. **131** (2004) 48;
G. Balossini *et al.*, Nucl. Phys. B **758** (2006) 227.

[3] G. Balossini *et al.*, Phys. Lett. B **663** (2008) 209.

[4] S. Jadach, W. Płaczek, B. F. L. Ward, Phys. Lett. B **390** (1997) 298.

[5] A. B. Arbuzov *et al.*, Eur. Phys. J. C **46** (2006) 689.

[6] F. A. Berends, R. Kleiss, Nucl. Phys. B **228** (1983) 537;
F. A. Berends, R. Kleiss, Nucl. Phys. B **186**, 22 (1981).

[7] R. Bonciani *et al.*, Nucl. Phys. B **701** (2004) 121;
R. Bonciani *et al.*, Nucl. Phys. B **716** (2005) 280;
A. A. Penin, Phys. Rev. Lett. **95** (2005) 010408;
A. A. Penin, Nucl. Phys. B **734** (2006) 185;
R. Bonciani *et al.*, Nucl. Phys. B **681** (2004) 261 (Erratum: Nucl. Phys. B **702** (2004) 364);
R. Bonciani, P. Mastrolia and E. Remiddi, Nucl. Phys. B **676** (2004) 399;
R. Bonciani, P. Mastrolia and E. Remiddi, Nucl. Phys. B **661** (2003) 289 (Erratum: Nucl. Phys. B **702** (2004) 359);
R. Bonciani and A. Ferroglia, Phys. Rev. D **72** (2005) 056004;
M. Czakon, J. Gluza and T. Riemann, Nucl. Phys. B **751** (2006) 1;
M. Czakon, J. Gluza and T. Riemann, Phys. Rev. D **71** (2005) 073009.

[8] S. Actis *et al.* [Working Group on Radiative Corrections and Monte Carlo Generators for Low Energies], Eur. Phys. J. C **66** (2010) 585

[9] C. M. Carloni Calame *et al.*, Nucl. Phys. Proc. Suppl. **225-227** (2012) 293;
C. M. Carloni Calame *et al.*, JHEP **1107** (2011) 126.

[10] L. Barzè *et al.*,Eur. Phys. J. C **71** (2011) 1680

[11] A. Anastasi *et al.*, Phys. Lett. B **750** (2015) 633;
D. Babusci *et al.* [KLOE-2 Collaboration], Phys. Lett. B **736** (2014) 459.

Measuring the leading-order hadronic contribution to the muon g-2 in the space-like region

Carlo M. Carloni Calame[1,a]

[1] *INFN, Sezione di Pavia, via A. Bassi 6, 27100 Pavia, Italy*

Abstract. A new experiment is proposed to measure the running of the electromagnetic coupling constant in the space-like region by scattering high-energy muons on atomic electrons of a low-Z target. The differential cross section of the elastic process $\mu e \to \mu e$ provides direct sensitivity to the leading-order hadronic contribution to the muon anomaly $a_\mu^{\rm HLO}$. It is argued that by using the 150-GeV muon beam available at the CERN North Area, with an average rate of $\sim 1.3 \times 10^7$ muon/s, a statistical uncertainty of $\sim 0.3\%$ can be achieved on $a_\mu^{\rm HLO}$ after two years of data taking. The direct measurement of $a_\mu^{\rm HLO}$ via μe scattering will provide an independent determination and consolidate the theoretical prediction for the muon g-2 in the Standard Model. It will allow therefore a firmer interpretation of the measurements of the future muon g-2 experiments at Fermilab and J-PARC.

1 Introduction

The material presented here is largely based on the recently published paper of Ref. [1].

The long-standing (3–4)σ discrepancy between the experimental value of the muon anomalous magnetic moment $a_\mu = (g-2)/2$ and the Standard Model (SM) prediction, $\Delta a_\mu(\mathrm{Exp} - \mathrm{SM}) \sim (28 \pm 8) \times 10^{-10}$ [2, 3], is considered as one of the most intriguing indications of physics beyond the SM. However, the accuracy of the SM prediction, 5×10^{-10}, is limited by strong interaction effects, which cannot be computed perturbatively at low energies. By using analyticity and unitarity, it was shown [4] that the leading-order (LO) hadronic contribution to the muon g-2, $a_\mu^{\rm HLO}$, can be computed via a dispersion integral of the hadron production cross section in e^+e^- annihilation at low-energy. With this technique, the present error on $a_\mu^{\rm HLO}$, $\sim 4 \times 10^{-10}$ (corresponding to a fractional accuracy of 0.6%), constitutes the main uncertainty of the SM prediction.

From the experimental side, the error achieved by the BNL E821 experiment, $\delta a_\mu^{\rm Exp} = 6.3 \times 10^{-10}$ (corresponding to 0.54 ppm) [6], is dominated by the available statistics. New experiments at Fermilab and J-PARC, aiming at measuring the muon g-2 to a precision of 1.6×10^{-10} (0.14 ppm), are in preparation [7, 8].

Together with a fourfold improved precision on the experimental side, an improvement on the LO hadronic contribution is highly desirable. Differently from the dispersive approach, in Ref. [1] it is proposed to determine $a_\mu^{\rm HLO}$ from a measurement of the effective electromagnetic coupling in the space-like region, where the vacuum polarization (VP) is a smooth function of the squared momentum transfer. The method in Ref. [1] exploits $\mu e \to \mu e$ scattering, while the main ideas were firstly proposed in Ref. [9] by using Bhabha scattering data. [1]

How the hadronic contribution to the running of α can be determined unambiguously through the t-channel μe elastic scattering, from which $a_\mu^{\rm HLO}$ can be obtained, is sketched in the next sections.

2 Theoretical framework

With the help of dispersion relations and the optical theorem, $a_\mu^{\rm HLO}$ is given by the well-known formula [4, 12]

$$a_\mu^{\rm HLO} = \left(\frac{\alpha m_\mu}{3\pi}\right)^2 \int_{4m_\pi^2}^{\infty} ds\, \frac{\hat{K}(s) R_{\rm had}(s)}{s^2}, \tag{1}$$

where $R_{\rm had}(s)$ is the ratio of the total $e^+e^- \to$ hadrons and the Born $e^+e^- \to \mu^+\mu^-$ cross sections, $\hat{K}(s)$ is a smooth function and m_μ (m_π) is the muon (pion) mass. It should be remarked that $R_{\rm had}(s)$ in the integrand function of Eq. (1) is highly fluctuating at low energy due to hadronic resonances. The dispersive integral in Eq. (1) is usually calculated by using the experimental value of $R_{\rm had}(s)$ up to a certain value of s [5, 13, 14] and by using perturbative QCD (pQCD) [15] in the high-energy tail.

For the calculation of $a_\mu^{\rm HLO}$, an alternative formula can also be exploited [9, 16], namely

$$a_\mu^{\rm HLO} = \frac{\alpha}{\pi} \int_0^1 dx\, (1-x)\, \Delta\alpha_{\rm had}[t(x)], \tag{2}$$

[a] e-mail: carlo.carloni.calame@pv.infn.it

[1] It is worth mentioning that a method to determine the running of α by using small-angle Bhabha scattering was described in [10] and applied to LEP data in [11].

where $\Delta\alpha_{\rm had}(t)$ is the hadronic contribution to the running of the fine-structure constant, evaluated at

$$t(x) = \frac{x^2 m_\mu^2}{x-1} < 0, \qquad (3)$$

the space-like (negative) squared four-momentum transfer. In contrast with the integrand function of Eq. (1), the integrand of Eq. (2) is smooth and free of resonances.

By measuring the running of α,

$$\alpha(t) = \frac{\alpha(0)}{1-\Delta\alpha(t)}, \qquad (4)$$

where $t = q^2 < 0$, the hadronic contribution $\Delta\alpha_{\rm had}(t)$ can be extracted by subtracting from $\Delta\alpha(t)$ the purely leptonic part $\Delta\alpha_{\rm lep}(t)$, which can be calculated in perturbation theory with high accuracy.

Fig. 1 (top) shows $\Delta\alpha_{\rm lep}$ and $\Delta\alpha_{\rm had}$ as a function of the variables x and t. The range $x \in (0,1)$ corresponds to $t \in (-\infty, 0)$, with $x = 0$ for $t = 0$. The integrand of Eq. (2), calculated with the routine hadr5n12 [17], which uses time-like hadroproduction data and perturbative QCD, is plotted in Fig. 1 (bottom). The peak of the integrand occurs at $x_{\rm peak} \simeq 0.914$ (corresponding to $t_{\rm peak} \simeq -0.108$ GeV^2) and $\Delta\alpha_{\rm had}(t_{\rm peak}) \simeq 7.86 \times 10^{-4}$.

3 Experimental proposal

Through Eq. (2), $a_\mu^{\rm HLO}$ can be determined by measuring the running of α in the space-like region with a muon beam of $E_\mu = 150$ GeV on a fixed electron target. The proposed measurement is similar to the one used for the determination of the pion form factor, as described in [18], and it is very appealing for the following reasons:

1. at tree-level, $\mu e \to \mu e$ is a t-channel process, making the dependence on t of the differential cross section, when including VP effects, proportional to $|\alpha(t)/\alpha(0)|^2$:

$$\frac{d\sigma}{dt} = \frac{d\sigma_0}{dt}\left|\frac{\alpha(t)}{\alpha(0)}\right|^2. \qquad (5)$$

The VP effect, in the leading photon t-channel exchange, is incorporated in the running of α and gives rise to the factor $|\alpha(t)/\alpha(0)|^2$. It is understood that for a high precision measurement also higher-order radiative corrections must be included;

2. given the incoming muon energy E_μ, in a fixed-target experiment the t variable is related to the energy of the scattered electron E_e^f or its angle θ_e^f. With $E_\mu = 150$ GeV, the angle θ_e^f spans the range (0–31.85) mrad for the electron energy E_e^f in the range (1–139.8) GeV (the low-energy cut at 1 GeV is arbitrary);

3. it turns out that -0.143 GeV$^2 < t < 0$ GeV^2 if $E_\mu =$ 150 GeV. This implies that the region of x extends up to 0.93, while the peak of the integrand function of Eq. (2) is at $x_{peak} = 0.914$, corresponding to an electron scattering angle of 1.5 mrad;

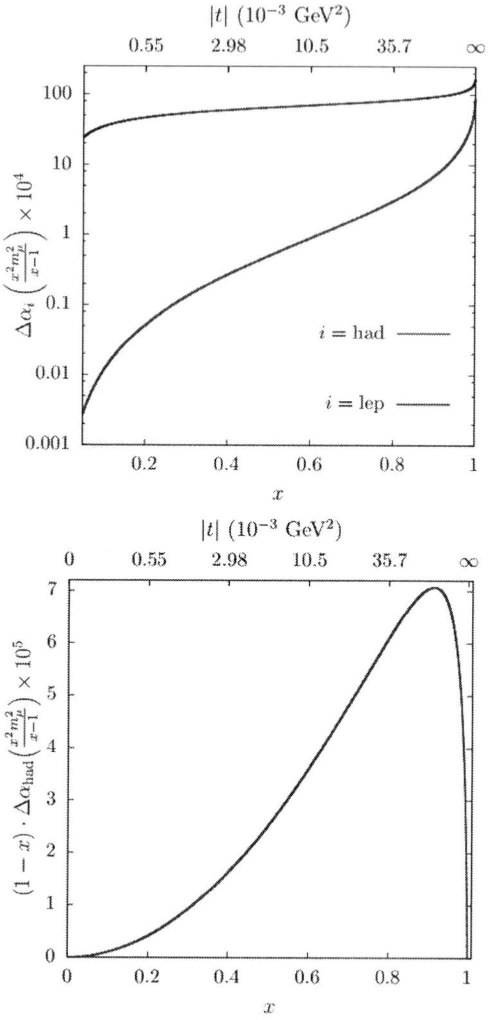

Figure 1. Top: $\Delta\alpha_{\rm had}[t(x)] \times 10^4$ (red) and, for comparison, $\Delta\alpha_{\rm lep}[t(x)] \times 10^4$ (blue), as a function of x and t (upper scale). Bottom: the integrand $(1-x)\Delta\alpha_{\rm had}[t(x)] \times 10^5$ as a function of x and t. The peak value is at $x_{\rm peak} \simeq 0.914$.

4. the angles of the scattered e^- and μ are correlated as shown in Fig. 2 (drawn for $E_\mu = 150$ GeV). This constraint is extremely important to select elastic scattering events, rejecting background events from radiative or inelastic processes and to minimize systematic effects in the determination of t. Note that for scattering angles of (2–3) mrad there is an ambiguity between the outgoing e^- and μ, as their angles and momenta are similar, to be resolved by means of μ/e discrimination;

5. the boosted kinematics allows the same detector to cover the whole acceptance. Many systematic errors, *e.g.* on the efficiency, will cancel out (at least at first order) in the relative ratios of event counts in the high and low q^2 regions (signal and normalization regions).

DOI: 10.1051/epjconf/201714201007

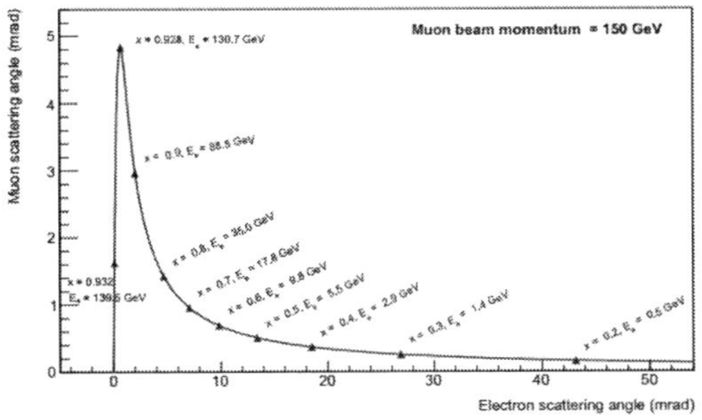

Figure 2. The relation between the μ and e^- scattering angles for 150 GeV incident muon beam. Blue triangles indicate reference values of the x variable and e^- energy.

Assuming a 150 GeV muon beam with an average intensity of $\sim 1.3 \times 10^7$ muons/s, presently available at CERN's North Area, incident on a target consisting of twenty Beryllium layers, each 3 cm thick (see Sect. 4), and two years of data taking with a running time of 2×10^7 s/yr, one can reach an integrated luminosity of about 1.5×10^7 nb^{-1}: taking into account the process cross section and with this luminosity, it is estimated with a simplified simulation that one can reach a statistical sensitivity of roughly 0.3% on the value of a_μ^{HLO}. The integrand in the region $x \in [0.93, 1]$, accounting for 13% of the a_μ^{HLO} integral, cannot be reached by the proposed experiment, but can be determined using time-like data and perturbative QCD, and/or lattice QCD results [19–21].

4 Considerations on the detector

In order to perform the measurement with the required precision, a dedicated detector is necessary. In this section, a possible and preliminary setup to measure the direction and momentum of the incident muon and the directions of the outgoing electron and muon is sketched.

The CERN 150-GeV muon beam M2 has the characteristics needed for such a measurement. The beam intensity provides the required event yield. Its time structure allows to tag the incident muon while keeping low the background related to incoming particles (*e.g.* electrons). The electron contamination is very small. The beam provides both positive and negative muons, which can be both exploited.

The target consists of atomic electrons. To reach the required statistics, it must contain an adequate amount of material to give a sufficient number of electron scattering centres. The target has to be made of a low-Z material to minimize the impact of multiple scattering and the background due to bremsstrahlung and pair production processes.

A promising idea, presently under study, is to use 20 identical modules, each consisting of a 3 cm thick layer of Be (or C) coupled to 2 Si stations located at a relative distance of one meter from each other and spaced by intermediate air gaps.

The arrangement provides both a distributed target with low-Z and the tracking system. As downstream particle identifiers it is planned to use a calorimeter for the electrons and a muon system for the muons (a filter plus active planes). This particle identifier system is required to solve the muon-electron ambiguity for electron scattering angles around (2–3) mrad.

Preliminary studies of such an apparatus, performed with the help of GEANT4, indicate that a tracking angular resolution for the outgoing particles of ~ 0.02 mrad can be reached using modern silicon strip detectors.

The detector acceptance covers the region of the signal, with the electron emitted at extremely forward angles and high energies, as well as the normalization region, where the electron has much lower energy (around 1 GeV) and an emission angle of some tens of mrad.

The incoming muons have to be tagged and their direction and momentum precisely measured. To this purpose, a detector similar to those used by COMPASS [22] or NA62 [23] can be employed.

5 Systematic uncertainties

Significant contributions of the hadronic VP to the $\mu e \to \mu e$ differential cross section are essentially restricted to electron scattering angles below 10 mrad, corresponding to electron energies above 10 GeV. The net effect of these contributions is to increase the cross section by a few per mille: a precise determination of a_μ^{HLO} requires not only high statistics, but also a high systematic accuracy, as the final goal of the experiment is equivalent to a determination of the differential cross section with $\sim$10 ppm systematic uncertainty at the peak of the integrand function.

Such an accuracy can be achieved if the efficiency is kept highly uniform over the entire q^2 range, including the normalization region, and over all the detector components. This motivates the choice of a purely angular measurement: an acceptance of tens of mrad can be covered with a single sensor of modern silicon detectors, positioned at a distance of about one meter from the target. It has to be stressed that particle identification is necessary to solve the electron-muon ambiguity in the region below 5 mrad. The wrong assignment probability can be measured with the data by using the rate of muon-muon and electron-electron events.

Another requirement for reaching very high accuracy is to measure all the relevant contributions to systematic uncertainties from the data themselves. An important effect, which distinguishes the normalization from the signal region, is multiple scattering, as the electron energy in the normalization region is as low as 1 GeV. Multiple scattering breaks the muon-electron two-body angular correlation, moving events out of the kinematic line in the 2D plot of Fig. 2. In addition, multiple scattering in general causes acoplanarity, while two-body events are planar, within the resolution. These facts allow effects to be modelled and measured using data. An additional handle on multiple scattering could be the inclusion of a thin layer in the apparatus, made of the same material as the main target modules. This possibility will be studied in detail with simulation.

The challenge of the proposed measurement is the feasibility of achieving a systematic uncertainty at the level of 10 ppm. This is the key point from the experimental side. In order to demonstrate that such a precision can be realistic, a very detailed optimization of the experimental apparatus is necessary. Tests with beams (electrons and muons), and with one or two modules of the detector, will be necessary and a crucial tool to understand if and to what extent the systematic uncertainties can be kept under control. They will provide a proof-of-concept of the proposed method.

From the theoretical side, the control of the systematic uncertainties requires the development of high-precision Monte Carlo tools, including all the relevant QED radiative corrections to reach the needed theoretical precision. To this aim, exact fixed order (NLO, NNLO) calculations, properly matched to leading-logarithmic corrections resummed up to all orders of perturbation theory, are mandatory to achieve the necessary theoretical accuracy on the relevant differential cross sections.

Tools to calculate Bhabha scattering exist, like for instance the `BabaYaga` event generator [24] (see also Ref. [25] and references therein), which implement exact NLO corrections matched with leading-logarithmic resummation, ensuring that the differential cross section is theoretically under control at the $O(10^{-4})$ level. The same algorithmic framework can be extended to $\mu e \to \mu e$ scattering and generalized to include exact diagrammatic NNLO corrections, which are not completely known yet for this process.

Work is in progress to extend the available Monte Carlo tools to $\mu e \to \mu e$ scattering and to quantify the achievable accuracy in the computation of the ratio of signal and normalization cross sections, by means of dedicated and realistic simulations.

6 Conclusions

A novel approach is under study to measure the running of α in the space-like region which can be used to determine a_μ^{HLO}, the leading hadronic contribution to the muon g-2. The proposed experiment exploits the process $\mu e \to \mu e$ by scattering high-energy muons on atomic electrons of a low-Z target. The experiment is primarily based on a precise measurement of the scattering angles of the two outgoing particles, from which the q^2 of the muon-electron interaction can be directly accessed.

An advantage of the muon beam is the possibility of employing a modular apparatus, with the target subdivided in subsequent layers. A low-Z solid target is preferred in order to provide the required event rate, limiting at the same time the effect of multiple scattering as well as of other types of muon interactions (pair production, bremsstrahlung and nuclear interactions).

The normalization of the cross section is provided by the very same $\mu e \to \mu e$ process in the low-q^2 region, where the effect of the hadronic corrections on $\alpha(t)$ is negligible. Such a simple and robust technique has the potential to keep systematic effects under control, aiming to reach a systematic uncertainty of the same order as the statistical one. For this purpose a preliminary detector layout has been described. By considering a beam of 150 GeV muons with an average intensity of $\sim 1.3 \times 10^7$ muon/s, currently available at the CERN North Area, a statistical uncertainty of $\sim 0.3\%$ can be achieved on a_μ^{HLO} in two years of data taking.

A test performed using a single detector module, exploiting the muon beam facility, can provide a validation of the proposed method and it is scheduled for the next future.

Acknowledgements

The author aknowledges the collaboration on this new proposed experiment with a group of enthusiastic colleagues: Giovanni Abbiendi, Marco Incagli, Umberto Marconi, Clara Matteuzzi, Guido Montagna, Oreste Nicrosini, Massimo Passera, Fulvio Piccinini, Roberto Tenchini, Luca Trentadue and Graziano Venanzoni.

The author is indebted with Dr. Giuseppe Mandaglio for the kind invitation and would like to warmly thank him, Prof. Giorgio Giardina and the Local Organizing Committee for the organization of a fruitful workshop.

References

[1] G. Abbiendi *et al.*, Eur. Phys. J. C **77** (2017) no.3, 139.

[2] T. Blum, A. Denig, I. Logashenko, E. de Rafael, B. Lee Roberts, T. Teubner and G. Venanzoni, arXiv:1311.2198 [hep-ph].

[3] F. Jegerlehner, EPJ Web Conf. **118** (2016) 01016.

[4] C. Bouchiat, L. Michel, J. Phys. Radium 22 (1961) 121;
L. Durand, Phys. Rev. **128** (1962) 441 [Erratum-ibid. **129** (1963) 2835];
M. Gourdin, E. De Rafael, Nucl. Phys. B **10** (1969) 667.

[5] F. Jegerlehner, A. Nyffeler, Phys. Rept. **477** (2009) 1.

[6] G. W. Bennett et al. (The g-2 Collaboration), Phys. Rev. **D73** (2006) 072003.

[7] J. Grange *et al.* [Muon g-2 Collaboration], arXiv:1501.06858 [physics.ins-det].

[8] N. Saito [J-PARC g-2/EDM Collaboration], AIP Conf. Proc. **1467** (2012) 45.

[9] C. M. Carloni Calame, M. Passera, L. Trentadue and G. Venanzoni, Phys. Lett. B **746** (2015) 325.

[10] A.B. Arbuzov, D. Haidt, C. Matteuzzi, M. Paganoni and L. Trentadue, Eur. Phys. J. C **34** (2004) 267.

[11] G. Abbiendi *et al.* [OPAL Collaboration], Eur. Phys. J. C **45** (2006) 1.

[12] F. Jegerlehner, "The anomalous magnetic moment of the muon", Springer Tracts Mod. Phys. **226**, 2008.

[13] M. Davier, A. Hoecker, B. Malaescu, Z. Zhang, Eur. Phys. J. C **71** (2011) 1515 [Erratum-ibid. C **72** (2012) 1874].

[14] K. Hagiwara, R. Liao, A.D. Martin, D. Nomura, T. Teubner, J. Phys. G **38** (2011) 085003.

[15] R.V. Harlander, M. Steinhauser, Comput. Phys. Commun. **153** (2003) 244.

[16] B.E. Lautrup, A. Peterman, E. de Rafael, Phys. Rept. **3** (1972) 193.

[17] S. Eidelman and F. Jegerlehner, Z. Phys. C **67** (1995) 585; F. Jegerlehner, Nucl. Phys. Proc. Suppl. **181-182** (2008) 135.

[18] S. R. Amendolia *et al.*, Phys. Lett. B **146** (1984) 116; S. R. Amendolia *et al.* [NA7 Collaboration], Nucl. Phys. B **277** (1986) 168.

[19] F. Jegerlehner, Nucl.Phys.Proc.Suppl. 181-182 (2008) 135-140.

[20] M. Marinkovic, P. Boyle, L. Del Debbio, A. Juettner, K. Maltman, A. Portelli, PoS LATTICE2014 (2015) 160, http://inspirehep.net/record/1345265.

[21] M. Golterman, K. Maltman, S. Peris, Phys. Rev. D90 (2014) 7, 074508.

[22] P. Abbon *et al.* [COMPASS Collaboration], Nucl. Instrum. Meth. A **577** (2007) 455.

[23] Mathieu Perrin-Terrin for the NA62 Collaboration, Proceed. 24th IWVD, Santa Fe, June 2015, pp.016.

[24] G. Balossini, C. M. Carloni Calame, G. Montagna, O. Nicrosini and F. Piccinini, Nucl. Phys. B **758** (2006) 227; C. M. Carloni Calame, Phys. Lett. B **520** (2001) 16; C. M. Carloni Calame, C. Lunardini, G. Montagna, O. Nicrosini and F. Piccinini, Nucl. Phys. B **584** (2000) 459.

[25] S. Actis *et al.* [Working Group on Radiative Corrections and Monte Carlo Generators for Low Energies Collaboration], Eur. Phys. J. C **66** (2010) 585.

Doubly-polarised pion photoproduction on the nucleon at MAMI

Susanna Costanza[1,2,a] for the A2 Collaboration

[1] *Dipartimento di Fisica, Università degli Studi di Pavia, I-27100 Pavia, Italy*
[2] *INFN, Sezione di Pavia, I-27100 Pavia, Italy*

Abstract. The A2 Collaboration at MAMI (Mainz) carried out new measurements of the helicity dependence of the total inclusive photo-absorption cross section and of the partial cross sections for several reaction channels on the proton and on the neutron in the photon energy region $200 < E_\gamma < 1500$ MeV.
The experiments were performed at the tagged photon beam facility of the MAMI accelerator in Mainz, using circularly and linearly polarised photons on longitudinally polarised proton, deuteron and ^{3}He targets. Hadronic reaction products were detected by the large acceptance Crystal Ball-TAPS spectrometer, complemented by plastic scintillators and vertex detectors for charged particle tracking and identification.
These new, high-quality doubly-polarised pion-photoproduction data sets provide a valuable input to the study of the nucleon structure and excitation spectrum by significantly constraining the electromagnetic multipole evaluation performed by the different available partial wave analysis models. Furthermore, the helicity dependent observables provide the main ingredient for the verification of the well-known Gerasimov-Drell-Hearn (GDH) sum rule, which relates the helicity-dependent photoasborption process to the main static nucleon properties (mass, charge, spin).

1 Introduction

The structure of the nucleus has been a central issue of nuclear and particle physics since many years, as well as the subject of intense research, both from the theoretical and experimental point of view. The study of the excited states of the nucleon is of particular interest: they are related to the fundamental properties of the strong interaction in the same way as the excitation spectra of atoms reflect the properties of the electromagnetic interaction.
Since Quantum Chromodynamics (QCD) is non-perturbative at this energy scale, we have to rely on phenomenological quark models to interpret the experimental results. Unfortunately, this approach has limited success: at present, while there is quite an agreement between the experimental findings and model predictions in the low mass region of the nucleon excitation spectrum, there is still a pletora of higher lying states predicted by theories and not yet observed ("missing resonances").
From the experimental point of view, the reason for this mismatch can be due to data analysis relying mostly on pion scattering, that will therefore miss states that couple only weakly to πN: it may be that the higher lying states couple preferentially to decay channels involving heavier mesons or decay via intermediate excited states [1].
In order to understand the nucleon excitation spectra, an alternative is represented by the use of electromagnetic-induced reactions, like meson photoproduction: in such a way, it is possible to explore multiple meson production states, like $\pi\pi$, $\pi\eta$, ..., and access resonances that decay mainly via intermediate excited states.
In particular, photoproduction reactions of single pseudoscalar mesons allow to measure a set of 16 polarisation observables, for every fixed value of energy and angle (W, θ). In addition to the unpolarised cross section σ, there are 3 single polarisation observables and 12 double polarisation observables, grouped in beam-target, beam-recoil nucleon and target-recoil polarisation classes (table 1). The measurement of 7 (8) properly chosen observables is necessary to yield a model-independent complete analysis [2, 3].

2 The A2 Collaboration at MAMI

Thanks to the use of polarised proton, deuteron and ^{3}He targets, the A2 Collaboration can carry out a comprehensive experimental program in order to give a better insight into the GDH sum rule on the proton and on the neutron, and to perform an accurate investigation of the properties of baryonic resonances through experiments devoted to the study of double-polarisation observables.

2.1 The detector

The experimental setup of the A2 Collaboration is located at the MAMI tagged photon beam facility in Mainz (Germany). The photon beam is produced via

[a] e-mail: susanna.costanza@pv.infn.it

 DOI: 10.1051/epjconf/201714201008

Table 1. Polarisation observables.

Photon polarisation		Target polarisation			Recoil nucleon polarisation			Target and recoil polarisation			
		X	Y	$Z_{(beam)}$	X′	Y′	Z′	X′ X	X′ Z	Z′ X	Z′ Z
unpolarised	σ	-	T	-	-	P	-	$T_{X'}$	$L_{X'}$	$T_{Z'}$	$L_{Z'}$
linear	-Σ	H	(-P)	-G	$O_{X'}$	(-T)	$O_{Z'}$	$(-L_{Z'})$	$(T_{Z'})$	$(L_{X'})$	$(-T_{X'})$
circular	-	F	-	-E	$C_{X'}$	-	$C_{Z'}$	-	-	-	-

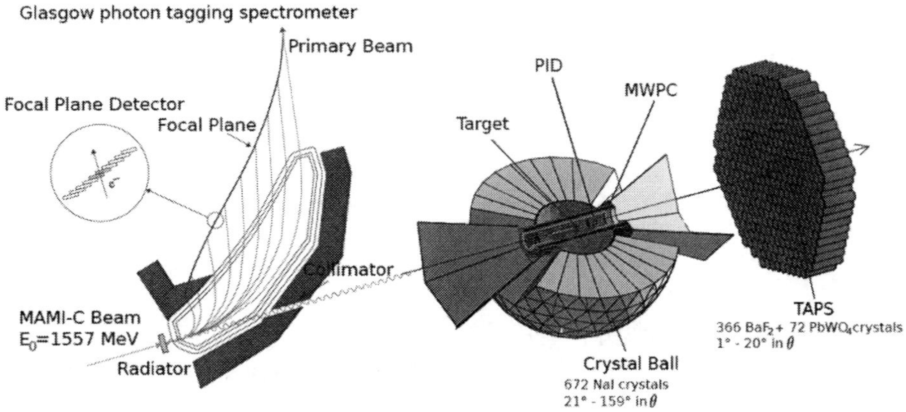

Figure 1. Schematic view of the A2 experimental setup.

bremsstrahlung of an electron beam of 1557 MeV provided by the MAMI accelerator, and polarised up to ~75-78%. The photons are then tagged using the Glasgow-Mainz magnetic spectrometer, which has an energy resolution of ~2 MeV.

The A2 apparatus can make use of different targets: for the experiments described in the following, a frozen-spin butanol (C_4H_9OH) [4] and deuterated butanol (C_4OD_{10}) targets, and a high pressure ^{3}He gas target have been used. Concerning the butanol target, the 2 cm long target cell was located in a ^{3}He/^{4}He dilution refrigerator and the target nucleons were polarised via Dynamic Nuclear Polarisation (DNP) [5] to an initial polarisation degree up to 90%. During the data taking, a magnetic field of 0.68 T together with the 25 mK temperature provided by the dilution refrigerator ensured long relaxation times (up to 2000 h) before repolarisation was required. Since only the hydrogen nuclei of the butanol can be polarised, a carbon foam target was used in a dedicated beamtime to study background contributions from unpolarised carbon nuclei.

In the case of the frozen-spin target using deuterated butanol, a starting value of target polarisation up to 70%, with a relaxation time of ~200 hours, was obtained, thanks to a new doping material [6]. Also in the case of gaseous ^{3}He an initial polarisation value up to 70% was obtained, via the metastability exchange optical pumping (MEOP [7]) method, with a total relaxation time of ~20 hours.

The target cell is sitting inside the central detector system and surrounded by the Particle IDentification (PID) detector, made of plastic scintillators. The PID is enclosed by the Multi-Wire Proportional Chambers (MWPCs), used to discriminate between charged and neutral particles. The MWPCs are complemented by the Crystal Ball (CB) NaI spectrometer, a large solid angle, highly segmented photon and hadron spectrometer. By combining the information provided by these three detectors, precise energy measurements, as well as angle and particle identification in the azimuthal (ϕ) and polar (θ) angular regions from 0° to 360° and from 21° to 159°, respectively, are obtained.

3 Double polarisation observables G and E

The double-polarisation observables G and E, as shown in table 1, require the polarisation of both the photon beam and the target.

They are derived from the differential cross section for pseudo-scalar meson photoproduction; in the case of elliptically polarised photons in combination with a longitudinally polarised target, the cross section is given by:

$$\frac{d\sigma}{d\Omega}(\theta,\phi) = \frac{d\sigma}{d\Omega_0}(\theta)[1 - P_{lin}\Sigma\cos(2(\alpha-\phi)) - P_z(-P_{lin}G\sin(2(\alpha-\phi)) + P_{circ}E]. \quad (1)$$

The E observable is extracted from eq. 1 by integrating over ϕ:

$$N_B\Big|_{\pm\alpha}^{\pm P_z}(\theta) = N_B(\theta)\cdot\left[1 - P_{circ}P_zE\right] \quad\rightarrow$$

$$E = \frac{\sigma^{1/2}-\sigma^{3/2}}{\sigma^{1/2}+\sigma^{3/2}} = \frac{N_B^{1/2}-N_B^{3/2}}{N_B^{1/2}+N_B^{3/2}}\cdot\frac{1}{d}\cdot\frac{1}{P_{circ}P_z}, \quad (2)$$

where $N_B^{1/2}$ and $N_B^{3/2}$ are the helicity-dependent count rates, d is the dilution factor and $P_{circ}P_z$ are the degrees of target and beam polarisation.
Since the experiment was performed with a butanol target, the count rates take into account the contribution of both the polarisable free protons in hydrogen nuclei and the unpolarised bound nucleons in carbon and oxygen nuclei. The amount of polarisable free protons in the data is specified by the dilution factor d ($d = N_{free}/(N_{free} + N_{bound})$), which was determined during a dedicated data taking using a carbon foam target. d was obtained as $d = 1 - s \cdot N_C/N_B$, where the coplanarity distribution of the carbon data (N_C) is scaled to the one of the butanol data (N_B). In the formula, s takes into account possible differences in acceptance, photon flux and target density during the carbon and butanol beamtimes.
In the same way as E, the observable G is obtained by integrating Eq. 1 over all possible helicity states:

$$N_B\Big|_{\pm\alpha}^{\pm P_z}(\theta,\phi) = N_B(\theta)\cdot[1 - P_{lin}\Sigma_B\cos(2(\alpha-\phi)) + dP_{lin}P_zG\sin(2(\alpha-\phi))]. \quad (3)$$

3.1 Experimental results

The double polarisation observables G and E were extracted from the data according to Eqs. 2 and 3 as a function of the beam energy E_γ for the entire angular range $-1 \leq \cos\theta_\pi \leq 1$.
Figures 2 and 3 show preliminary results for the photoproduction reaction $\vec{\gamma}\vec{p} \to p\pi^0$. The results are compared to data from the CBELSA/TAPS collaboration ([8] for the G observable and [9] for E) and to different partial wave analyses: the agreement with the existing data and mostly with the different PWA models, especially with the BnGa2014-02 PWA solution, is shown for both the double polarisation observables.

4 The GDH sum rule

Since sum rules connect information from all energies to fundamental parameters of the interaction models of interest, they are important tools to study the excited states of the nucleons.
One of these sum rules is the well known Gerasimov-Drell-Hearn (GDH) sum rule [13, 14], which relates the nucleon anomalous magnetic moment (AMM) κ, the spin S and the mass M of a nucleon to the integral over the weighted helicity difference of the total absorption cross section for circularly polarised photons on a longitudinally polarised nucleon target:

$$I_{GDH} = \int_{\nu_{th}}^{\infty}\frac{\sigma_p - \sigma_a}{\nu}d\nu = 4\pi^2\kappa^2\frac{e^2}{M^2}S, \quad (4)$$

where ν is the photon energy, and σ_p and σ_a denote the total absorption cross section for parallel and antiparallel orientation of photon and particle spins, respectively. The lower limit of the integral, ν_{th}, corresponds to pion production and photodisintegration threshold for a nucleonic and nuclear target, respectively.
The GDH sum rule is derived from very general fundamental physical principles, like Lorentz invariance, the optical and the low energy theorems. Therefore, its verification is of particular interest in the understanding of our knowledge of the $\gamma - N$ interaction, as well as of the physics of strongly interacting systems and of the existing photo-reaction models. Furthermore, by studying the helicity dependent $N\pi$ channels, it is possible to test multipole models, to access new observables and to study the baryonic resonances.
Thanks to the use of polarised proton, deuteron and ^{3}He targets, the A2 Collaboration can study all the $\gamma N \to N\pi(\pi)$ partial channels, as well as the total cross sections, for the proton and the neutron. It can also give a better insight into the GDH sum rule and perform an accurate investigation of the baryonic resonance properties.

4.1 Experimental results

The experiments were carried out by using circularly polarised photons obtained via bremsstrahlung of longitudinally polarised electrons of about 75%.
Different targets have been used: for the proton studies, the same frozen-spin butanol data collected for the E and G double polarisation observables were used.
To compensate the lack of free neutron targets, polarised deuteron and high pressure ^{3}He gas targets were used as polarised neutron sources, as described in section 2.1.

4.1.1 Results on the proton and deuteron

Fig. 4 (blue circles) shows the preliminary results of the helicity dependence of the total inclusive cross section $\Delta\sigma_{tot} = (\sigma_a - \sigma_p)$ on the proton (a) and on the neutron (b). The A2 results are compared to those obtained by the GDH Collaboration [15, 16] (red circles), showing a good agreement within the errors.
The results for the partial channels $\gamma p \to \pi^0 X$ and $\gamma d \to \pi^0 X$ are depicted in Fig. 5: also in these plots, the A2 data (blue circles) are compared with good agreement to the GDH Collaboration results. Since the GDH data were collected only in the Δ-resonance region, it is therefore clear that the A2 recent results can provide a useful contribution in an energy region where no other data are available.
Our experimental results are also compared to the predictions of the MAID multipole analysis: the discrepancy with the data is due to the fact that the model is a free nucleon estimation, therefore it does not take into account the nuclear effects inside the deuteron target (Fig. 5b).

4.1.2 Results on ^{3}He

Figure 6 shows the unpolarised and helicity-dependent total cross sections for the partial channels $\gamma^3He \to \pi^0 X$ and $\gamma^3He \to \pi^\pm X$ in the energy region $200 < E_\gamma < 500$ MeV [17].
Here the data are compared to the prediction of the Fix-Arenhövel (FA) model (red line), which is an extention of

 DOI: 10.1051/epjconf/201714201008

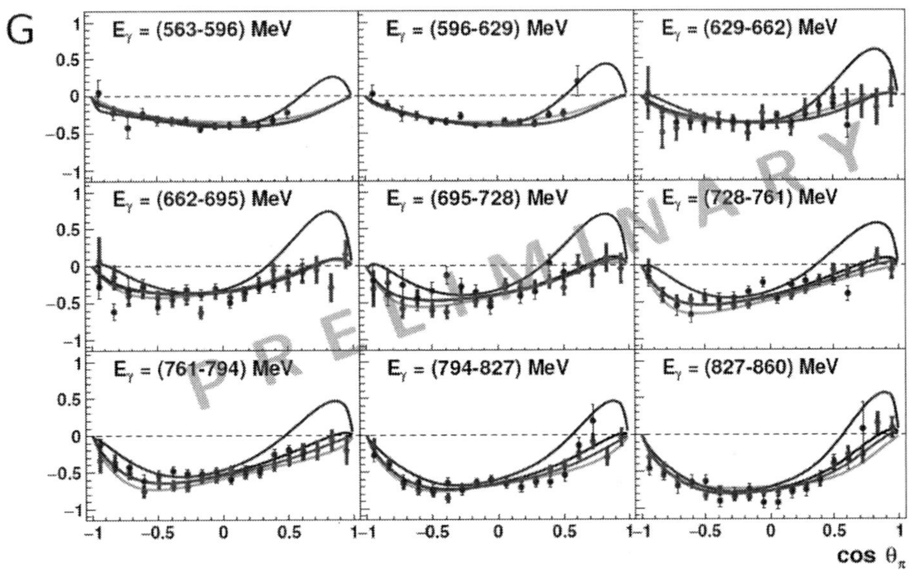

Figure 2. Results for G in π^0-photoproduction for selected energy bins, from 560 MeV to 860 MeV. The experimental results from A2 (blue circles) are compared to the CBELSA/TAPS data (green circles) [8], to PWA fits (BnGa_2014_02 [10], red line and BnGa_2014_01, black line) and PWA predictions from MAID2007 [11] (green line) and SAID-CM12 [12] (blue line).

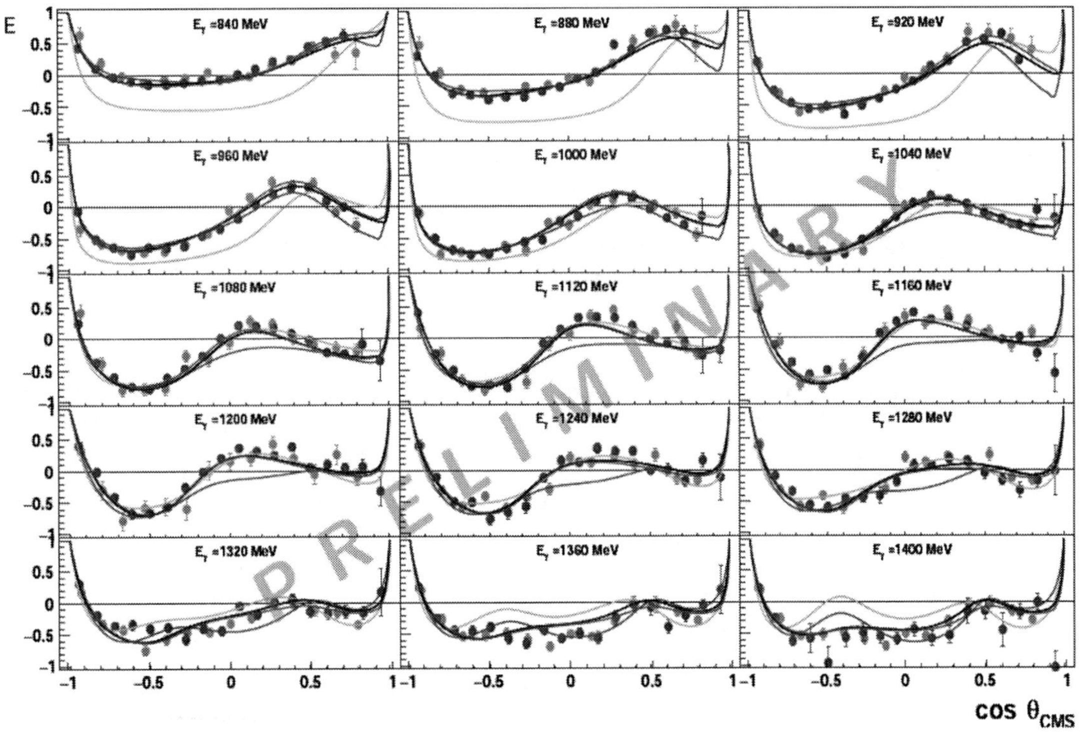

Figure 3. Results for E in π^0-photoproduction for selected energy bins from 840 MeV to 1400 MeV. The experimental results from A2 (blue circles) are compared to the CBELSA/TAPS data (green circles) [9], to PWA fits (BnGa_2014_02 [10], red line and BnGa_2014_01, black line) and PWA predictions from MAID2007 [11] (green line) and SAID-CM12 [12] (blue line).

 DOI: 10.1051/epjconf/201714201008

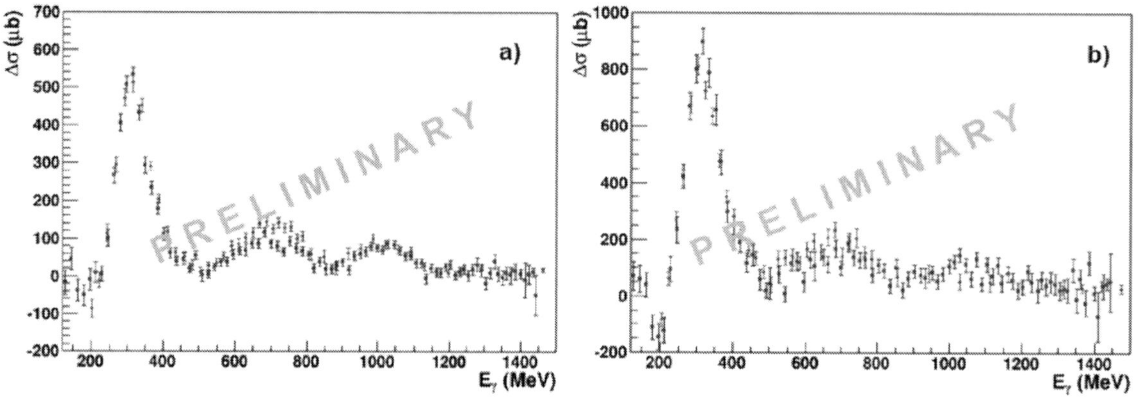

Figure 4. Helicity dependent total inclusive cross section for the $\gamma p \to X$ (a) and $\gamma d \to X$ channels (b). The A2 results (blue circles) are compared to the results from the GDH Collaboration (red circles) [15, 16].

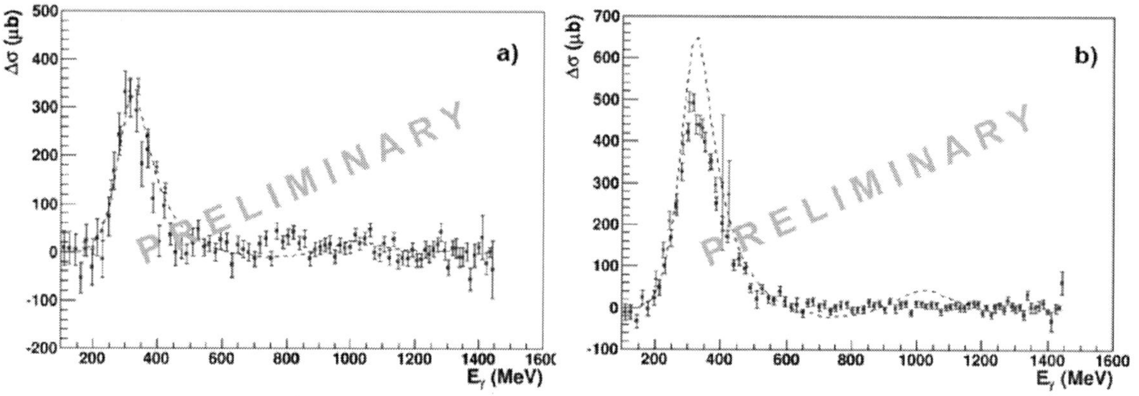

Figure 5. Helicity dependent total cross section for the semi-exclusive channels $\gamma p \to \pi^0 X$ (a) and $\gamma d \to \pi^0 X$ channels (b). The A2 results (blue circles) are compared to the results from the GDH Collaboration (red circles) [15, 16] and to the MAID model (dashed line).

the model on the deuteron [18]. In addition, the A2 results are compared to a simple plane-wave impulse approximation (PWIA) model (blue line), where the cross sections are evaluated as an incoherent sum of quasi-free single nucleon contributions, determined using the MAID multipole analysis and the momentum distribution of the nucleons inside ^{3}He as parametrised in [19]. The difference between the two models, i.e. the role of nuclear effects, results in damping and broadening the peak corresponding to the Δ resonance excitation.
In the unpolarised case, for the neutral partial channel the FA model is in good agreement with our data for $E_\gamma \geq$ 250 MeV, while it underestimates them at lower photon energies, where the coherent $\gamma^3He \to \pi^0\, ^3He$ reactions is expected to play a dominant role. On the contrary, the FA model describes the measured cross section of the charged channel less well and for $E_\gamma \geq 350$ MeV the PWIA model does better.
As in the unpolarised case, in the helicity dependent cross section case, the FA model describes the $\gamma^3He \to \pi^0 X$ data at higher photon energies while there is less agreement with the experimental data of the charged partial channel. The PWIA model, instead, reproduces the data at higher E_γ values for both reactions reasonably well.
Furthermore, to understand the origin of the observed discrepancies between the experimental models and the data, an analysis of both the unpolarised and polarised differential cross sections for the $\gamma^3He \to \pi^0 X$ and $\gamma^3He \to \pi^\pm X$ has been performed: the results can be found in [20].

5 Conclusions

Double polarised pion photoproduction experiments are a useful tool for the understanding of the nucleon excitation spectra.
The A2 Collaboration at MAMI has performed the first simultaneous measurement of the double polarisation observables E and G, by using a longitudinally polarised electron beam incident on a diamond radiator and a longitudinally polarised frozen-spin target. The preliminary

DOI: 10.1051/epjconf/201714201008

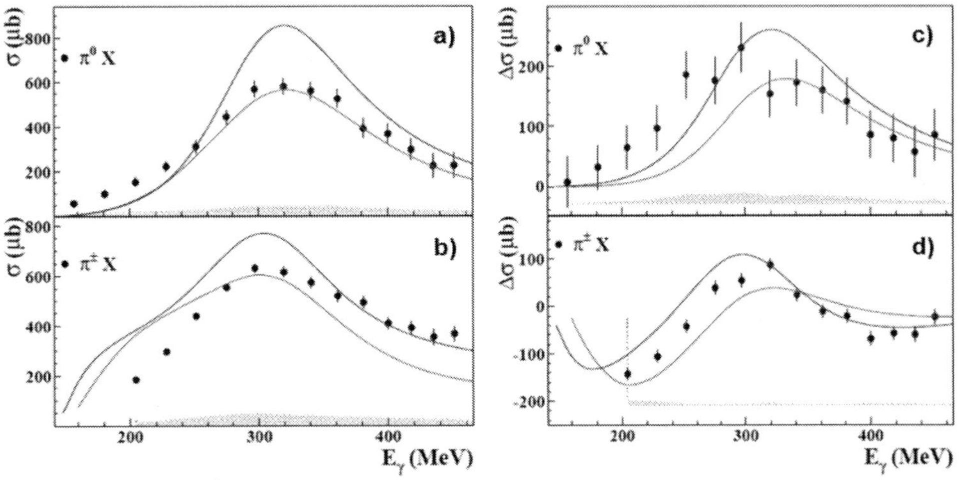

Figure 6. Unpolarised (left) and polarised (right) total cross section for $\gamma^3He \rightarrow \pi^0X$ (a and c) and $\gamma^3He \rightarrow \pi^{\pm}X$ (b and d) channels. The error bars are statistical and the hatched bands show the systematic uncertainties. The measured cross sections are compared to the FA model (red line) and to our PWIA model (blue line).

results available up to E_γ = 1500 MeV are in good agreement with the existing data from the CBELSA/TAPS Collaboration and provide an original contribution in the Δ(1232) resonance region.
The experimental data obtained from these experiments can be used also to study the Gerasimov-Drell-Hearn sum rule. Furthermore, thanks to the use of polarised deuteron and ^{3}He targets, the A2 Collaboration has carried out a comprehensive experimental program to study the GDH sum rule not only on the proton, but also on the neutron and on ^{3}He.
The new precise results on double-polarisation measurements of the total and differential cross sections for the partial $\gamma N \rightarrow \pi X$ channels on the proton and on the neutron, obtained by the A2 Collaboration, are compared to the existing model predictions and to the few, available results. These new data are significantly increasing the available statistics, especially on the neutron, thus providing an important testing ground for all existing models.
Moreover, the results obtained on ^{3}He give information not only on the GDH integral on the neutron, but also on the ^{3}He nuclear structure and allow an investigation of the nucleon properties inside this nucleus.
These unprecedented data have proved the feasibility of the use of polarised ^{3}He gas target to check the GDH sum rule on the neutron.

References

[1] B. Krusche and S. Schadmand, arXiv:nucl-ex/0306023v1
[2] I. S. Barker, A. Donnachie, J. K. Storrow, Nucl. Phys. **B95**, 347 (1975)
[3] W.-T. Chiang and F. Tabakin, Phys. Rev. C **55**, 2054 (1997)
[4] D. Goertz et al., Proceedings of PSTP205, PoS(PSTP2015)009 (2016)
[5] D.G. Crabb and W. Meyer, Annu. Rev. Nucl. Part. Sci. **47**, 67 (1997)
[6] S. Goertx et al., Nucl. Instr. Meth. A **526**, 43 (2004)
[7] F. Colegrove et al., Phys. Rev. Lett. **132**, 2561 (1963)
[8] A. Thiel et al., Phys. Rev. Lett. **109**, 102001 (2012)
[9] M. Gottschall et a., Phys. Rev. Lett. **112**, 012003 (2014)
[10] E. Gutz et al., Eur. Phys. J. A **50**, 74 (2014)
[11] D. Drechsel, S.S. Kamalov and L. Tiator, Eur. Phys. J. A **34**, 69 (2007)
[12] R.L. Workman, M.W. Paris, W.J. Briscoe, I.I. Strakovsky, Phys. Rev. C **86**, 015202 (2012)
[13] S. B. Gerasimov, Sov. J. Nucl. Phys. **2**, 430 (1966)
[14] S. D. Drell and A. C. Hearn, Phys. Rev. **96**, 1428 (1966)
[15] J. Ahrens et al.,Phys. Lett. B **672**, 328 (2009)
[16] H. Dutz et al., Phys. Rev. Lett. **94**, 162001 (2005)
[17] P. Aguar Bartolomè et al., Phys. Lett. B **723** (2013) 71-77
[18] H. Arenhövel, A. Fix, Phys. Rev. C **72** (2005) 064004
[19] R. Schiavilla, C. Pandharipande, Nucl. Phys. A **449** (1986) 219
[20] S. Costanza et al., Eur. Phys. J. A **50** 173 (2014)

 DOI: 10.1051/epjconf/201714201009

Dark Searches and $\gamma\gamma$ Physics at KLOE

Francesca Curciarello on behalf of the KLOE-2 Collaboration[1,a]

[1] *Laboratori Nazionali di Frascati, 00044, Frascati, Italy*

Abstract. The search for a dark sector mediated by a new gauge boson, the dark photon, is motivated by many astrophysical anomalies and by the $g-2$ discrepancy. The KLOE experiment, operating at the e^+e^- DAΦNE collider in Frascati, searched for a visibly-decaying dark photon by investigating the ϕ-Dalitz decay into the η meson, the dark photon production from continuum, and the Higgsstrahlung process. The KLOE-2 run started in November 2014, after the upgrade of both, DAΦNE and the KLOE apparatus. In particular, two high electron and positron tagger stations were installed in the DAΦNE layout to study $\gamma\gamma$ interactions at 1 GeV. Progress status of the project is given.

1 Introduction

One of the unsolved problems in modern physics is represented by the dark matter (DM). As for visible matter the only way to gain insights into the DM particle nature is to make it interact with baryonic matter. For this reason the hypothesis of a new interaction, dubbed as dark force, between DM and Standard Model (SM) particles, has motivated a great world-wide experimental effort, mostly regarding recent puzzling astrophysical and terrestrial anomalies [6–12]. The dark force should be carried by a new light (<1 GeV) boson messenger [1] with a very weak coupling defined as the ratio of effective dark and fine structure coupling constants, $\varepsilon^2 = \alpha'/\alpha \ll 1$ [1–5]. Moreover, the dark photon should get mass by means of a Higgs-like mechanism which implies the existence of at least one additional scalar particle, the dark Higgs h'. Dark photon search is also strongly motivated since it would give a positive one-loop contribution to the calculated value of the muon magnetic moment anomaly, a_μ, thus explaining the observed discrepancy with the experimental value, for dark photon masses of 10–100 MeV and coupling constant ε of about 10^{-3} [13]. In this paper, we present a review of all KLOE results on visibly-decaying dark photon, obtained by analysing data of the first KLOE run (2000-2006).

A new run, KLOE-2, has started in November 2014 with an upgraded DAΦNE and KLOE detector. One of the main goal of KLOE-2 is the precision measurement of the π^0 width, $\Gamma_{\pi^0\to\gamma\gamma}$, which is considered a strong test of low-energy QCD. The $\Gamma_{\pi^0\to\gamma\gamma}$ is predicted with 1.4% of precision ($\Gamma^{\rm Th.}_{\pi^0\to\gamma\gamma} = 8.09 \pm 0.11$eV [14, 15]) while the world average experimental value is known with 6% precision ($\Gamma^{\rm PDG}_{\pi^0\to\gamma\gamma} = 7.74 \pm 0.48$eV [16]), dominated by the discrepancy between the most sensitive measurements, the direct one and that obtained with the Primakoff conversion [17]. A way to achieve the precision needed to test QCD predictions is to study the π^0 production through $\gamma\gamma$ fusion in the $e^+e^- \to e^+e^-\gamma^\star\gamma^\star \to e^+e^-\pi^0$ reaction. This purpose could be achieved by KLOE-2 thanks to the installation, in both arms of the DAΦNE layout, of two tagger stations in order to tag high energy scattered electron and positrons coming from $\gamma\gamma$ interactions. About 10^4 events are expected at KLOE-2 with an integrated luminosity of 5 fb^{-1} [18].

KLOE-2 aims also to perform the first measurement of the $F_{\pi^0\gamma^\star\gamma}$ form factor at low $Q^2 \leq 0.1$ GeV^2 in the space-like region. This measurement can have an impact on the value and precision of the contribution of one-neutral pion exchange to the hadronic light-by-light scattering, $a_\mu^{\rm LbyL;\pi^0}$, a term of the muon anomaly, g-2 [18].

Progress and operation status of the $\gamma\gamma$ taggers are reviewed in Sec. 5.

2 DAΦNE collider and KLOE detector

DAΦNE is an e$^+$e$^-$ collider running at the energy $\sqrt{s} = m_\phi = 1.0195$ GeV located at Laboratori Nazionali di Frascati of INFN. It consists of a linear accelerator, a damping ring, nearly 180 m of transfer lines, and two storage rings that intersect at two points.

The KLOE detector is made up of a large cylindrical drift chamber (DC) [19], surrounded by a lead scintillating fiber electromagnetic calorimeter (EMC) [20]. A superconducting coil around the EMC provides a 0.52 T magnetic field. The calorimeter is divided into a barrel and two end–caps and covers 98% of the solid angle. The modules are read out at both ends by 4880 photo–multipliers. Energy and time resolutions are $\sigma_E/E = 5.7\%/\sqrt{E(\rm GeV)}$ and $\sigma_t = 57\,\text{ps} / \sqrt{E(\rm GeV)} \oplus 100\,\text{ps}$, respectively. The all-stereo drift chamber, $4\,m$ in diameter and $3.3\,m$ long, has a mechanical structure of carbon fiber-epoxy composite

[a] e-mail: Francesca.Curciarello@lnf.infn.it

[1] Referred to also as U, A', γ' or simply dark photon.

and operates with a light gas mixture (90% helium, 10% isobutane). The position resolutions are $\sigma_{xy} \sim 150\,\mu$m and $\sigma_z \sim 2$ mm. The momentum resolution $\sigma_{p_\perp}/p_\perp$ is better than 0.4% for large angle tracks. Vertices are reconstructed with a spatial resolution of ~ 3 mm.

To allow a better vertex reconstruction near the interaction point (IP), an higher acceptance to low p_t tracks and to increase the tightness of the detector, KLOE has been upgraded with a inner tracker [21] (four layers of triple GEM) and two calorimeters, the CCALT [22], made of LYSO crystals, and the QCALT [22], surrounding the DAΦNE quadrupoles and made of tungsten and plastic scintillator tiles. Moreover, in both arms of the DAΦNE layout two couple of tagger stations have been installed to study $\gamma\gamma$ physics. The low energy tagger (LET) [23], has been installed at 1 m from the IP in order to tag e^+e^- of $160 < E < 400$ MeV, and it is made up of LYSO crystals. The high energy tagger (HET) [24] is a plastic scintillator hodoscope whose energy acceptance is expected from $420 < E < 490$ MeV and has been installed at 11 m from the IP just after the DAΦNE bending dipoles.

3 Dark forces

The KLOE Collaboration searched for dark photon signature by investigating three processes and six different channels. In Fig. 1 all limits set by KLOE are reported in comparison with all other constraints in the 0-1 GeV mass range. In the next sections all the analyses performed to extract the KLOE limits are described.

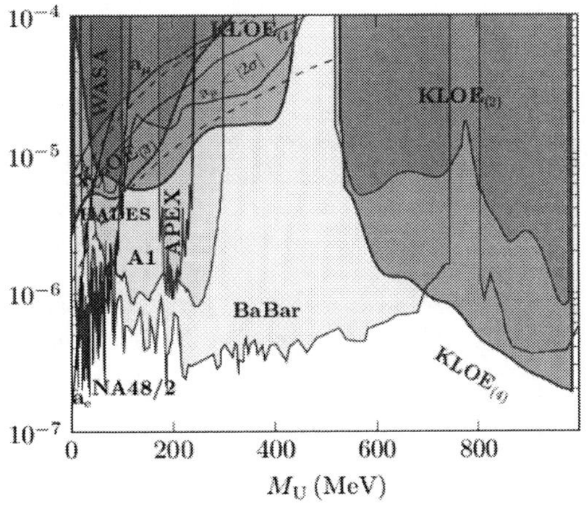

Figure 1. 90% CL exclusion plot for ε^2 as a function of the U-boson mass. The limits from the A1 [25] and APEX [26] fixed-target experiments, the limits from the ϕ Dalitz dacay (KLOE$_{(1)}$) [27, 28] and $e^+e^- \to U\gamma$, $U \to \mu^+\mu^-$, e^+e^-, $\pi^+\pi^-$ (KLOE$_{(2)}$, KLOE$_{(3)}$ and KLOE$_{(4)}$ respectively) [29–31], the WASA [32], HADES [33], BaBar [34] and NA48/2 [35] limits are shown. The solid lines are the limits from the muon and electron anomaly [13]. The gray line shows the U boson parameters that could explain the discrepancy between SM prediction and the experimental value of muon anomaly, a_μ, with a 2σ error band (gray dashed lines) [13].

3.1 ϕ-Dalitz decay

The dark photon is expected to be produced in vector to pseudoscalar meson decays with a rate ε^2 times suppressed with respect to the ordinary transitions [36], producing a peak in the invariant mass distribution of the electron-positron pair over the continuum Dalitz background.

The KLOE Collaboration set a constraint on the U-boson coupling ε^2, by exploiting the $\phi \to \eta e^+e^-$ decay, where the η meson is tagged by $\pi^+\pi^-\pi^0$ [27] and $3\pi^0$ decays [28]. The first analysis used a data sample of 1.5 fb^{-1} of integrated luminosity. The limit on the number of U-boson events has been set by using the Confidence Level Signal (CL$_S$) technique [37–39]. This first upper limit (UL) has been then updated, improving sample statistics and background rejection (2% of background contamination), and combined with a new limit derived by tagging the η meson by its neutral decay into $3\pi^0$ [28]. For this new analysis, 30577 events are selected from a data sample of 1.7 fb^{-1} with 3% of residual background. For each channel, the irreducible background is extracted directly by a fit to data (see red line in Fig. 2) excluding for each U-boson mass hypothesis the signal region used for the upper limit evaluation (5 MeV centred around m_U). A

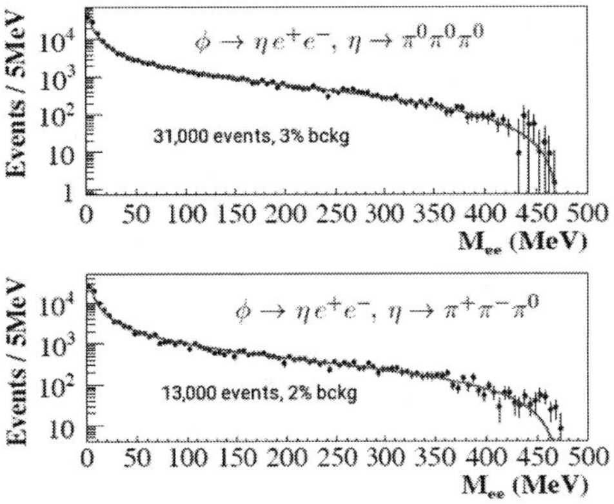

Figure 2. M_{ee} spectrum for the Dalitz decay $\phi \to \eta e^+e^-$, with $\eta \to \pi^0\pi^0\pi^0$ (top) and $\eta \to \pi^+\pi^-\pi^0$ (bottom), red line is a fit to the distribution.

combined UL on the parameter ε^2 at 90% CL has been derived by using the Vector Meson Dominance expectation for the transition form factor slope ($b_{\phi\eta} \sim 1$ GeV^2) resulting in $\varepsilon^2 < 1.7\times10^{-5}$ for $30 < M_U < 400$ MeV, and $\varepsilon^2 < 8.0\times10^{-6}$ for the sub-region $50 < M_U < 210$ MeV. The above final combined limit is shown in Fig. 1 and dubbed as KLOE$_{(1)}$. This limit, published in 2013 [28] rules out a wide range of U-boson parameters that could explain the a_μ discrepancy in the hypothesis of visibly-decaying dark photon.

3.2 Uγ events

Radiative U-boson production in $e^+e^- \to U\gamma$, $U \to l^+l^-$, $l = e, \mu, \pi$ events is a very sensitive process, independent of the details of the dark Higgs sector. The U boson

DOI: 10.1051/epjconf/201714201009

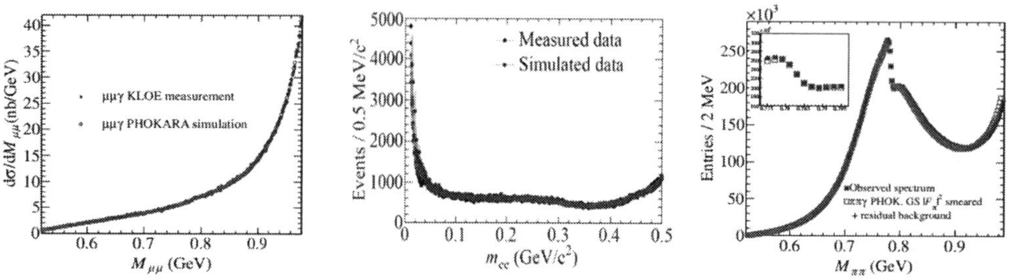

Figure 3. Invariant mass distributions at the end of the analysis selections for the $e^+e^- \to U\gamma$, $U \to \mu^+\mu^+$, e^+e^-, $\pi^+\pi^+$ processes. From left to right: $M_{\mu\mu}$, M_{ee} and $M_{\pi\pi}$ in comparison with expectation from PHOKHARA ($M_{\mu\mu}$ and $M_{\pi\pi}$) and BabaYaga (M_{ee}) Monte Carlo code.

should appear as a resonant peak in the invariant mass of lepton or hadron pairs. KLOE investigated both leptonic and hadron decays into pions. Particularly, the hadronic channel allows to increase the sensitivity in the $\rho - \omega$ resonance region because of the U dominant branching fraction into hadrons. The searches for U $\to \mu^+\mu^-$, $\pi^+\pi^-$ exploited a statistics corresponding to an integrated luminosity of 239.3 pb^{-1} and 1.93 fb^{-1} respectively, and selected events with a small angle Initial State Radiation (ISR) photon [40] and 2 charged tracks with acceptance between 50° and 130°. The application of kinematical cuts and the small angle event selection allowed to reduce the background coming from Final State Radiation (FSR) and ϕ-resonant processes and to increase sensitivity [41] on the dark photon decay. To approach the dielectron mass threshold, the search for U $\to e^+e^-$ has been performed by applying a large angle event selection for both ISR photon and charged leptons ($55° < \theta_{e,\gamma} < 125°$) to a data sample of about 1.5 fb^{-1}. Kinematical cuts have been used to remove FSR and resonant backgrounds. To avoid contamination from γ conversion processes in the beam pipe, we asked for events entirely contained within the vacuum pipe ($\rho_{PCA} < 1$ cm, and $|z_{PCA}| < 6$ cm). At the end of the analysis selection the residual background is less then 1%. No significant dark photon signature has been observed and limits at 90% CL have been extracted for all processes on the number of U events by means of the CL_S technique [37–39]. The expected backgrounds have been estimated by a fit to side bands for electron and pion decay channels while for the muon channel a PHOKHARA [42] MC generation has been used (see Fig. 3). The limits on U events have then been converted in limits on ε^2 by using the formula reported in Refs. [29–31]. The resulting exclusion limits are shown in Fig. 1 with all other existing limits in the region 0–1000 MeV [25–35]. The $e^+e^-\gamma$ limit excludes some of the remaining $g-2$ favored region.

4 Higgsstrahlung

The KLOE Collaboration investigated also the Higgsstrahulung process, sensitive to the dark coupling constant α_D and less suppressed with respect previous processes, with an expected cross section up to 1 pb at KLOE. The invisible scenario was considered, where the dark Higgs is lighter then the U boson and escapes detection, in the energy range between the dimuon mass threshold and 1 GeV. In this case, the expected signal is a muon pair from the U-boson decay plus missing energy. The analysis has been performed on two data samples of 1.65 fb^{-1} (collected on the ϕ peak) and 0.2 fb^{-1} collected at E_{cm} = 1000 MeV (off-peak sample) which is not affected by resonant backgrounds (see bottom panel of Fig. 4). In the on-peak analysis the huge background coming from kaon decays (see top panel of Fig. 4) has been reduced thanks to cuts on the quality of the vertex fit. No signal signature has been observed and a Bayesian limit on the number of signal events at 90% CL has been evaluated, bin-by-bin, for the on-peak and off-peak sample separately. Results have been translated in terms of $\alpha_D \times \varepsilon^2$ by using the integrated luminosity information, the signal efficiency, the dark Higgsstrahlung cross section and the branching fraction of the U $\to \mu^+\mu^-$ decay [43]. The combined upper limits [44] projected in the $M_{\mu\mu}$ and M_{miss} directions, for different dark Higgs mass hypotheses and slightly smoothed, are shown in Fig. 5. Values of the order of $10^{-9} \div 10^{-8}$ in $\alpha_D \times \varepsilon^2$ are excluded at 90% CL for a large range of the dark photon and dark Higgs masses. These limits translate in $\varepsilon \sim 10^{-3} - 10^{-4}$ for $\alpha_D = \alpha$ and are in agreement and complementary with BaBar and Belle results [45, 46] as they refer to the same process in a different final state and phase space region.

5 $\gamma\gamma$ physics at KLOE

One of the distinctive features of the KLOE-2 experiment is the possibility to investigate the two photon physics at very low transferred momentum in the space-like region. To distinguish the signal from the huge background coming from ϕ decays and reach the level of precision needed to test the strong QCD dynamic, the study of the interaction between two $\gamma's$ requires to tag the electron and the positron coming from the $e^+e^- \to e^+e^-\gamma^\star\gamma^\star \to e^+e^-X$ process. The leptons are scattered at very small polar angles and X is, a pseudoscalar meson produced by the fusion of the two quasi-real photons. In case of π^0 or η production the photons from the decay have to be detected in the central KLOE detector in association with the HET

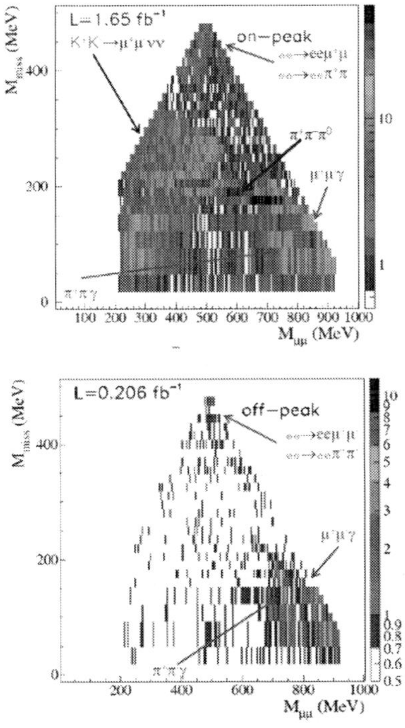

Figure 4. Top panel: $M_{\rm miss}$ Vs $M_{\mu\mu}$ for on-peak sample (1.65 fb^{-1} integrated luminosity). Bottom panel: $M_{\rm miss}$ Vs $M_{\mu\mu}$ for the off-peak sample, (0.2 fb^{-1} integrated luminosity).

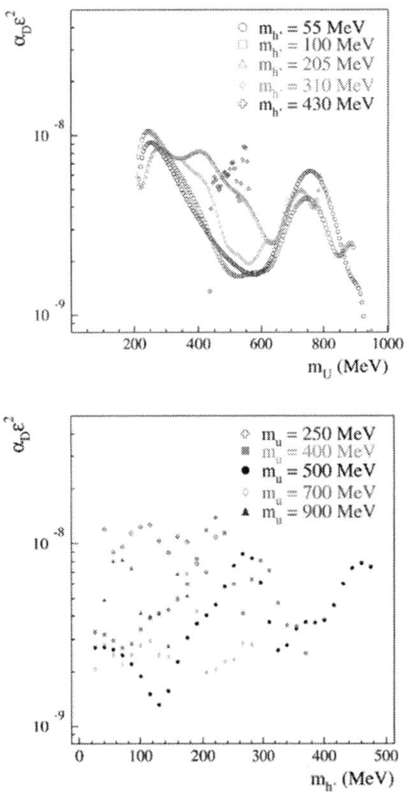

Figure 5. Combined 90% CL upper limits in $\alpha_{\rm D}\times\varepsilon^2$ as a function of $M_{\mu\mu}$ for different values of $m_{h'}$ (top panel) and as a function of $M_{\rm miss}$ for different values of $m_{\rm U}$ (bottom panel).

signal. For this reason two HET stations have been designed and installed just after the bending dipoles (acting like spectrometers) of DAΦNE and are currently operated during the KLOE-2 data taking.

In this section, the most relevant features of the KLOE-2 tagging system operation and the requirements used to classify double arm and single arm events will be presented.

5.1 The HET detector

The HET is a position detector able to measure the deviation of leptons from their main orbit in DAΦNE. The HET stations are placed at the exit of the dipole magnets, 11 m away from the IP, on both the positron and the electron arm. The sensitive area is made up of a set of 28 plastic scintillators (EJ-228), designed for ultra-fast timing and ultra-fast counting applications, with dimensions $(3 \times 5 \times 6)$ mm^3 each. One additional scintillator, of dimensions $(3 \times 50 \times 6)$ mm^3, whose acceptance covers all the others, is used to reinforce evidence for real particles crossing the detector. The light emitted by each of the 28 scintillators is read out through a plastic light guide by a high quantum efficiency photomultiplier (35% for wavelengths between 300 and 400 nm) which matches well the scintillator spectrum. The 28 scintillators are placed at different distances from the beam-line, in such a way that the measurement of the distance between beam and particle can be performed simply knowing which scintillator has been fired. A dedicated DAQ electronic board, based on a Xilinx Virtex-5 FPGA, has been developed for this detector. It provides a MultiHit TDC with a time resolution of 550(1) ps and the possibility to clearly identify the correct bunch crossing ($\Delta T_{\rm bunch}$ = 2.7ns).

5.2 HET operation

The two HET tagger stations have been operated since the very beginning of the KLOE-2 data taking in November 2014. Since KLOE and HET acquisition systems are asynchronous, we use the DAΦNE radiofrequency signal issued once per machine turn, every 325 ns, that is independently recorded by the HET and KLOE, to associate their data. There is a strong evidence that the HET counting rate is almost entirely due to Bhabha scattering events. The hit time structure, indeed, closely reproduces the DAΦNE bunch structure, as can be seen in Fig. 6, while the rate timeline strictly follows the luminosity timeline as measured by the KLOE central detector (KLOEcd). Fig. 7 shows the timelines of the counting rate of the HET stations. The measurement (blu points) are compared with the expectation (red curve) which can be described by the

DOI: 10.1051/epjconf/201714201009

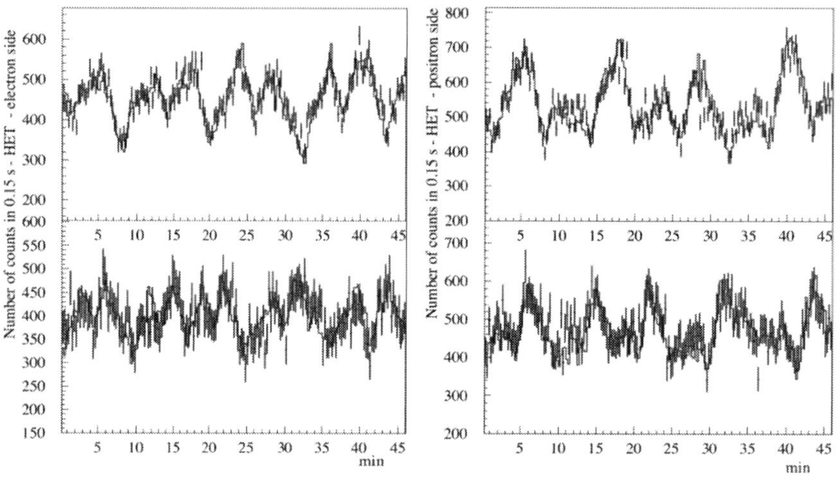

Figure 7. Timelines of the counting rate of the HET stations. The measurements (points with error bars) are compared with the expectation assuming stable DAΦNE operation and a dependence from the Touschek contribution and the luminosity, as obtained from no-collision runs, and the KLOE central detector, respectively. Upper plots are 2015 data and lower are data taken in year 2016. Left plots refer to the electron arm while right plots to the positron side.

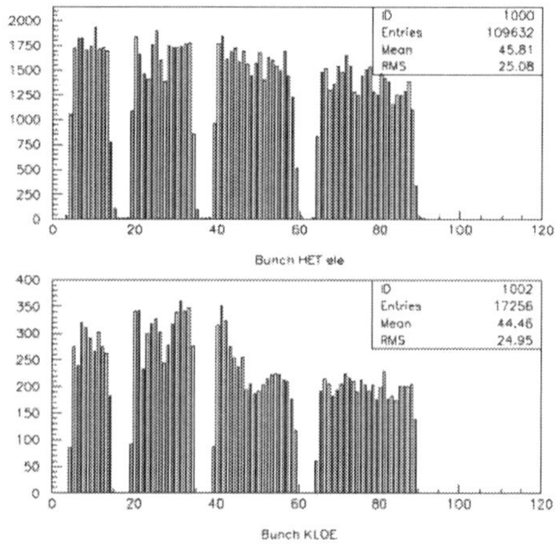

Figure 6. The DAΦNE bunch structure as measured by the electron HET station (upper panel) and by KLOE central detector (lower panel).

following relation:

$$R_{\mathrm{HET}_{e,p}} = \frac{R_{\mathrm{trig}}}{\mathrm{kHz}}(\alpha_{\mathrm{L}_{e,p}} \frac{Lumi}{0.2\mathrm{nb}^{-1}\mathrm{s}^{-1}} + \beta_{\mathrm{e,p}} \frac{I^2_{\mathrm{e,p}}}{A^2}). \quad (1)$$

R_{trig} is the KLOEcd trigger rate, *Lumi* is the instantaneous luminosity value , $I^2_{\mathrm{e,p}}$ represents the electron or the positron beam currents squared. $\alpha_{\mathrm{L}_{e,p}}$ and $\beta_{\mathrm{e,p}}$ are coefficients (in Hz), normalized to a luminosity of 0.2 $\mathrm{nb}^{-1}\mathrm{s}^{-1}$ and 1 A squared, respectively. Since the HET acquisition system transfers data to the KLOE event-building process only when a valid trigger is asserted by the asynchronous data acquisition system of the KLOEcd, the recorded HET count rate is proportional to the KLOEcd trigger rate and the coefficients of eq. 1 are normalized to the instantaneous trigger rate (in kHz) to take into account such a dependence. In absence of collisions the timeline rate closely follows the approximate I^2 dependence expected for intra-bunch scattering events (Touschek particles). This can be seen in the upper plots of Fig. 8 where the timelines of the counting rate of the HET stations, from a special run operating DAΦNE in no-collision mode in year 2016, are shown. Blue points are measurements while the expectation is represented by the red curve. In the bottom plots the corresponding timelines for the positron and electron circulating currents are reported. The total HET rate is dominated by Bhabha scattering events and is at level of 500-600 kHz. The high counting rates and the fast and reliable feedbacks on machine operation make the HET an ideal luminometer detector as well.

5.3 π^0 search

Data are classified as single arm (SA) double arm (DA) events. The DA data sample has been selected by requiring a time delay between the two stations $\Delta_{\mathrm{e,p}}$ of 18.9 ns corresponding to $|\Delta_{\mathrm{e,p}}| < 7$ bunches. This data sample contains the HET-electron * HET-positron coincidences which are expected, from resolution studies, in a window of ± 1 bunch and a control sample, large enough to characterize the coincidences, with $2 < \Delta_{\mathrm{e,p}} < 7$ bunches. The SA data sample has been selected by choosing:

- events in time with the KLOE trigger, that is a time difference between trigger and clusters $-3 \leq \Delta T_{\mathrm{Trig-clu}} \leq 8$ bunches;
- events in time with a bunch giving two clusters in KLOE with a cluster energy $20 < E_{\mathrm{clu}} < 300$ MeV, with a time delay between the HET station and KLOE within 4 bunches ($|\Delta T_{\mathrm{HET-KLOE}}| < 4$ bunches).

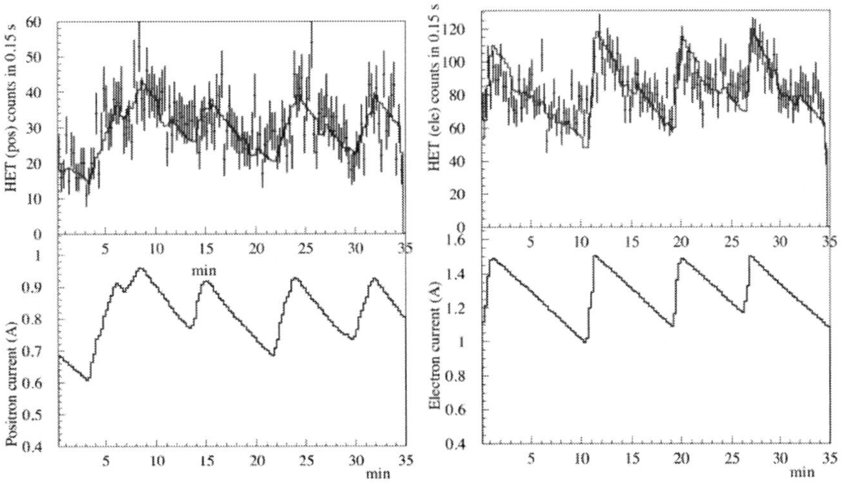

Figure 8. Timelines of the counting rate of the HET stations, from a special run operating DAΦNE in no-collision mode in year 2016. The measurements (points with error bars) are compared with the expectation assuming stable DAΦNE operation and a dependence from the Touschek contribution varying with the square of the circulating current. Bottom panel: timelines of the circulating currents for the same run. The additional peaks in the count rate are due to the KLOE trigger, needed to record HET data. Left plots refer to positron, right plots refer to electrons.

A fine inter-calibration of HET and KLOE TDCs based on bunch structure seen by the KLOE EMC and HET has been applied run by run.

An integrated luminosity of about 500 pb^{-1} has been processed so far, recording information on the hit in the tagger, trigger, DAΦNE operational parameters, clusters and tracks reconstructed in the KLOE central detector. The pre-selected DA sample statistics consists of about 10^8 fully-reconstructed events. The number of $\gamma\gamma$ expected events is predicted by using the *EKHARA* MC generator [47], taking into account the KLOE EMC resolution, in combination with the BDSIM package [48] (GEANT4 application), which allows to track leptons along the machine optics. About 650 events are expected from π^0 decays almost at rest at the current statistics, taking into account detector and trigger efficiency (65%). The analysis of the π^0 candidates requires:

(a) the coincidence between the taggers hits: $|\Delta_{ep}| < 2$ bunches and with the KLOE trigger; two KLOE clusters associated with the same bunch with $\Delta T_{\mathrm{KLOE_{clu}-HET}} \leq 4$ bunches and with energy $E_\gamma < 300$ MeV (see blue curve in Fig. 9)

(b) $E_\gamma > 20$ MeV (see black curve in Fig. 9)

(c) $30 < E_\gamma < 135$ MeV (see green curve in Fig. 9)

(d) $P_{\pi^0} < 90$ MeV; $\cos\alpha_{\gamma\gamma} < -0.8$;
$80 < M_{\gamma\gamma} < 230$ MeV; $|\Delta T - \Delta R/c| < 1.1$ ns (see red curve in Fig. 9)

Figure 9 shows the distribution of the total energy of the photon pairs at different stages of the π^0 selection, the distribution in red is obtained by applying the constraints suggested by EKHARA and BDSIM simulation (see point (d) above) taking into account of the detector resolution. The background evaluation is performed on a run-by-run basis. We select as background the events out of the HET coincidence window which we use in this study together with HET events that can not be in time with the KLOE trigger. The sample is dominated by random coincidences between HET and KLOE detector even at the end of the π^0 selection. Techniques to separate the random coincidences from π^0 signal are still in progress.

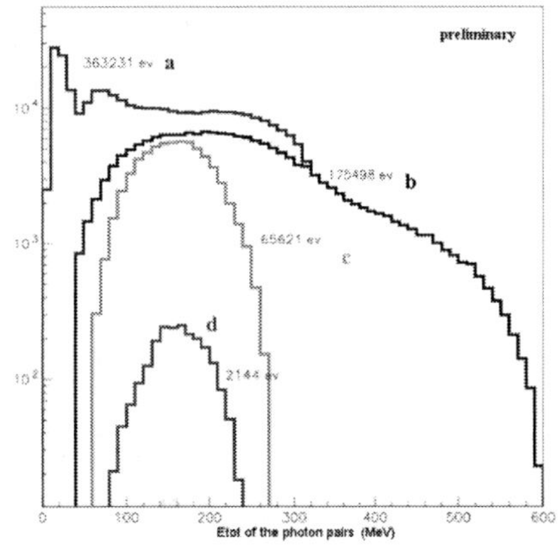

Figure 9. Distributions of the total energy of the photon pairs for samples satisfying different criteria as described in the text. The distribution in red is obtained from events within the ranges indicated by EKHARA and BDSIM simulations.

Conclusions

The KLOE Collaboration investigated the existence of the dark photon in the prompt and visibly-decay hypothesis, by means of the $\phi-$Dalitz decay, continuum processes and Higgsstrahlung. No signal has been found and stringent limits have been extracted below 1 GeV, constraining the $g-2$ favored region. A new run, with the KLOE-2 upgraded detector has started in November 2014 to record 5 fb^−1 within March 2018. The expected larger new statistical sample and the KLOE-2 inner tracking detector, which allows a better vertex reconstruction and a higher invariant mass resolution, will improve the current sensitivity on dark photon signal of a factor of about 2. Moreover, new searches for invisible dark photon decays into light DM states and for a leptophobic B boson are also planned. KLOE-2 aims also to perform the precision measurement (1% level) of the π^0 width which allows to test the strong QCD dynamics. For this reason two high energy taggers have been installed in the DAΦNE layout and are currently operated during the KLOE-2 data taking. The two HET stations are completely noiseless and the timelines of the counting rate for electron and positron stations show only two visible contributions from luminosity and from Touschek The total rate is dominated by the Bhabha scattering and is at the level of 500-600 kHz. These characteristics allow to have fast and reliable feed-backs on the machine operation making the HET a luminometer detector as well. We have pre-filtered candidates of single-π^0 production from $\gamma\gamma$ scattering. A total integrated luminosity of about 500 pb^{-1} of DA and SA data are being currently analyzed.

References

[1] B. Holdom, Phys. Lett. B **166**, 196 (1985)
[2] C. Boehm and P. Fayet, Nucl. Phys. B **683**, 219 (2004)
[3] P. Fayet, Phys. Rev. D **75**, 115017 (2007)
[4] Y. Mambrini, J. Cosmol. Astropart. Phys. **1009**, 022 (2010)
[5] M. Pospelov, A. Ritz and M.B. Voloshin, Phys. Lett. B **662**, 53 (2008)
[6] O. Adriani et al., Nature **458**, 607 (2009)
[7] M. Aguilar et al., Phys. Rev. Lett. **110**, 141102 (2013)
[8] P. Jean et al., Astronomy Astrophysics **407**, L55 (2003)
[9] J. Chang et al., Nature **456**, 362 (2008)
[10] F. Aharonian et al., Phys. Rev. Lett. **101**, 261104 (2008)
[11] A. A. Abdo et al., Phys. Rev. Lett. **102**, 181101 (2009)
[12] R. Barnabei et al., Eur. Phys. J. C **56**, 333 (2008)
[13] M. Pospelov, Phys. Rev. D **80**, 095002 (2009)
[14] K. Kampf, B. Moussallam, Phys. Rev. D **79**, 076005 (2009)
[15] J. Bijnens, K. Kampf, Nucl. Phys. B, Proc. Suppl. 207-208, 220-223 (2010)
[16] K. Nakamura, J. Phys. G **37**, 075021 (2010) (controllare quello 2012)
[17] H. Primakoff, Phys. Rev. **81**, 899 (1951)
[18] D. Babusci et al., Eur. Phys. J. C **72**, 1917 (2012)
[19] M. Adinolfi, et al., Nucl. Instrum. Meth. A **488**, 51 (2002)
[20] M. Adinolfi, et al., Nucl. Instrum. Meth. A **482**, 364 (2002)
[21] Bencivenni G., Domenici D., Nucl. Instrum. Meth. A, **581**, 581 (2007); Di Cicco A. and Morello G., Acta Phys. Pol. B, **46**, 73 (2015)
[22] Happacher F. and Martini M., Acta Phys. Pol. B, **46**, 87 (2015)
[23] Babusci D. et al., Acta Phys. Pol. B, **46**, 87 (2015)
[24] Babusci D. et al., Acta Phys. Pol. B, **46**, 81 (2015)
[25] H. Merkel et al., Phys. Rev. Lett. **112**, 221802 (2004)
[26] S. Abrahamyan et al., Phys. Rev. Lett. **107**, 191804 (2011)
[27] F. Archilli et al. Phys. Lett. B **706**, 251 (2012)
[28] D. Babusci et al., Phys. Lett. B **720**, 111 (2013)
[29] D. Babusci et al., Phys. Lett. B . **736**, 459 (2014)
[30] A. Anastasi et al., Phys. Lett. B **750**, 633 (2015)
[31] A. Anastasi et al., Phys. Lett .B **757**, 356 (2016)
[32] P. Adlarson et al., Phys. Lett. B **726**, 187 (2013)
[33] G. Agakishiev et al., Phys. Lett. B **731**, 265 (2014)
[34] J. P. Lees, Phys. Rev. Lett. **113**, 191804 (2014)
[35] J.R. Batley et al., Phys. Lett. B **746**, (2015)
[36] M. Reece and L. T. Wang, JHEP **0907**, 051 (2009)
[37] G. J. Feldman and R. D. Cousins, Phys. Rev. D **57**, 3873 (1998)
[38] T. Junk, Nucl. Instr. Meth. A **434**, 435 (1999)
[39] A. L. Read, J. Phys. G: Nucl. Part. Phys. **28**, 2693 (2002)
[40] D. Babusci et al., Phys. Lett. B **720**, 336 (2013)
[41] L. Barzè et al., Eur. Phys. J. C **71**, 1680 (2011)
[42] H. Czyż, A. Grzelinsk, J. H. Kühn and G. Rodrigo Eur. Phys. J. C **39**, 411 (2005)
[43] B. Batell et al., Phys. Rev. D **79**, 11508 (2009)
[44] A. Anastatsi et al., Phys. Lett. B **747**, 365 (2015)
[45] J. P. Lees et al., Phys. Rev. Lett. **108**, 211801 (2012)
[46] I. Jaegle et al., Phys. Rev. Lett. **114**, 211801 (2015)
[47] H. Czyz, S. Ivashyn, Comput. Phys. Commun. **182**, 1338 (2011)
[48] I. Agapov, G.A. Blair, S. Malton, L. Deacon, Nucl. Instrum. Meth. A **606**, 708 (2009)

EPJ Web of Conferences **142**, 01010 (2017) DOI: 10.1051/epjconf/201714201010
Advances in Dark Matter and Particle Physics

Monte Carlo generators for hadron physics: updates on PHOKHARA and EKHARA generators*

Henryk Czyż[1], Patrycja Kisza[1], and Szymon Tracz[1]

[1] *Institute of Physics, University of Silesia, PL-40007 Katowice, Poland.*

Abstract. A short review of current upgrades of the PHOKHARA and EKHARA generators is presented together with a report on the work in progress. The upgrades are based on a newly constructed model of the $\chi_{c_i} - \gamma^* - \gamma^*$ and $\chi_{c_i} - J/\psi^* - \gamma^*$ form factors. Within this model predictions were made for the electronic widths of χ_{c_1} and χ_{c_2} and, based on the event generators results, cross sections $e^+e^- \to \chi_{c_1}(\to J/\psi(\to \mu^+\mu^-)\gamma)$, $e^+e^- \to e^+e^-\chi_{c_1}$ and $e^+e^- \to e^+e^-\chi_{c_1}(\to J/\psi(\to \mu^+\mu^-)\gamma)$.

1 Introduction

The event generators PHOKHARA and EKHARA were developed to help measurements of the $e^+e^- \to$ hadrons + photons and $e^+e^- \to e^+e^-$ + hadrons cross sections. From their first publications [1, 2] the groups developing the codes put an effort to guarantee technical precision of the simulations at the level of 0.01%. In the same time models of hadron-photon interactions were developed and radiative corrections calculated and implemented into the event generator PHOKHARA to guarantee the physical accuracy at the level required by the experimental groups. Till now the radiative corrections were not implemented into the distributed version of the generator EKHARA, even if they were calculated. Problems with the low efficiency of the code for the processes $e^+e^- \to e^+e^-$+hadrons+photons where not yet solved even if the works towards the solution are well advanced.

For the processes simulated by the event generator PHOKHARA, the radiative corrections to the initial state radiation (ISR) were calculated in [3] and implemented in [1, 4]. The final state radiation (FSR) requires not only calculation of the radiative corrections, but also modeling of the hadron-photon interactions. This was studied only for the most important hadronic final states: charged pion and kaon pair production [5–8] and proton-anti-proton pair production [9]. The latter mode was at first implemented with ISR corrections only [10]. For other hadronic final states: three- [11] and four- [12, 13] pion modes and lambda pair production with their subsequent decays [14], only ISR corrections are implemented.

The $e^+e^- \to \mu^+\mu^-\gamma$ reaction is a separate subject of the studies as the final state does not involve hadrons. It serves as one of the luminosity monitoring tools. For this process the radiative corrections were gradually added [4, 7] and finally the complete corrections were calculated and implemented into the event generator in [15].

A special care was devoted to the treatment of narrow resonances [16], when modeling the radiative return processes [8]. New physical and technical problems arise in the vicinity of narrow resonances, which were solved in the event generator implementations. Lately, in [17], a possibility of simulation of the hadronic processes in a scan mode was also added. In this mode only ISR corrections are implemented with the exception of the proton-anti-proton pair production, where the Coulomb factor models approximately FSR corrections.

The first version of the EKHARA generator [2] was developed for the background studies at KLOE for the pion form factor measurements. Only later it was upgraded for studies of the $\gamma^* - \gamma^*$ processes starting with the reaction $e^+e^- \to e^+e^-\pi^0$ [18] and adding subsequently [19] better modeling of the $\pi - \gamma^* - \gamma^*$ form factor and possibility of the generation of the reactions $e^+e^- \to e^+e^-\eta(\eta')$. In [19] also a new model of the $\eta(\eta') - \gamma^* - \gamma^*$ form factors was developed to give predictions in agreement with experimental data.

In this proceedings we report on the latest progress on the upgrades of both event generators EKHARA and PHOKHARA. Results of the studies [20] on a direct production of $\chi_{c_1}(1P)$ and and $\chi_{c_2}(1P)$, J^{++}, $J = 1, 2$ charmonia in electron-positron annihilation are reported in Section 2. In Section 3 we report on the studies [21] on the prospects of the detailed investigations of the χ_{c_i}, $i = 0, 1, 2$ properties at BELLE2, using measurements of the reactions $e^+e^- \to e^+e^-\chi_{c_i}$ and $e^+e^- \to e^+e^-\chi_{c_i}(\to J/\psi(\to \mu^+\mu^-)\gamma)$. In Section 4 we give a short summary and sketch the near future developments in both event generators.

*Work supported in part by the Polish National Science Centre, grant number DEC-2012/07/B/ST2/03867 and German Research Foundation DFG under Contract No. Collaborative Research Center CRC-1044.

Table 1. Predictions of the electronic widths of the χ_{c_1} and χ_{c_2} charmonia within recently published models.

	[20]	[27]	[26]	[28]
$\Gamma(\chi_{c_1} \to e^+e^-)$ [eV]	0.43	0.046	0.367	0.1
$\Gamma(\chi_{c_2} \to e^+e^-)$ [eV]	4.25	0.037	0.137	-

2 PHOKHARA generator: direct χ_{c_1} and χ_{c_2} production in e^+e^- annihilation

The production of the quarkonium states with even charge conjugation in e^+e^- annihilation is suppressed, as two photon exchange is necessary to produce them. Yet, with the luminosity of the nowadays colliders, the loop induced processes, leading to direct production of such states are within experimental reach. This possibility was investigated in detail in [20]. The best place to search for such processes is the BES-III experiment even if in the original proposal [22] this possibility was not investigated. The reaction $e^+e^- \to \chi_{c_i}(\to J/\psi(\to \mu^+\mu^-)\gamma)$, $i = 1,2$ is especially suited for such a measurement as the muons and photon are well identified within the BES-III detector. The χ_{c_0} production rate is proportional to the electron mass and thus beyond reach of any near future experiment. For any realistic prediction of the cross section a model of the amplitudes $\chi_{c_i} \to J/\psi^*\gamma^*$, $\chi_{c_i} \to \gamma^*\gamma^*$ and $\psi(2S) \to \chi^*_{c_i}\gamma^*$ is needed with parameters describing well the observed decay widths $\Gamma(\chi_{c_i} \to J/\psi\gamma)$, $\Gamma(\chi_{c_i} \to \gamma\gamma)$ and $\Gamma(\psi(2S) \to \chi_{c_i}\gamma)$. Several models were built to meet this requirement [20, 23–28]. All of them are able to give predictions for these widths in agreement with the experimental data [29]. Yet, when using these models to predict the electronic widths of the χ_{c_1} and χ_{c_2} charmonia the models give predictions, which vary within one order of magnitude. The situation involving only recent papers is summarised in Table 1. The reason for these differences is clear: the predictions of the electronic widths involve the calculation of loop integrals and thus the shape of the form factors is very much important. While the widths, discussed above, fix only couplings at some well defined scales. It was shown also in [20] that the phases between different contributions are crucial for the size of the predicted electronic widths. The phases come from QED interactions and thus are almost fully predicted. Moreover the interferences give sizable contributions both to the electronic widths and the cross sections.

The non-reducible background for the χ_{c_1} and χ_{c_2} production in the reactions $e^+e^- \to \chi_{c_i}(\to J/\psi(\to \mu^+\mu^-)\gamma)$, $i = 1,2$ is the J/ψ production in radiative return with its subsequent decay to $\mu^+\mu^-$. This was discussed in [20], where it was shown that within the model developed there the main effect comes from the interference between the background and signal diagrams. We show it in Fig. 1 for χ_{c_1} and in Fig. 2 for χ_{c_2}. We assume here a beam resolution of 1 MeV for each beam and Gaussian distribution of energies within a beam. For χ_{c_1} there is also visible, but small, contribution coming from Z^0 exchange. This result

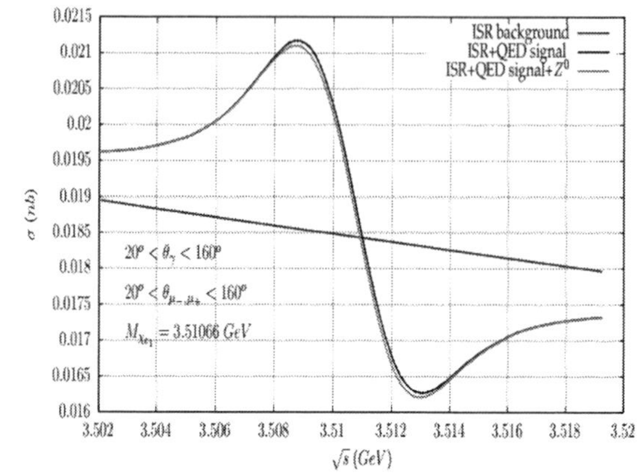

Figure 1. The cross section $e^+e^- \to \mu^+\mu^-\gamma$, see text for details.

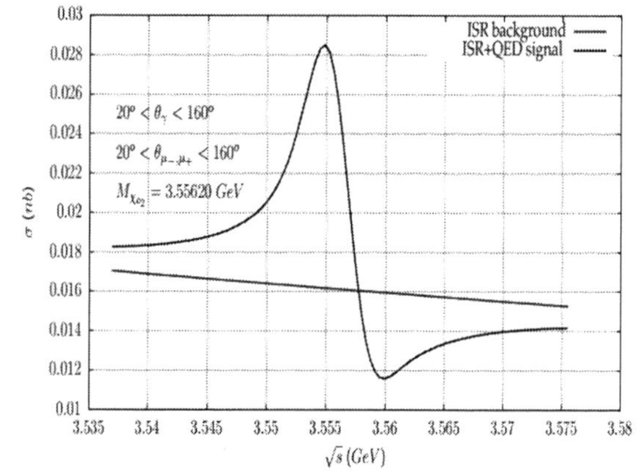

Figure 2. The cross section $e^+e^- \to \mu^+\mu^-\gamma$, see text for details.

is also model dependent. It can be tested experimentally with a scan experiment with energies chosen around the χ_{c_1} and χ_{c_2} masses.

The angular distributions of the muons are however more affected in regions, where the interference is not maximal. In Fig. 3 it is shown for χ_{c_1} and in Fig. 4 for χ_{c_2}. The energy of the experiment in both cases was chosen to be equal to the mass of the χ_{c_i}, $(i = 1,2)$ charmonium. In principle the studies of angular distributions would allow for detailed studies of the $\chi_{c_i} - J/\psi - \gamma$ amplitudes, but it would require collection of a big sample of events. Within the model used here [20] only one of the five allowed amplitudes gives contribution to the $\chi_{c_2} - J/\psi - \gamma$ amplitude and only one specific combination of the allowed amplitudes gives contribution to the $\chi_{c_1} - J/\psi - \gamma$ amplitude. Thus the predictions shown here are again specific to one of the models. Yet, one has to say that only in [20] the production of the χ_{c_1} and χ_{c_2} states in e^+e^- annihilation was studied, while other groups gave only predictions for the electronic widths. As the interference effects with the non-reducible background are crucial for these studies, the predictions of the electronic widths are not sufficient for the

DOI: 10.1051/epjconf/201714201010

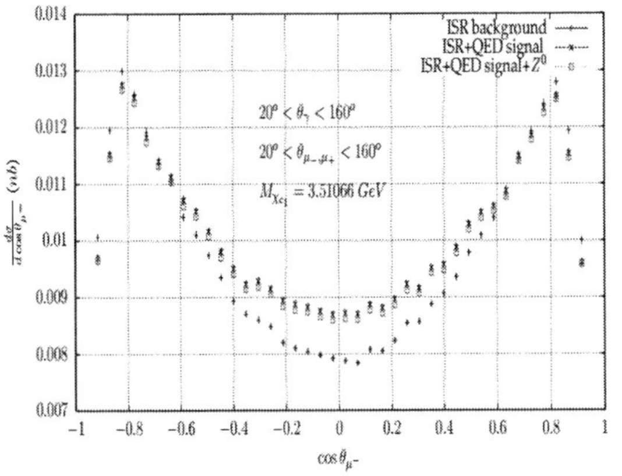

Figure 3. The differential (in μ^- polar angle)cross section $e^+e^- \to \mu^+\mu^-\gamma$, see text for details.

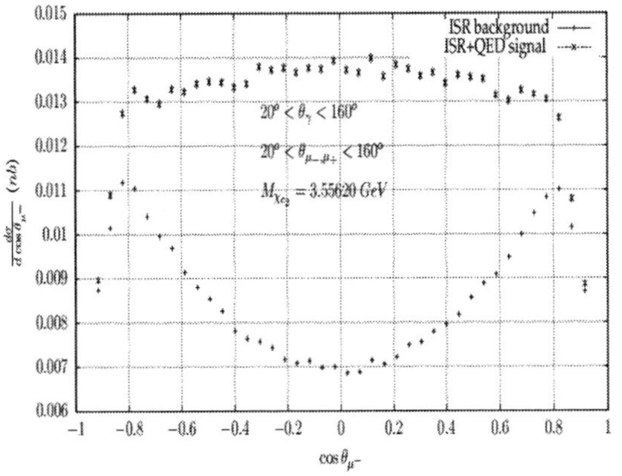

Figure 4. The differential (in μ^- polar angle) cross section $e^+e^- \to \mu^+\mu^-\gamma$, see text for details.

extraction of the model parameters from the experimental data.

3 EKHARA generator: χ_{c_i}, $i = 0, 1, 2$ production in $\gamma^* - \gamma^*$ processes

It is clear from the previous section that a measurement of the $\chi_{c_i} - \gamma^* - \gamma^*$ form factors is crucial to test the models in detail. In [21] it was advocated that such measurement will be possible at BELLE2 experiment [30]. Here we report on this possibility showing also a complementary material not presented in [21]. The process which was advocated to be used is $e^+e^- \to e^+e^-\chi_{c_i}(\to J/\psi(\to \mu^+\mu^-)\gamma)$. Measurement of the muons and photon four-momenta will allow for a construction of the χ_{c_i} and J/ψ invariant masses and thus for a clear identification of the final state. In the presented plots we assume that the invariant mass of the $\mu^+\mu^-$ pair is within ten J/ψ widths from the mass of the J/ψ and the invariant mass of the $\mu^+\mu^-\gamma$ particles is within ten $\chi_{c_i}, (i = 0, 1, 2)$ widths from its mass. We show the distributions of the event rates in the laboratory frame of the

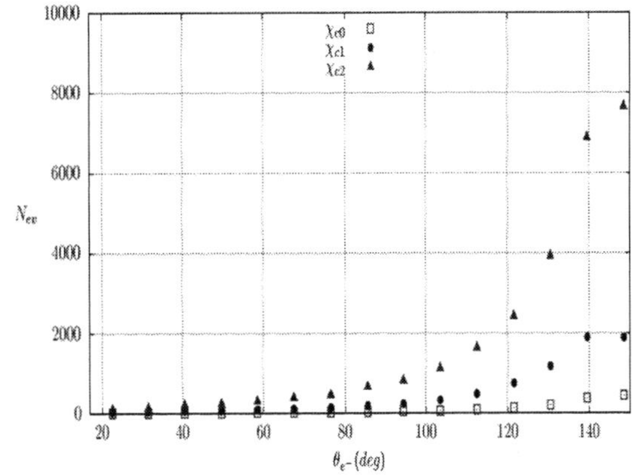

Figure 5. The differential (in e^- polar angle) cross section $e^+e^- \to e^+e^-J/\psi(\to \mu^+\mu^-)\gamma$, see text for details.

BELLE2 experiment assuming the asymmetric beams of 4 and 7 GeV with the half crossing angle of 41.5 mrad. We assume also that the particles (μ^+, μ^-, photon and electron or positron) can be detected and their four momenta measured if their polar angles are between 17° and 150° [31]. With the single tag events, when the final electron momenta is also measured, it will allow to cover the range of the χ_{c_i}–γ^*–γ^* transition form factor up to about $-25\ \mathrm{GeV}^2$ for one of the invariants [21]. This corresponds to the polar angle distribution of the observed electron shown in Fig. 5. The number of events corresponds there to the integrated luminosity of 50 ab^{-1}.

The accumulated luminosity of 50 ab^{-1} will allow also for a measurement of the $\chi_{c_i} - \gamma^* - \gamma^*$ transition form factors with both invariants substantially different from zero. In these case one has to measure both four momenta of the final electron and the final positron. The range of the γ^* invariant masses which can be covered spans up to about -10 GeV^2 [21]. It corresponds to the polar angles distribution of the final electron and final positron shown in Fig. 6 for χ_{c_1} and in Fig. 7 for χ_{c_2}.

4 Conclusions and near future developments

In [20] a model of the $\chi_{c_i} - \gamma^* - \gamma^*$ and $\chi_{c_i} - J/\psi^* - \gamma^*$ $(i = 1, 2)$ transition form factors was developed. In [21] it was extended to cover also the case of χ_{c_0}. It allowed [20] to predict the electronic widths of χ_{c_1} and χ_{c_2} and to show how to study the direct production of these states at BES-III experiment. In [21] it was shown that at BELLE2 experiment detailed studies of the $\chi_{c_i} - \gamma^* - \gamma^*$ transition form factors will be possible. The amplitudes predicted within the developed model were implemented into PHOKHARA9.2 (http://ific.uv.es/~rodrigo/phokhara/) and EKHARA2.2 (http://www.us.edu.pl/~ekhara) event generators. Hopefully the generators will serve well the experimental community, helping in performing the measurements discussed here.

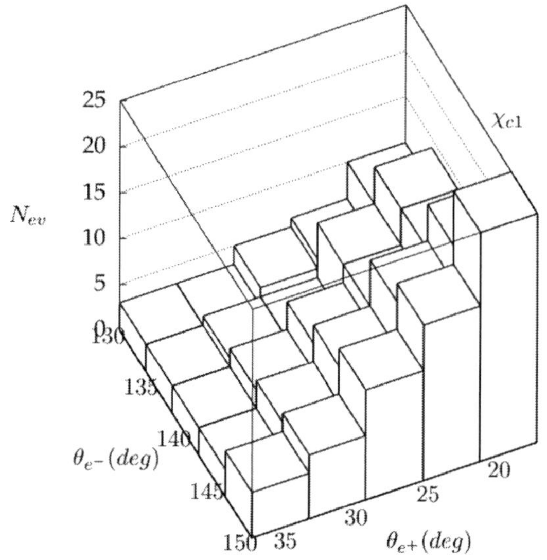

Figure 6. The differential (in e^- and e^+ polar angles) cross section $e^+e^- \to e^+e^-\chi_{c_1}(\to J/\psi(\to \mu^+\mu^-)\gamma)$, see text for details.

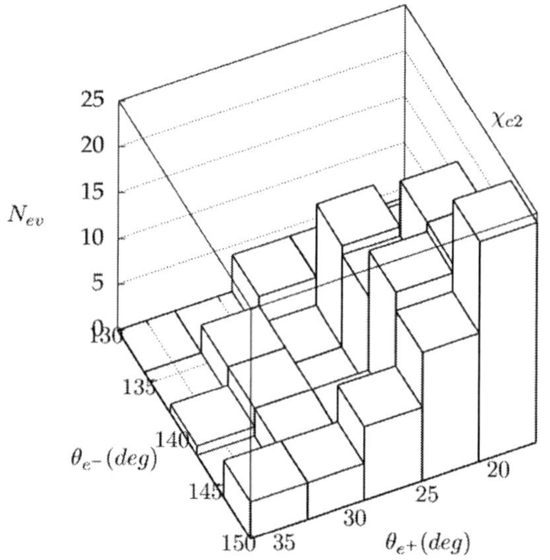

Figure 7. The differential (in e^- and e^+ polar angles) cross section $e^+e^- \to e^+e^-\chi_{c_2}(\to J/\psi(\to \mu^+\mu^-)\gamma)$, see text for details.

The generators will be further extended in the near future: in PHOKHARA generator the radiative corrections for the reaction $e^+e^- \to \pi^+\pi^-\gamma$ will be completed, while in the EKHARA generator the radiative corrections will be included for the first time. Both works are well advanced.

References

[1] G. Rodrigo, H. Czyz, J.H. Kühn, M. Szopa, Eur.Phys.J. **C24**, 71 (2002), hep-ph/0112184

[2] H. Czyz, E. Nowak-Kubat, Phys. Lett. **B634**, 493 (2006), hep-ph/0601169

[3] G. Rodrigo, A. Gehrmann-De Ridder, M. Guilleaume, J.H. Kuhn, Eur. Phys. J. **C22**, 81 (2001), hep-ph/0106132

[4] H. Czyz, A. Grzelinska, J.H. Kuhn, G. Rodrigo, Eur. Phys. J. **C27**, 563 (2003), hep-ph/0212225

[5] H. Czyz, A. Grzelinska, J.H. Kuhn, G. Rodrigo, Eur. Phys. J. **C33**, 333 (2004), hep-ph/0308312

[6] H. Czyz, A. Grzelinska, J.H. Kuhn, Phys. Lett. **B611**, 116 (2005), hep-ph/0412239

[7] H. Czyz, A. Grzelinska, J.H. Kuhn, G. Rodrigo, Eur. Phys. J. **C39**, 411 (2005), hep-ph/0404078

[8] H. Czyz, A. Grzelinska, J.H. Kuhn, Phys. Rev. **D81**, 094014 (2010), 1002.0279

[9] H. Czyz, J.H. Kühn, S. Tracz, Phys. Rev. **D90**, 114021 (2014), 1407.7995

[10] H. Czyz, J.H. Kuhn, E. Nowak, G. Rodrigo, Eur. Phys. J. **C35**, 527 (2004), hep-ph/0403062

[11] H. Czyz, A. Grzelinska, J.H. Kuhn, G. Rodrigo, Eur. Phys. J. **C47**, 617 (2006), hep-ph/0512180

[12] H. Czyz, J.H. Kuhn, Eur. Phys. J. **C18**, 497 (2001), hep-ph/0008262

[13] H. Czyz, J.H. Kuhn, A. Wapienik, Phys. Rev. **D77**, 114005 (2008), 0804.0359

[14] H. Czyz, A. Grzelinska, J.H. Kuhn, Phys. Rev. **D75**, 074026 (2007), hep-ph/0702122

[15] F. Campanario, H. Czyż, J. Gluza, M. Gunia, T. Riemann, G. Rodrigo, V. Yundin, JHEP **02**, 114 (2014), 1312.3610

[16] H. Czyz, J.H. Kuhn, Phys. Rev. **D80**, 034035 (2009), 0904.0515

[17] H. Czyż, M. Gunia, J.H. Kühn, JHEP **08**, 110 (2013), 1306.1985

[18] H. Czyz, S. Ivashyn, Comput. Phys. Commun. **182**, 1338 (2011), 1009.1881

[19] H. Czyz, S. Ivashyn, A. Korchin, O. Shekhovtsova, Phys. Rev. **D85**, 094010 (2012), 1202.1171

[20] H. Czyz, J.H. Kühn, S. Tracz, Phys. Rev. **D94**, 034033 (2016), 1605.06803

[21] H. Czyż, P. Kisza (2016), 1612.07509

[22] D.M. Asner et al., Int. J. Mod. Phys. **A24**, S1 (2009), 0809.1869

[23] J.H. Kühn, J. Kaplan, E.G.O. Safiani, Nucl.Phys. **B157**, 125 (1979)

[24] R. Barbieri, R. Gatto, R. Kogerler, Phys. Lett. **B60**, 183 (1976)

[25] J. Kaplan, J.H. Kühn, Phys. Lett. **B78**, 252 (1978)

[26] D. Yang, S. Zhao, Eur. Phys. J. **C72**, 1996 (2012), 1203.3389

[27] N. Kivel, M. Vanderhaeghen, JHEP **02**, 032 (2016), 1509.07375

[28] A. Denig, F.K. Guo, C. Hanhart, A.V. Nefediev, Phys. Lett. **B736**, 221 (2014), 1405.3404

[29] C. Patrignani et al. (Particle Data Group), Chin. Phys. **C40**, 100001 (2016)

[30] T. Aushev et al. (2010), 1002.5012

[31] T. Abe et al. (Belle-II) (2010), 1011.0352

The HPS experiment at JLab

Marzio De Napoli[1,a] for the HPS Collaboration

[1] *INFN - Sezione di Catania, Via Santa Sofia 64, 95123 Catania, Italy*

Abstract. Many Beyond-Standard-Model theories predict a new massive gauge boson, such as a "dark" or "heavy photon". The heavy photon is expected to mix with the Standard Model photon through kinetic mixing and therefore have a small coupling to electric charge. The Heavy Photon Search (HPS) experiment is searching for a heavy photon at the Thomas Jefferson National Accelerator Facility (JLab), in the mass range 20-500 MeV/c^2. In particular HPS looks for the e^+e^- decay channel of heavy photons radiated by electron Bremsstrahlung, employing both an invariant mass search and detached vertexing techniques. The experiment employs a compact forward spectrometer comprising silicon microstrip detectors for vertexing and tracking and an electromagnetic calorimeter for particle identification and triggering. HPS took data successfully in 2015 and 2016 at 1.05 GeV and 2.3 GeV beam energies, respectively. First results are expected to be presented soon.

1 Introduction

The existence of an additional U(1) gauge symmetry in nature is predicted by several beyond the Standard Model theories [1, 2]. The associated gauge boson A', aka "dark" or "heavy photon", can kinematically mix with the photon through quantum loops of heavy particles that carry both the standard model hypercharge and the dark force equivalent [3, 4]. This kinetic mixing results in an effective interaction $\epsilon e A'_\mu J^\mu_{EM}$ of the A' to the electromagnetic current J^μ_{EM}, suppressed relative to the electron charge e by a factor $\epsilon \leq 10^{-3}$. Since heavy photons couple to electrons, they can be produced by electron Bremsstrahlung off heavy nuclear targets and can decay to e^+e^- pairs.

HPS is an experiment to search for A' in fixed target electroproduction at Hall B of the JLab. The high duty-cycle, high-quality and high-intensity CEBAF electron beam is crucial to generate large integrated luminosities while keeping beam-related background under control. The electron beam, of intensity between 50 nA and 400 nA, impinging on 0.15% - 0.25% radiation length tungsten foils, can produce heavy photons through a process analogous to Bremsstrahlung, and the A' can then decay to e^+e^- pairs (Fig. 1). Since the coupling of heavy photons to e^+e^- pairs is much weaker than the electromagnetic coupling, heavy photon production is buried under a background of pairs from virtual photons. By reconstructing the momenta of the e^+e^- pairs, HPS is searching for a narrow resonance in the invariant mass spectrum over the quantum electrodynamics (QED) background (the traditional "bump-hunting"). The main QED background reactions are the radiative trident production and the Bethe-Heitler process (Fig. 1). Although the Bethe-Heitler process has much larger total cross-section than the radiative one, it can be significantly reduced with appropriate kinematic cuts. Instead, the radiative trident process is an irreducible background, with the same kinematics as the signal in a small invariant-mass window centered on the A' mass. The signal is characterized by a very-forward emitted A' that carries most of the beam energy E. Therefore, the A' decay products are boosted forward with typical polar angles $\sim M_{A'}/E$, of the order of few degrees. Moreover, depending on $M_{A'}$ and ϵ, A' can be long-lived, with a distinct secondary decay vertex detached from the target. HPS is also using this signature to distinguish the signal from the QED e^+e^- pairs promptly produced in the target.

2 The HPS Experiment

The HPS detector was designed to match the experimental requirements dictated by the search strategies and the signal kinematics discussed above. Resonance search requires excellent track reconstruction for electrons and positrons with a good angular and momentum resolution, which is most easily accomplished with charged particle spectrometry. Detached vertexing requires the tracking detectors to be placed just downstream of the target to minimize extrapolation errors and optimize vertex resolution. Detectors must be also placed as close to the beam direction as possible to cover with high acceptance the forward angular region.

The proximity of the tracking detectors to the beam dictates stringent requirements on acceptable beam size, stability, and halo. At the same time it necessitates protection collimation and real time monitoring and circuitry to protect against errant beam motion. During the data taking in 2015 and 2016 high quality beam was delivered and

[a] e-mail: marzio.denapoli@ct.infn.it

DOI: 10.1051/epjconf/201714201011

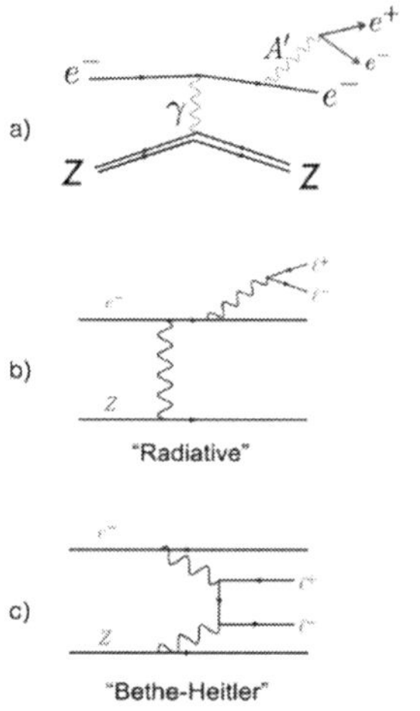

Figure 1. (*a*) Feynman diagram for radiative A' electroproduction and A' decay to e^+e^- pairs. Diagrams of radiative (*b*) and Bethe-Heitler (*c*) QED reactions, the primary background for the A' search, are also shown.

the beam feedback and fast shut down systems described in [5] worked in maintaining the beam position stable and protecting the silicon sensors from errant beam exposure.
The HPS apparatus is split into symmetric upper and lower halves since the detectors must avoid a region of intense electron flux caused by the small angle multiple coulomb scattering in the target and the horizontal dispersion in the magnetic field of the analyzing magnet of the electrons which have radiated energy in the target. The resulting "dead zone" has an angular range of θ_y <15 mrad (y is the vertical dimension). In addition, the beam is transported in vacuum through the tracker to minimize beam-gas interaction backgrounds. The detector setup is shown in Fig. 2. It is made of a $\sim$ 1m long Silicon Vertex Tracker (SVT) inside an analyzing dipole magnet to reconstruct charged particle trajectories and vertexes, and a fast lead-tungstate electromagnetic calorimeter (ECal) to measure particle energies, identify the e^+e^- pairs, and generate the trigger of the experiment.
The SVT is made of 6 layers of pairs of closely-spaced silicon microstrip sensors mounted back-to-back to form a module. A small stereo angle between the two layers (50 or 100 mrad) provides three-dimensional tracking and vertexing. The SVT material budget in the tracking volume is minimized to reduce multiple scattering and optimize both the invariant mass and vertex resolutions that directly affect the experimental sensitivity. The sensors have an active area of 98.33×38.34 mm^2, they are 320 $\pm$ 20μm thick and have a sense (readout) pitch of 30 (60) μm. To achieve large acceptance and good vertexing resolution, the first layer of silicon detectors is placed just 10 cm downstream of the target with the sensor edges only 500 μm above and below the beam. Sensors are readout by the APV25 chip [6] with a custom DAQ system based on the Advanced Telecom Communications Architecture (ATCA) technology.
The ECal [7] is made of 442 $PbWO_4$ crystals, 16 cm thick, with a 13.3x13.3 mm^2 (16x16mm^2) front (rear) face. In each half of the detector, crystals are arranged in five rows of 46 crystals, except the one closest to the beam that has only 37 crystals. The light from each crystal, approximately 120 photons/MeV, is read out by large-area Avalanche PhotoDiodes (APD) glued to the back face. Signals from the APDs are read and amplified using custom-made pre-amplifiers before being sent to the JLab Flash ADC boards for digitization and processing. The on-board FPGA processes the digitized samples to provide information about the energy and the timing associated with each hit. This information, available during data-taking, is exploited by an advanced triggering algorithm matched to the foreseen A' signal.
The data collected in 2015 and 2016 demonstrated excellent performances of both the detectors. The SVT and Ecal primary goals of providing an high quality vertexing and tracking and a fast trigger in a large background environment were fully achieved. We anticipate first results from the brief 2015 1.1 GeV run in early 2017.

References

[1] Feldman D., Kors B. and Nath P., Phys. Rev. D **75**, 023503 (2007)

[2] Andreas S., Goodsell M. and Ringwald A., Phys. Rev. D **87**, 025007 (2013)

[3] Galison P. and Manohar A., Phys. Lett. B **136**, 279 (1984)

[4] Holdom B., Phys. Lett. B **166**, 196 (1984)

[5] N. Baltzell et al., arXiv:1612.07821

[6] French M., Jones L., Morrissey Q., Neviani A., Turchetta R. et al., Nucl. Instr. and Meth. A **466**, 359-65 (2001)

[7] Balossino I. et al., Nucl. Instr. and Meth. A **854**, 89-99 (2017)

DOI: 10.1051/epjconf/201714201011

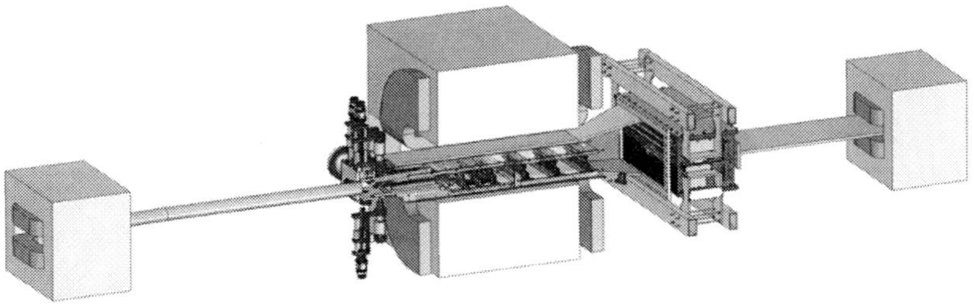

Figure 2. Layout of the HPS detector setup. From left to right: the first dipole magnet deflects the beam toward the target placed at the entrance of the second analyzing dipole magnet, the SVT inside the analyzing magnet, the ECal and the third dipole magnet to steer back the beam toward the beam dump.

EPJ Web of Conferences **142**, 01012 (2017) DOI: 10.1051/epjconf/201714201012
Advances in Dark Matter and Particle Physics

Problems with the sources of the observed gravitational waves and their resolution

A.D. Dolgov[1,2,a]

[1] *NSU, Novosibirsk, 630090, Russia*
[2] *ITEP, Moscow, 117218, Russia*

Abstract. Recent direct registration of gravitational waves by LIGO and astronomical observations of the universe at redshifts 5-10 demonstrate that the standard astrophysics and cosmology are in tension with the data. The origin of the source of the GW150914 event, which presumably is a binary of coalescing black holes with masses about 30 solar masses, each with zero spin, as well as the densely populated universe at z= 5-10 by superheavy black holes, blight galaxies, supernovae, and dust does not fit the standard astrophysical picture. It is shown here that the model of primordial black hole (PBH) formation, suggested in 1993, nicely explains all these and more puzzles, including those in contemporary universe, such as MACHOs and the mass spectrum of the observed solar mass black holes.. The mass spectrum and density of PBH is predicted. The scenario may possibly lead to abundant antimatter in the universe and even in the Galaxy.

1 Introduction

The standard cosmological ΛCDM model (here CMD stands for Cold Dark Matter and Λ for vacuum-like dark energy, or, what is the same, Λ-term) very well describes gross features of the universe such as the spectrum of perturbations at large scales, features of CMB (especially the shape of the angular fluctuation spectrum), baryogenesis, big bang nucleosynthesis, *etc* at expense of a few parameters. However many ingredients of the standard cosmology are absent in the minimal standard model (MSM) of particle physics, in particular, dark matter, dark energy, baryogenesis, and vacuum energy. To be more precise, the situation with vacuum energy is opposite: there is too much vacuum energy in the MSM. For example the energy of the gluon and quark condensate established by quantum chromodynamics are roughly 45 orders of magnitude higher than the observed magnitude of the dark energy.

So we have to conclude that new physics beyond the frameworks of MSM is a necessity. Still, except for the vacuum energy problem, the new physics may be almost the old one with introduction of some new fields of particles, while it is quite possible that for the solution of the vacuum energy problem a revolutionary modification of the existing theory would be necessary.

In addition to these well known good old problems the last several years revealed many features which look surprising and even completely mysterious in the frameworks of the ΛCDM model. These recent discoveries are reviewed in what follows. It is argued that all them can be explained by formation of heavy primordial black holes (PBHs) through the mechanisms suggested and discussed in detail in papers [1]. This mechanism allows for formation of PBHs with the masses in the range from a fraction of the solar mass, say, up to $10^4 M_\odot$ or even higher. Usually only the PBHs with rather low masses $\sim 10^{20}$ g were considered.

Moreover, the puzzling properties of the sources of the gravitational waves recently discovered by LIGO [2] are explained by the same scenario of heavy PBH formation.

The content of the talk is the following:
1. GW observation by LIGO.
2. Problems with the GW sources.
3. Solution of the problems and predictions.
4. Dense population of the universe at $z \sim 10$ by the objects which could not be there. They include in particularly supermassive BHs, early supernovae and gamma-bursters, evolved chemistry and dust in high z universe.
5. Problems in present day universe: MACHOs, PBH dark matter, supermassive BHs in large galaxies and even in almost empty space.

The talk is based on several our papers [3–7], where the relevant references can be found.

2 Direct discovery of gravitations waves

On February 11, LIGO (Laser Interferometer Gravitational wave Observatory) collaborations announced discovery of gravitational waves from a coalescing binary systems of black holes [2]. Two more events were reported shortly, see below Table. 1.

The shape of the signal is in perfect agreement with the theory of BH interactions in the strong (Schwarzschild) sefl-fields, so it can be considered as a first direct proof of BH existence and a persuasive confirmation of General

[a] e-mail: dolgov@fe.infn.it

DOI: 10.1051/epjconf/201714201012

Relativity for a large deviation of geometry from the flat Minkowsky one. All the previous tests were about weak fields only.

This discovery opened a new era of gravitational waves telescopes which will presumably allow to observe several (many) such catastrophic events per year. The anticipated increase of the LIGO sensitivity by factor three will give enhance the number of the registered events 27-fold. With the expected onset of operation of VIRGO (Italy) and KAGRA (The Kamioka Gravitational Wave Detector, Japan) the direction to the sources can be reliably established, so the sources can be studied by optical and other electromagnetic telescopes. New discoveries are imminent.

In Table I, copied from ref. [2], the properties of the two coalescing black holes, which produced the powerful burst of gravitational waves, are presented. The mass and spin of the final BH, and the total energy radiated in gravitational waves are estimated by the fits to numerical simulations of binary black hole mergers. The estimated total energy radiated in gravitational waves is $(3.0 \pm 0.5)M_\odot$ and the peak of gravitational-wave luminosity is $3.0^{+0.5}_{-0.4} \times 10^{56}$ erg/sec equivalent to $200M_\odot$/sec, more than whole radiation power of the visible universe. Rotational energy (outside the BH) is about $0.3M_\odot$. It may be in principle extracted.

TABLE I. Source parameters for GW150914. We report median values with 90% credible intervals that include statistical errors, and systematic errors from averaging the results of different waveform models. Masses are given in the source frame; to convert to the detector frame multiply by $(1+z)$ [90]. The source redshift assumes standard cosmology [91].

Primary black hole mass	$36^{+5}_{-4}M_\odot$
Secondary black hole mass	$29^{+4}_{-4}M_\odot$
Final black hole mass	$62^{+4}_{-4}M_\odot$
Final black hole spin	$0.67^{+0.05}_{-0.07}$
Luminosity distance	410^{+160}_{-180} Mpc
Source redshift z	$0.09^{+0.03}_{-0.04}$

Three GW event observed to the present time are described in Fig. 1. The last event is questionable while the first two look reliable enough. The essential difference between GW150914 and GW151226 is the following: the masses of the sources in the first case are very high, they are are 36 and 29 solar masses, while in the second case they are $14M_\odot$ and $7M_\odot$, which look astrophysically normal. The spins of both progenitors in the first case are low, compatible with zero. In the second case one of the black holes has noticeable spin $a > 0.2$. The formation mechanism of the initial back holes in the first case demands some unknown, unusual astrophysical processes, while each of the companions on the second pair may be created through the usual astrophysical channel. However, the formation of the binaries in both cases is not well understood, see below.

There are essentially three problems in the standard theory:
1. Origin of heavy BHs ($\sim 30M_\odot$).
2. Low spins of the coalescing BHs.
3. Formation of BH binaries from original stellar binaries.

The first problem is a heavy BH origin. Such BHs are believed to be created by massive star collapse, though a convincing theory is still lacking. To form so heavy BHs, the progenitors should have $M > 100M_\odot$ and a low metal abundance to avoid too much mass loss during the evolution. Such heavy stars might be present in young star-forming galaxies but they are not yet observed in sufficiently high number.

Another problem is the low value of the BH spins in GW150914. It strongly constrains astrophysical BH formation from close binary systems. However, the dynamical formation of double massive low-spin BHs in dense stellar clusters is not excluded. The second reliable LIGO detection, GW151226, turned out to be closer to the standard binary BH system.

Last but not the least is the problem of formation of BH binaries. Stellar binaries were formed from common interstellar gas clouds and are quite frequent in galaxies. If BH is created through stellar collapse, a small non-sphericity of collapse results in a huge velocity of the BH and the binary is destroyed. An indirect evidence for that is presented by large velocities of pulsars in the Galaxy. Their velocities are about 1000 km/sec, while the average star velocities are only 200 - 300 km/sec. Moreover, the BH formation from PopIII stars and subsequent formation of BH binaries with $\sim (30 + 30)M_\odot$ is analyzed in the literature and is found to be negligible.

All these problems are solved if the observed sources of GWs are the binaries of primordial black holes (PBH). Here a model of PBH formation is presented which naturally reproduces the puzzling properties of GW150914, the rate of binary BH merging events inferred from the first LIGO science run, and provides seeds for early supermassive BH formation. These PBHs are created at rest and so their mutual capture to form a binary is not inhibited after they loose their relative velocity due e.g. to dynamical friction. The spins of the original BHs are naturally zero because rotational perturbations are absent in the early universe.

In addition, the mechanism explains an avalanche of mysteries discovered recently and may provide all or a large fraction of cosmological DM in the form of PBHs with rather wide mass spectrum.

3 Mechanism of massive PBH formation.

The model of an early BH creation is based on the supersymmetric (Affleck-Dine) scenario for baryogenesis [8], modified by introduction of a general renormalizable coupling to the inflaton field, see below, Eq. (6). It was suggested in 1993 [1] and discussed in more details in several our papers applied to an explanation of existence of the observed "old" objected in the young universe. As a byproduct the model may lead to an abundant antimatter objects in the universe and, in particular, in the Galaxy.

The basic ingredient of the Affleck-Dine (AD) scenario is a scalar field χ with non-zero baryonic number, $B \neq 0$. As is well known, scalar baryons must exist in supersymmetric theories. Generically the potential of such field has the so called flat directions along which the potential does not rise. We can take as toy model example the potential of the form:

$$U_\lambda(\chi) = \lambda|\chi|^4 (1 - \cos 4\theta), \tag{1}$$

where θ is the phase of the complex field $\chi = |\chi|\exp(i\theta)$. Such bosons, χ, may condense along flat directions of this potential accumulating large baryonic number. To be more precise, the baryonic number is accumulated not in the large amplitude of χ but in its angular momentum associated with variation of $\theta(t)$, see eq. (4).

In addition to the quartic potential there may exist quadratic mass term, $m^2\chi^2 + m^{*2}\chi^{*2}$, which also could have flat directions generally different from those in quartic poteetial:

$$U_m(\chi) = m^2|\chi|^2[1 - \cos(2\theta + 2\alpha),], \tag{2}$$

where $\chi = |\chi|\exp(i\theta)$ and $m = |m|e^{i\alpha}$. If $\alpha \neq 0$, C and CP are broken. In GUT SUSY baryonic number is naturally non-conserved. In out model this non-conservation is induced by non-invariance of $U(\chi)$ w.r.t. the phase rotation of χ.

Initially (after inflation) χ was naturally away from the origin due to rising quantum fluctuations of light fields. After inflation terminated, χ started to evolve down to the equilibrium point, $\chi = 0$, according to the its equation of motion, which for homogenous field formally coincides with the equation of motion for point-like body in the Newtonian mechanics:

$$\ddot{\chi} + 3H\dot{\chi} + U'(\chi) = 0. \tag{3}$$

The second term in this equation, induced by the universe expansion, is called the Hubble friction term. In the mechanical analogy it is equivalent to liquid friction.

The baryonic number of χ is equivalent to the angular momentum of χ-rotation in two dimensional complex plane $[Re\,\chi, Im\,\chi]$:

$$B_\chi = \dot{\theta}|\chi|^2. \tag{4}$$

Later the decays of χ transferred its baryonic number to baryonic number of quarks in B-conserving process. The Affleck-Dine baryogenesis could lead to the cosmological baryon asymmetry of order of unity, much larger than the observed value $\beta = N_B/N_\gamma \sim 10^{-9}$.

When inflation terminated and the Hubble friction drastically dropped down, the field χ started to move to the origin, $\chi = 0$, along the flat direction of the quartic potential. At sufficiently small χ the quadratic mass term started to dominate and χ moved from the quartic valley to the quadratic one gaining nonzero and typically large angular momentum. This process occurred if the quartic and quadratic flat directions are different. This surely happens if $\alpha \neq 0$ but may also happen even with $\alpha = 0$ if initially χ was in a different quartic valley from that of the quadratic $U_m(\chi)$ (2).

If the CP-odd phase α is small but non-vanishing, both baryonic and antibaryonic regions are possible with dominance of one of them. Matter and antimatter domain may exist but globally $B \neq 0$.

In ref. [1] a general renormalizable coupling of χ to the inflaton field Φ was introduced:

$$\delta U = g|\chi|^2(\Phi - \Phi_1)^2, \tag{5}$$

where g is a dimensionless coupling constant and Φ_1 is a value of the inflaton which it passed during inflation. This potential looks as a very special one, but it is not so. It contains three renormalizable contributions: quartic, cubic, and quadratic terms. The only mild tuning of this potential is the value of Φ_1.

With the new term and with an account of the so called Coleman-Weinberg contribution [9], arising from one-loop radiation corrections, the total potential governing the evolution of χ takes the form:

$$\begin{aligned} U &= g|\chi|^2(\Phi - \Phi_1)^2 + \lambda|\chi|^4 \ln\left(\frac{|\chi|^2}{\sigma^2}\right) \\ &+ \lambda_1\left(\chi^4 + h.c.\right) + (m^2\chi^2 + h.c.). \end{aligned} \tag{6}$$

The shape of this potential as a function of $|\chi|$ for different values of Φ is presented in Fig. 2.

When Φ is close to Φ_1, the window to the flat directions is open but only for a relatively short period. At this stage cosmologically small but possibly astronomically large bubbles with high values of χ could be created. Later these high χ regions would create bubbles with very high baryonic number density occupying a small fraction of the universe volume, while the rest of the universe would have the normal baryon asymmetry $\beta \approx 6 \cdot 10^{-10}$, created in the dominant part of the universe with small χ. The process when χ reaches large value but with low probability can be called phase transition of 3/2 order, since in a sense such a behavior is between first and second order phase transitions.

The density contrast between the bubbles with high baryonic density (high B bubbles or HBB) and the low B average cosmological background is initially small, since before the QCD phase transition quarks populating the cosmological plasma are essentially massless, so the initial density perturbations are predominantly isocurvature. However, after the phase transition to the quark confinement phase quarks turned into heavy nucleons and the density contrast between HBB and the rest of the universe became large. Note that the contribution of HBBs into the total cosmological energy/mass density can be higher than the contribution of the normal free baryons.

The high density contrast, which appeared after the QCD phase transition, could lead to an early formation of compact stellar-type objects and possibly to a comparable amount of anti-objects, such that the bulk of baryons and (equal) antibaryons are in the form of compact stellar-like objects or PBH, plus the sub-dominant observed homogeneous baryonic background. The amount of antimatter may be comparable or even larger than of the known baryons, but such "compact" (anti)baryonic objects do not contradict any existing observations [5, 7].

DOI: 10.1051/epjconf/201714201012

The distributions of high baryon density bubbles over length and mass have log-normal form [1]:

$$\frac{dN}{dM} = C_M \exp\left[-\gamma \ln^2(M/M_0)\right] \qquad (7)$$

where C_M, γ, and M_0 are constant parameters. The spectrum is practically model independent, it is basically determined by inflation.

Adjusting parameters of the mass spectrum (7) using available astronomical data and constraints, we can predict the density of PBHs produced by the considered here mechanism. The prediction as a function of the black hole mass together with the existing observational bounds is presented in Fig. 3.

4 Problems in contemporary and near contemporary universe.

4.1 Dense population of $z = 5 - 10$ universe

Astronomical data accumulated during the last few years have revealed that the early, $z \sim 10$, universe is unexpectedly dense, populated by the evolved objects which demand much more time for their creation than was available at that high redshifts. Among them there are bright but too young galaxies, QSO/supermassive BHs, and gamma-bursters (supernovae). Moreover, the early universe contains much more dust than can be reasonably expected. A more detailed review and the list of literature is presented in ref. [4]. Here we discuss only the most striking creatures.

About 40 quasars with $z > 6$ are already known, each quasar containing BH with $M \sim 10^9 M_\odot$. Such black holes, when the Universe was less than one billion years old, present substantial challenges to theories of the formation and growth of black holes and the coevolution of black holes and galaxies. Even the origin of supermassive black holes in contemporary universe, which had 14 Gyr for their creation, is difficult to explain.

Very recently a new monster was discovered "An ultraluminous quasar with a twelve billion solar mass black hole at redshift 6.30" [10]. There is already a serious problem with formation of ten times lighter and less luminous quasars, which is multifold deepened with this new "creature". A formation of the new one with $M \approx 10^{10} M_\odot$ is absolutely impossible in the standard approach.

The premature appearance of supermassive black holes at $z \sim 6$ emerged as a great surprise. It is very difficult to understand how $10^9 M_\odot$ black holes appeared so quickly after the big bang without invoking non-standard accretion physics and the formation of massive seeds, both of which are not observed in the local Universe [11]. On the other hand, such massive seeds are exactly the objects which are supplied by the HBBs discussed in the previous section.

Several galaxies have been observed at high redshifts, with natural gravitational lens "telescopes. Their creation was also unexpectedly early. E.g. a galaxy at $z \approx 9.6$ was discovered [12], which was created when the universe was about 0.5 Gyr old. Moreover a galaxy at $z \approx 11$ has been observed [13], which was formed earlier than the universe age was 0.41 Gyr (or even shorter with larger H advocated by the recent traditional astronomical measurements).

An observation of not so young but extremely luminous galaxy is reported [14]. Its luminosity is thousand times higher than that of the Milky way: $L = 3 \cdot 10^{14} L_\odot$, while the age is only ~ 1.3 Gyr. The galactic seeds, or embryonic black holes, might be bigger than thought possible. According to one of the authors of the discovery P. Eisenhardt: "How do you get an elephant? One way is start with a baby elephant." However, the origin of the baby elephant" is mysterious. The seed black hole should be already billions of solar masses, when our universe was only a tenth of its present age of 13.8 billion years. There is no way to create such a seed with standard mechanisms.

The medium around the observed early quasars contains considerable amount of "metals" (elements heavier than He). According to the standard picture, only elements up to ^{4}He and traces of Li, Be, B were formed by the big bang nucleosynthesis, while heavier elements were created much later by stellar nucleosynthesis and dispersed in the interstellar space by supernova explosions. It already demands non-negligible time.

Much later molecular dust could form. Nevertheless it was discovered recently that the young universe at $z > 6$ is quite dusty [15]. Dusty galaxies show up at redshifts corresponding to a Universe which is only about 500 Myr old. The highest redshift such object, HFLS3, lies at z=6.34 and numerous other sources have been found [16].

Hence, prior to or simultaneously with the QSO formation a rapid star formation should take place. These stars should evolve to a large number of supernovae enriching interstellar space by metals through their explosions which later make molecules and dust. (We all are dust from SN explosions, but probably at much later time.)

Another possibility is a non-standard BBN due to very high baryonic density, which allows for formation of heavy elements beyond lithium.

Observations of high redshift gamma ray bursters (GBR) also indicate a high abundance of supernova at large redshifts. The highest redshift of the observed GBR is 9.4 and there are a few more GBRs with smaller but still high redshifts. The necessary star formation rate for explanation of these early GBRs is at odds with the canonical star formation theory.

All such early supernovae can be HBBs which were not massive enough to make PBHs but some compact stellar like objects which is also possible according to the discussed in sec. 3 scenario.

4.2 Back to the future.

Similar and possibly related mysteries exist in contemporary and near-contemporary universe.

An accumulation of quasars is discovered in a narrow spot in the sky [17]. at redshift $z \approx 2$. According to the authors of the paper: We discovered a physical association of four quasars embedded in a giant nebula. Quasars are rare

DOI: 10.1051/epjconf/201714201012

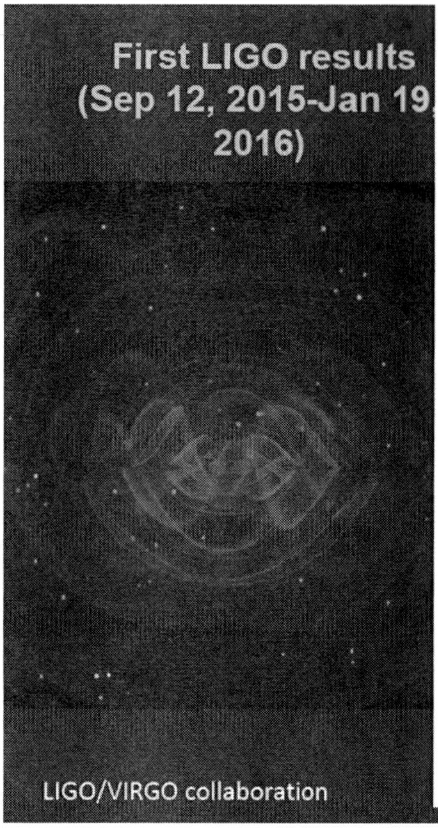

Event	GW150914	GW151226	LVT151012
Signal-to-noise ratio ρ	23.7	13.0	9.7
False alarm rate FAR/yr^{-1}	$< 6.0 \times 10^{-7}$	$< 6.0 \times 10^{-7}$	0.37
p-value	7.5×10^{-8}	7.5×10^{-8}	0.045
Significance	$> 5.3\sigma$	$> 5.3\sigma$	1.7σ
Primary mass $m_1^{\text{source}}/\text{M}_\odot$	$36.2^{+5.2}_{-3.8}$	$14.2^{+8.3}_{-3.7}$	23^{+18}_{-6}
Secondary mass $m_2^{\text{source}}/\text{M}_\odot$	$29.1^{+3.7}_{-4.4}$	$7.5^{+2.3}_{-2.3}$	13^{+4}_{-5}
Chirp mass $\mathcal{M}^{\text{source}}/\text{M}_\odot$	$28.1^{+1.8}_{-1.5}$	$8.9^{+0.3}_{-0.3}$	$15.1^{+1.4}_{-1.1}$
Total mass $M^{\text{source}}/\text{M}_\odot$	$65.3^{+4.1}_{-3.4}$	$21.8^{+5.9}_{-1.7}$	37^{+13}_{-4}
Effective inspiral spin χ_{eff}	$-0.06^{+0.14}_{-0.14}$	$0.21^{+0.20}_{-0.10}$	$0.0^{+0.3}_{-0.2}$
Final mass $M_f^{\text{source}}/\text{M}_\odot$	$62.3^{+3.7}_{-3.1}$	$20.8^{+6.1}_{-1.7}$	35^{+14}_{-4}
Final spin a_f	$0.68^{+0.05}_{-0.06}$	$0.74^{+0.06}_{-0.06}$	$0.66^{+0.09}_{-0.10}$
Radiated energy $E_{\text{rad}}/(\text{M}_\odot c^2)$	$3.0^{+0.5}_{-0.4}$	$1.0^{+0.1}_{-0.2}$	$1.5^{+0.3}_{-0.4}$
Peak luminosity $\ell_{\text{peak}}/(\text{erg s}^{-1})$	$3.6^{+0.5}_{-0.4} \times 10^{56}$	$3.3^{+0.8}_{-1.6} \times 10^{56}$	$3.1^{+0.8}_{-1.8} \times 10^{56}$

Figure 1. Three registered gravitational wave events.

objects separated by cosmological distances, so the chance of finding a quadruple quasar is $\sim 10^{-7}$. It implies that the most massive structures in the distant universe have a tremendous supply ($\sim 10^{11} M_\odot$) of cool dense ($n \approx 1/\text{cm}^3$) gas, in conflict with current cosmological simulations."

Every large galaxy and some smaller ones contain central supermassive BHs whose masses are larger than $10^9 M_\odot$ in giant elliptical and compact lenticular galaxies and $\sim 10^6 M_\odot$ in spiral galaxies like Milky Way. The origin of these superheavy BHs is not understood. Moreover, SHBs are observed in several small galaxies, where is no material to make a supermassive BH.

It is intriguing if the type of the galaxy is determined by the mass of the original black hole seed.

The mass of BH is typically 0.1% of the mass of the stellar bulge of galaxy but some galaxies may have huge BH: e.g. NGC 1277 has the central BH of $1.7 \times 10^{10} M_\odot$, or 60% of its bulge mass [18]. This fact creates serious problems for the standard scenario of formation of central supermassive BHs by accretion of matter in the central part of a galaxy.

Report of the similar observations can be found in ref. [19]. The authors conclude that although supermassive black holes correlate well with their host galaxies, there is an emerging view that outliers exist. Henize 2-10, NGC 4889, and NGC1277 are examples of supermassive BHs at least an order of magnitude more massive than their host

DOI: 10.1051/epjconf/201714201012

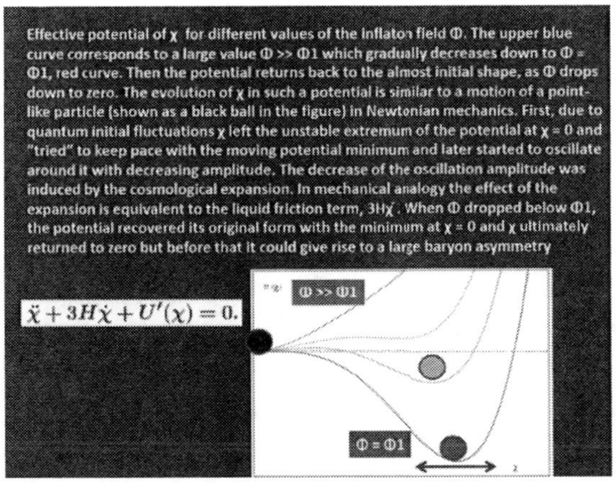

Figure 2. Evolution of the potential (6) as a function of χ for different values of Φ.

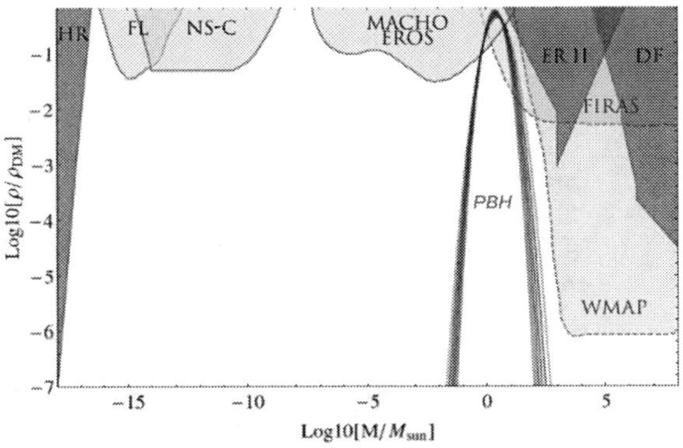

Figure 3. Constraints on PBH fraction in DM, $f = \rho_{\rm PBH}/\rho_{\rm DM}$, where the PBH mass distribution is taken as $\rho_{\rm PBH}(M) = M^2 dN/dM$. The existing constraints (extragalactic γ-rays from evaporation (HR), femtolensing of γ-ray bursts (F), neutron-star capture constraints (NS-C), MACHO, EROS, OGLE microlensing (MACHO, EROS) survival of star cluster in Eridanus II (E), dynamical friction on halo objects (DF), and accretion effects (WMAP, FIRAS)). The PBH distribution is shown for ADBD parameters $\mu = 10^{-43}$ Mpc^{-1}, $M_0 = \gamma + 0.1 \times \gamma^2 - 0.2 \times \gamma^3$ with $\gamma = 0.75 - 1.1$ (red solid lines), and $\gamma = 0.6 - 0.9$ (blue solid lines).

galaxy suggests. The dynamical effects of such ultramassive central black holes is unclear.

Several more observations of too heavy black holes in poor galaxies are presented in ref. [4] but the most striking one is given in the recent publication [20], where a supermassive BH was observed in practically empty space

Thus the inverted picture is more plausible: first a supermassive black hole was formed and attracted matter serving as a seed for subsequent galaxy formation [1, 21].

4.3 Old stars, black holes, and MACHOS in the Milky Way

Some more, at first sight unrelated, but probably the same kind problems are demonstrated by the contemporary universe. There are stars in the Milky Way, older than the Galaxy and even older than the universe (more than two sigma) and even one very old rocky planet. The mass distribution of black holes in the Galaxy and abundant MACHOs are also at odds with the standard astrophysics.

The new recently developed precise measurements allowed to determine age of several stars in the Milky Way with unprecedented accuracy. The results showed that quite a few stars are consuderably older than it was earlier expected.

Employing thorium and uranium abundances in comparison with each other and with several stable elements the age of metal-poor, halo star BD+17° 3248 was estimated as 13.8 ± 4 Gyr[22]. For comparison the age of inner halo of the Galaxy 11.4 ± 0.7 Gyr [23].

The age of a star in the galactic halo, HE 1523-0901, was estimated to be about 13.2 Gyr [24]. First time many different chronometers, such as the U/Th, U/Ir, Th/Eu, and Th/Os ratios to measure the star age have been employed.

Even more striking, the metal deficient high velocity subgiant in the solar neighborhood HD 140283 has the age 14.46 ± 0.31 Gyr [25]. The central value exceeds the universe age by two standard deviations, if $H = 67.3$ km/sec/Mpc and $t_U = 13.8$ Gyr. The excess is even bigger for $H = 74$ km/sec/Mpc, when $t_U = 12.5$ Gyr.

A discovery of a surprisingly old planet was recently announced [26] Its age is estimated as $10.6^{+1.5}_{-1.3}$ Gyr. (The age of the Earth: 4.54 Gyr.) A supenova explosion must precede formation of this planet.

The considered in this report scenario of HBB formation may explain these striking discoveries, because if the initial chemical content of a star is different from the canonical one, it may look older than it is in reality.

Some more mysteries are revealed by the black holes observed in the Galaxy. It was found that the BH masses are concentrated in the narrow range $(7.8 \pm 1.2) M_\odot$ [27].

This result agrees with another paper where a peak around $8 M_\odot$, a paucity of sources with masses below $5 M_\odot$, and a sharp drop-off above $10 M_\odot$ are observed [28].

These features are not expected in the standard model, but may fit our model with the log-normal mass spectrum of PBHs.

HBB can be also found by the effect of gravitational microlensing which may be caused by both visible and invisible stars. These objects are called Machos for Massive Astrophysical Compact Halo Objects. The observational situation with them is reviewed in ref. [5], where it is concluded that this population of the old invisible stars that evaded detection can well be HBBs discussed here. Those stars should be older than any kind of the oldest standard stellar populations. They will meet both criteria: they should be very weak and their cloud should have such a high velocity dispersion as needed. There is an intriguing possibility that some of those HBBs can be antistars in the Galaxy and in the galactic halo.

5 Conclusion

1. Supersymmetric baryogenesis could lead to abundant formation of PBHs and compact stellar-like objects in the early universe after the QCD phase transition at $z \sim 10^{12}$ or $t \gtrsim 10^{-5}$ sec.
2. These objects have log-normal mass spectrum.
3. Adjusting the spectrum parameters is possible to explain the peculiar features of the sources of the gravitational waves observed by LIGO.
4. The considered mechanism solves the numerous mysteries of $z \sim 10$ universe: abundant population of supermassive black holes, early created gamma-bursters and supernovae, early bright galaxies, and evolved chemistry including dust.
5. There is persuasive data in favor of the inverted picture of galaxy formation, when first a supermassive BH seeds are formed and later they accrete surrounding matter forming galaxies.
6. An existence of supermassive black holes observed in all large and some small galaxies and even in almost empty environment is naturally explained.
7. "Older than the universe" stars may exist.
8. Existence and high density of invisible "stars" (machos) can be understood.
9. Some noticeable fraction of dark matter or even all of it can be made of PBHs.
10. Large amount of astronomical data the data strongly demand abundant cosmological population of PBH with wide mass spectrum.

Testable predictions:
A. Rate and masses of the BH sources of the coming GW events.
B. Possible existence of antimatter in our neighborhood, even in the Galaxy.

Acknowledgement

This work was supported by the Grant of President of Russian Federation for the leading scientific schools of the Russian Federation, NSh-9022.2016.2.

References

[1] A. Dolgov and J. Silk, Phys.Rev. **D47**, 4244 (1993); A.D. Dolgov, M. Kawasaki, and N. Kevlishvili, Nucl. Phys. **B807**, 229 (2009).

[2] B. P. Abbott *et al*, Phys. Rev. Lett. **116**, 061102 (2016).

[3] S. Blinnikov, A. Dolgov, N.K. Porajko, and K. Postnov, JCAP **1611**, 036 (2016).

[4] A.D. Dolgov, Int. J. Mod. Phys. **A31**, 1645029 (2016); A.D. Dolgov, arXiv:1605.06749, Conference: C16-02-16.1.

[5] S.I. Blinnikov, A.D. Dolgov, and K.A. Postnov, Phys. Rev. **D92**, 023516 (2015).

[6] A.D. Dolgov and S.I. Biinnikov, Phys. Rev. **D89** 2, 021301 (2014).

[7] C. Bambi and A. D. Dolgov, Nuclear Physics **B 784**, 132 (2007).

[8] I. Affleck and M. Dine, Nucl. Phys. B **249**, 361 (1985).

[9] S.R. Coleman and E. Weinberg, Phys. Rev. **D7**, 1888 (1973).

[10] Xue-Bing Wu et al, Nature **518**, 512 (2015).

[11] F. Melia, Astron. J. **147**, 120 (2014).

[12] W. Zheng, *et al*, Nature **489**, 406 (2012).

[13] D. Coe *et al* Astrophys. J. **762**, 32 (2013).

[14] Chao-Wei Tsai, P.R.M. Eisenhardt *et al*, arXiv:1410.1751, Ap.J. in press.

[15] M. Mancini *et al* 1505.01841.

[16] L. Mattsson, arXiv:1505.04758v1.

[17] J.F. Hennawi *et al*, Science, **348**, 779 (2015).

[18] R. C. E., Gebhardt *et al*, Nature, **491**, 729 (2012).

[19] F. Khan, K. Holley-Bockelmann, P. Berczik, arXiv:1405.6425v1.

[20] J. J. Condon, J. Darling, Y. Y. Kovalev, L. Petrov, arXiv:1606.04067.

[21] R. C. E. van den Bosch *et al*, Nature, **491**, 729 (2012).

[22] J.J. Cowan *et al*, Ap.J. **572**, 861 (2002).

[23] J. Kalirai, Nature **486**, 90 (2012).

[24] A. Frebe, N. Christlieb, J.E. Norris, C. Thom Astrophys. J. **660**, L117 (2007).

[25] H. E. Bond *et l* Astrophys. J. Lett. **765**, L12 (2013)/

[26] X. Dumusque, *et al* Ap. J., **789**, 154 (2014).

[27] F. Ozel *et al* Astrophys. J. **725**, 1918 (2010).

[28] L. Kreidberg *et al* Astrophys. J. **757**, 36 (2012).

EPJ Web of Conferences **142**, 01013 (2017) DOI: 10.1051/epjconf/201714201013
Advances in Dark Matter and Particle Physics

Study of e^+e^- annihilation to hadrons at low energies at BABAR

V. P. Druzhinin[1,2,⋆]
for the BABAR Collaboration

[1] *Novosibirsk State University, 630090 Novosibirsk, Russia*
[2] *Budker Institute of Nuclear Physics, 630090 Novosibirsk, Russia*

Abstract. New results of study of e^+e^- annihilation to hadrons at low energies at BABAR are presented. The the cross sections for the $\pi^+\pi^-\pi^0\pi^0$, $\pi^+\pi^-\eta$ final states and several exclusive hadronic modes with kaons are measured using the initial-state-radiation technique.

1 Introduction

During the last decade the BABAR experiment [1, 2] at the SLAC National Accelerator Laboratory performs the intensive study of exclusive processes of e^+e^- annihilation into hadrons at low energies using the initial-state radiation (ISR) technique. The main goal of this study is to measure exclusive cross sections for all possible final hadronic states below 2 GeV and calculate the total cross section, which used in Standard Model (SM) calculations of the muon anomalous magnetic moment $a_\mu = (g_\mu - 2)/2$ and the running electromagnetic constant. The different SM contributions to a_μ taken from PDG review [3] are listed in Table 1 together with the experimental result [4]. The leading order (LO) hadronic vacuum polarization contribution ($a_\mu^{\text{had, LO}}$), which is calculated using hadronic cross section data gives a largest contribution into the error of the Standard Model prediction. Currently, about 3.5 sigmas difference is observed between experiment and the SM calculation. New experiments are planned, which will improve accuracy of a_μ measurement by a factor of about 4 [5].

The low energy region, below 2 GeV, give dominant contribution into $a_\mu^{\text{had, LO}}$. The total cross section in this region is calculated as a sum of exclusive channels. Above 2 GeV the total hadronic cross section is measured inclusively. Recently the accurate inclusive measurement was performed in two energy points below 2 GeV (1.84 and 1.94 GeV) in the KEDR experiment [6]. So, the comparison of inclusive and exclusive data and perturbative QCD calculation [7] may be performed in the energy interval between 1.8 and 2.0 GeV. The exclusive data are incomplete near 2 GeV. For example, recently two previously unmeasured processes $e^+e^- \to \omega\pi^0\eta$ and $e^+e^- \to \pi^+\pi^-\pi^0\eta$ were studied in the SND experiment [8, 9], which cross sections are about 5% of the total hadronic cross section near 2 GeV. The BABAR goal is to measure all significant exclusive channels below 2 GeV and perform the comparison with inclusive measurement and pQCD prediction.

Table 1. Contributions to the SM prediction for a_μ (10^{-10}) [3] and comparison with experiment [4].

QED	11658471.895	± 0.008
LO hadronic vacuum polarization	692.3	± 4.2
NLO hadronic vacuum polarization	−9.84	± 0.06
Hadronic light-by-light	10.5	± 2.6
Electroweak	15.36	± 0.10
Theory	11659180.3	± 4.9
Experiment (E821@BNL)	11659209.1	± 5.4 ± 3.3
Experiment − Theory	28.8	± 8.0

In this paper, we will discuss BABAR analyses and preliminary results on the processes $e^+e^- \to \pi^+\pi^-\pi^0\pi^0$, $\pi^+\pi^-\eta$, $K\bar{K}\pi$, $K\bar{K}2\pi$, $K_SK_L\eta$, and $K_SK^+\pi^-\eta$. All measurements are based on data sample with an integrated luminosity of about 470 fb^{-1} collected at center-of-mass energy near 10.6 GeV and use the initial state radiation technique.

2 Initial state radiation method

The mass spectrum of the hadronic system f in the ISR process $e^+e^- \to f\gamma$ is related to the cross section (σ_0) of the nonradiative process $e^+e^- \to f$:

$$\frac{d\sigma(s,m,\theta_\gamma)}{dmd\cos\theta_\gamma} = \frac{2m}{s}W(s,x,\theta_\gamma)\sigma_0(m), \quad (1)$$

where $\sqrt{s}$ is the e^+e^- center-of-mass energy, $x = 2E_\gamma/\sqrt{s} = 1 - m^2/s$, and E_γ and θ_γ is the energy and polar angle of the ISR photon. The function $W(s,x,\theta_\gamma)$ is calculated in QED with a high, better than 0.5%, accuracy.

The ISR photon is emitted predominantly along the e^+e^- collision axis. The produced hadronic system is

⋆e-mail: druzhinin@inp.nsk.su

 DOI: 10.1051/epjconf/201714201013

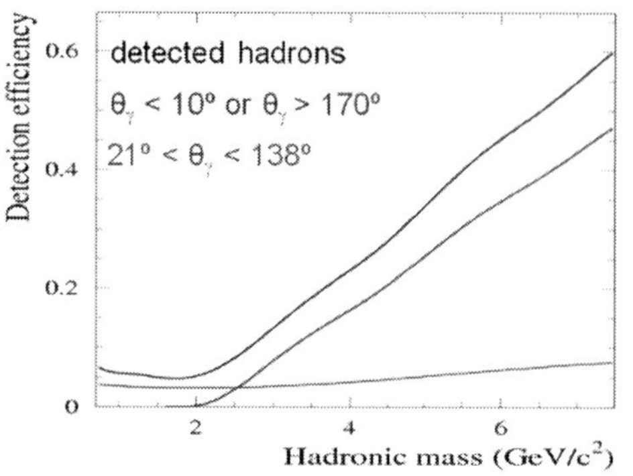

Figure 1. The detection efficiency for $e^+e^- \to 4\pi\gamma$ events obtained using MC simulation as a function of 4π invariant mass. The black curve represents the efficiency with requirement that all final hadrons are detected, the blue and red lines are the efficiencies with additional requirements that the ISR photon is emitted at small and large angles, respectively.

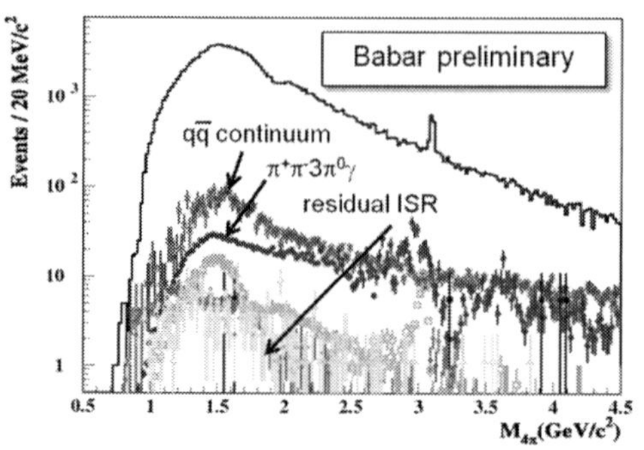

Figure 2. The $\pi^+\pi^-2\pi^0$ mass spectrum for selected data events (black histogram). The mass spectra for background events of different ISR and non-ISR processes calculated using MC simulation are shown by the color histograms.

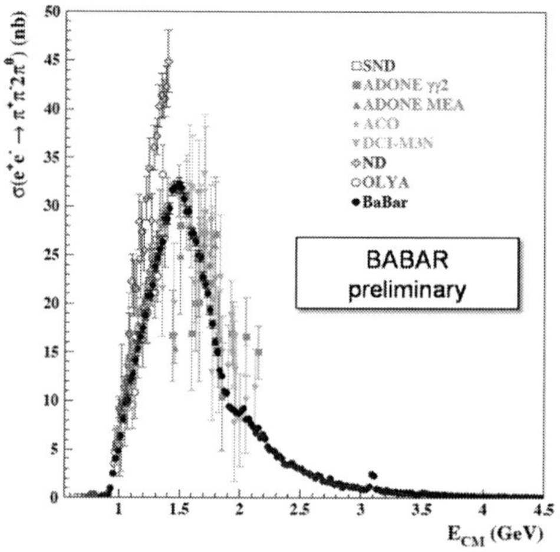

Figure 3. The measured $e^+e^- \to \pi^+\pi^-2\pi^0$ cross section in comparison with previous measurements. The errors are statistical.

boosted against the ISR photon. Figure 1 shows the dependence of the detection efficiency on the hadronic mass for the $e^+e^- \to 4\pi\gamma$ reaction. The black curve represents the efficiency with requirement that all final hadrons are detected. The blue curve is the efficiency with additional requirement that the ISR photon is emitted at small angle, while the red curve is the efficiency with requirement that the ISR photon is emitted at large angle and detected. Due to limited detector acceptance the mass region below 2 GeV can be studied only with detected ISR photon.

The fully exclusive approach is used for low-energy ISR measurements at BABAR. It is required that an event contains a photon with the center-of-mass energy higher 3 GeV and that all final hadrons are detected and identified. Large-angle ISR forces the hadronic system into the detector fiducial volume.

The advantage of the large-angle ISR method over conventional e^+e^- measurements is a weak dependence of the detection efficiency on dynamics of the hadronic system. This allow to measure the total cross section for a multibody final state without detail study of its dynamics with relatively small model uncertainty.

The second advantage is a weak dependence of the efficiency on hadron invariant mass. Measurement in a wide energy range, from threshold up to 4-5 GeV, can be performed with the same selection criteria.

A kinematic fit using requirements of energy and momentum balance improves hadronic mass resolution and provides background suppression.

3 Measurement of the $e^+e^- \to \pi^+\pi^-\pi^0\pi^0$ cross section

The $e^+e^- \to \pi^+\pi^-\pi^0\pi^0$ cross section is one of the least known cross section important for $(g-2)_\mu$.

We select events with all final particles detected and perform the kinematic fit with requirements of energy and momentum balance and two π^0-mass constraints. The requirement on the χ^2 of the kinematic fit ($\chi^2_{4\pi\gamma} < 30$) is applied to select signal events.

The background from the ISR processes with 3 and 5 photons in the final state, $\pi^+\pi^-\pi^0\gamma$, $\pi^+\pi^-\pi^0\eta\gamma$, and $\pi^+\pi^-\eta\eta\gamma$, is additionally suppressed by rejecting events with kinematic fits consistent with those hypotheses; e.g., require $\chi^2_{3\pi\gamma} > 25$. The process with kaons ($K^+K^-\eta\gamma$, $K^+K^-\pi^0\pi^0\gamma$, ...) are suppressed by requiring none of the charged particles to be identified as a kaon.

The largest remaining background is from non-ISR continuum hadronic events, arising from the misidentification of a photon from π^0 decay as an ISR photon. This background is subtracted using MC simulation, which is normalized to reproduce π^0 peak seen in data mass distribution for γ_{ISR} plus any other gamma combinations.

Residual ISR background is subtracted using MC simulation normalized to the existing measurements (rate and shape). The largest ISR background process $e^+e^- \to \pi^+\pi^-3\pi^0\gamma$ is not well measured. The $\pi^+\pi^-3\pi^0\gamma$ mass

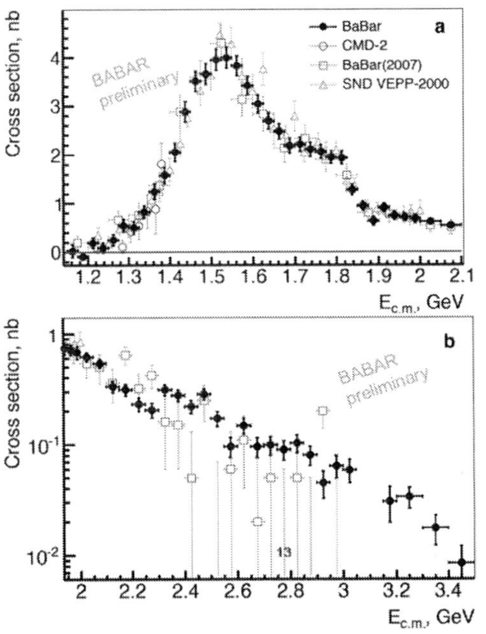

Figure 4. The measured $e^+e^- \to \pi^+\pi^-\eta$ cross section in comparison with previous measurements below 2.1 GeV (a) and between 1.97 and 3.5 GeV (b) on a log scale. The errors are statistical.

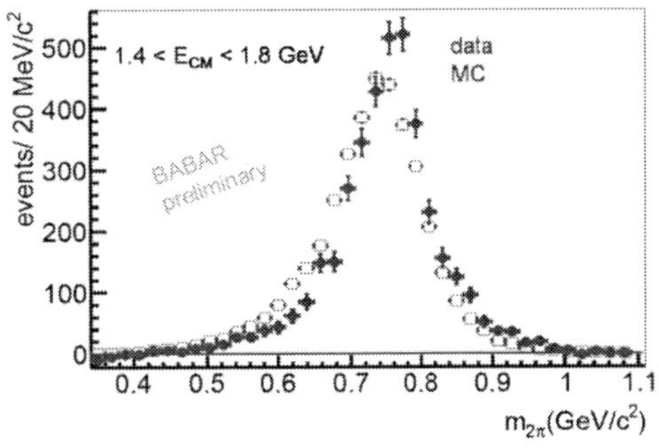

Figure 5. The $\pi^+\pi^-$ mass distribution for selected data events from the $\pi^+\pi^-\eta$ invariant mass range 1.4–1.8 GeV in comparison with the same distribution in simulation.

spectrum used to reweight simulation is extracted from BABAR data.

The data $\pi^+\pi^-2\pi^0$ mass spectrum together with the spectra for background events of different ISR and non-ISR processes is shown in Fig. 2. The background is relatively small, about 3% at 1.5 GeV, and about 15% near 4.5 GeV.

The measured $e^+e^- \to \pi^+\pi^-2\pi^0$ cross section in comparison with previous measurements is shown in Fig. 3, The BABAR results are most precise, especially above 1.4 GeV, and covers a wider energy region. The systematic uncertainty of our measurement is 3.1% in the energy region between 1.2 and 2.7 GeV.

The obtained cross section data is used to calculate the contribution from the $e^+e^- \to \pi^+\pi^-2\pi^0$ into $a_\mu^{\mathrm{had,\,LO}}$. For the energy region below 1.8 GeV it is equal to $(17.9 \pm 0.1_{\mathrm{stat}} \pm 0.6_{\mathrm{syst}}) \times 10^{-10}$. This result can be compared with the previous calculation [10] based on the preliminary BABAR data on $e^+e^- \to \pi^+\pi^-2\pi^0$ from 2007 [11]: $(18.0 \pm 0.1_{\mathrm{stat}} \pm 1.2_{\mathrm{syst}}) \times 10^{-10}$. Our new measurement improves accuracy of this contribution by a factor of 2.

4 Measurement of the $e^+e^- \to \pi^+\pi^-\eta$ cross section

The reaction $e^+e^-\!\to\pi^+\pi^-\eta$, which is expected to proceed via the $\rho\eta$ intermediate state, is important for spectroscopy of excited ρ-like states.

The process is studied in the $\eta \to \gamma\gamma$ decay mode. Event selection is based on χ^2 of the kinematic fit using requirements of energy and momentum balance. The number of signal events is determined from the fit to the two-photon mass distribution for the η-candidate. The fit removes background from the processes not containing η. Peaked background from ISR ($e^+e^- \to K^+K^-\eta\gamma$, $\pi^+\pi^-\pi^0\eta\gamma, \ldots$) and non-ISR processes ($e^+e^- \to \pi^+\pi^-\pi^0\eta$, $\ldots$) is subtracted using MC simulation normalized as described in the previous section.

The measured $e^+e^- \to \pi^+\pi^-\eta$ cross section is shown in Fig. 4a below 2.1 GeV, and between 1.97 and 3.5 GeV on a log scale in Fig. 4b. The BABAR results in the $\eta \to \gamma\gamma$ decay mode agrees well with the previous measurements, but is more precise and covers wider energy range. The systematic uncertainty near the cross section maximum, 1.35-1.80 GeV, is 4.5%.

Figure 5 represent the $\pi^+\pi^-$ mass distribution for data $\pi^+\pi^-\eta$ events in comparison with the same distribution in simulation. The simulation uses a model with the $\rho(770)\eta$ intermediate state. We confirm the SND observation [12] that the $\rho(770)\eta$ mechanism is dominant, but does not fully describes the $\pi^+\pi^-$ mass distribution observed in data. The observed shift between the data and simulation distributions may be result of the interference with other mechanism, for example, $\rho(1450)\eta$.

The cross section has complex energy dependence. To fit this dependence we use the vector meson dominance (VMD) model with one, two, and three excited states. In Model 1, the cross section data are fitted in the energy range 1.15 – 1.70 GeV with two resonances, $\rho(770)$ and $\rho(1450)$, with the relative phase between their amplitudes equal to π. The fit result is shown in Fig. 6. It is seen that Model 1 cannot reproduce the structure in the cross section near 1.8 GeV. In Models 2 and 3 we include an additional contribution from the $\rho(1700)$ resonance with phases $\phi_{\rho(1700)} = \pi$ and 0, respectively. Both models describe the data below 1.90 GeV reasonably well. There is an additional argument in favor of Model 3: the fitted value of the $\rho(770)$ amplitude better agrees with the its estimation from the $\rho(770) \to \eta\gamma$ decay width. Above 1.90 GeV the fitting curves for both the models lie below the data. Model 4 is Model 3 with a fourth resonance ρ''' added. The fitted energy range is extended up to 2.2 GeV. The fit result is shown in Fig. 6. It is seen that the model suc-

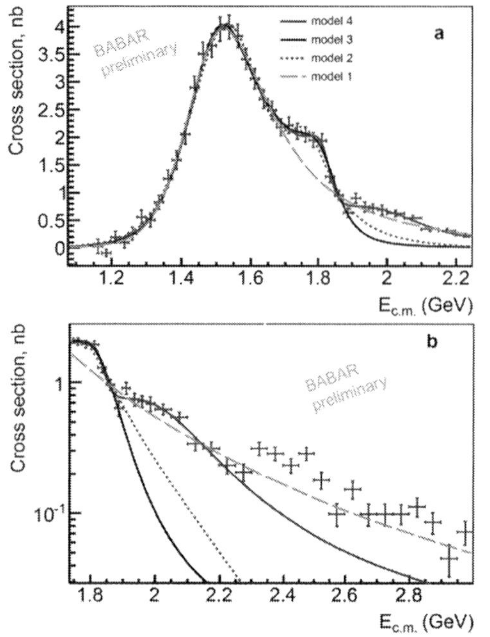

Figure 6. The measured $e^+e^- \to \pi^+pi^-\eta$ cross section fitted with the four models described in the text.

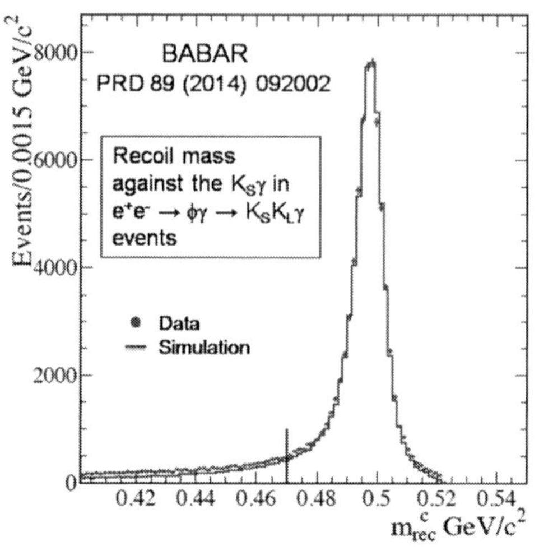

Figure 7. The distribution of the mass recoiling against $K_S\gamma$ system in selected $\phi\gamma$ events in data and simulation.

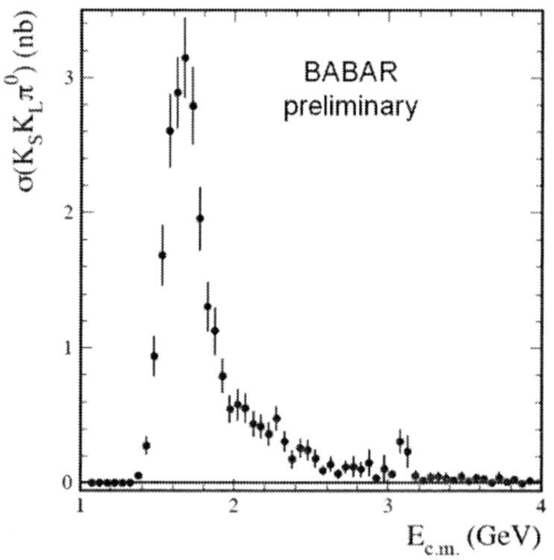

Figure 8. The $e^+e^- \to K_SK_L\pi^0$ cross section measured by BABAR.

cessfully describes the cross-section data up to 2.3 GeV. It is interesting that the cross section data from 1.9 to 3.0 GeV is reasonably well described by the simpler model with only one excited resonance.

Under CVC hypothesis, our data on the $\pi^+\pi^-\eta$ cross section can be used to predict the branching fraction for the corresponding τ decay. The CVC prediction based on BABAR data

$$B(\tau^- \to \pi^-\pi^0\eta\nu_\tau) = (0.162 \pm 0.008)\%, \qquad (2)$$

is in good agreement with, but more precise than, the same value based on the SND data [12] $(0.156 \pm 0.011)\%$ and is about 1.8σ higher than the current experimental result for the branching fraction from PDG table [3]. The difference between the CVS prediction and the PDG value, about 15% of the branching fraction, is too large to be explained by isospin-breaking corrections. The PDG value is dominated by the Belle measurement [13] $B(\tau^- \to \pi^-\pi^0\eta\nu_\tau) = (0.135\pm0.003\pm0.007)\%$ The difference between the CVC-based prediction and Belle result is 2.4σ.

5 Measurement of the $e^+e^- \to K_SK_L\pi^0$ cross section

In this analysis, the K_S is reconstructed via its decay into the $\pi^+\pi^-$ pair. The K_L candidate is identified as an isolated cluster in the electromagnetic calorimeter with the energy deposition larger than 0.2 GeV. The K_L detection efficiency is measured using $e^+e^- \to \phi\gamma$ events, which can be selected without the requirement of detected K_L meson. Figure 7 shows the spectrum of the mass recoiling against the $K_S\gamma$ system in selected $\phi\gamma$ events [14].

Figure 8 represents the first measurement of the $e^+e^- \to K_SK_L\pi^0$ cross section. The systematic uncertainty of this measurement is 10% near the cross-section peak and increases up to 30% at 3.0 GeV.

The dominant intermediate state (more than 95% events) in this reaction is $K^*(892)K$. The $K_2^*(1430)K$ and $\phi\pi^0$ intermediate states are also seen.

Two other charge combinations of the $e^+e^- \to K\bar{K}\pi$ reaction were measured by BABAR previously [15]. So, we can calculate the total $e^+e^- \to K\bar{K}\pi$ cross section without any assumptions about isospin relations between charge modes. The red points in Fig. 9 represent $e^+e^- \to K^+K^-\pi^0$ cross section, the blue points are the sum of the cross sections for $K^+K^-\pi^0$ and $K_SK_L\pi^0$, and the black points are the sum of all three charge modes. The $e^+e^- \to$

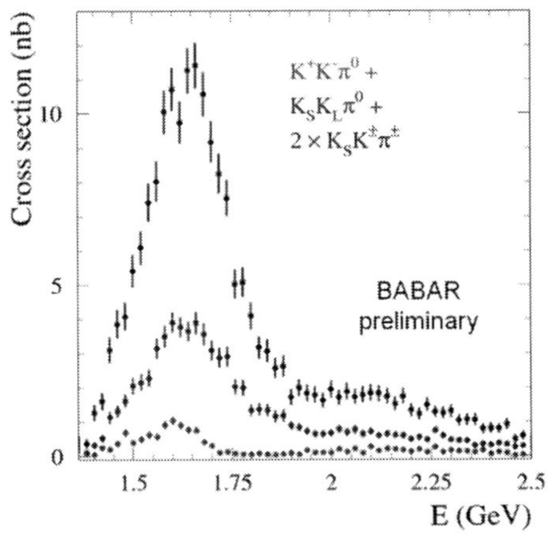

Figure 9. The total $e^+e^- \to K\bar{K}\pi$ cross section obtained as a sum of the cross sections measured in three charge modes: $\sigma(K_SK_L\pi^0) + \sigma(K^+K^-\pi^0) + 2\sigma(K_SK^\pm\pi^\mp)$. The red points represent $K^+K^-\pi^0$ cross section. The blue points are the sum of $K^+K^-\pi^0$ and $K_SK_L\pi^0$ final states.

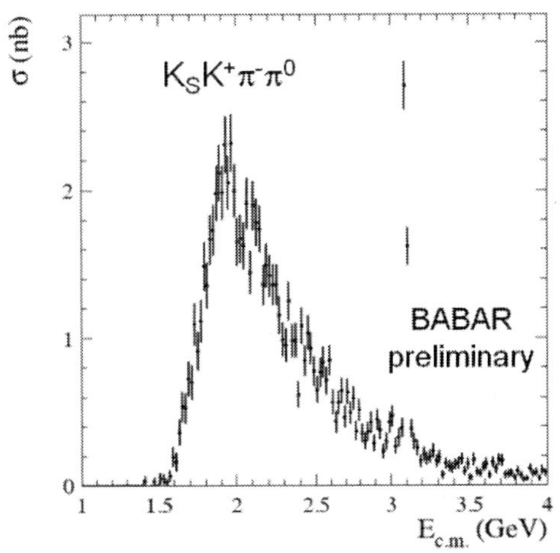

Figure 11. The $e^+e^- \to K_SK^\pm\pi^\mp\pi^0$ cross section measured by BABAR.

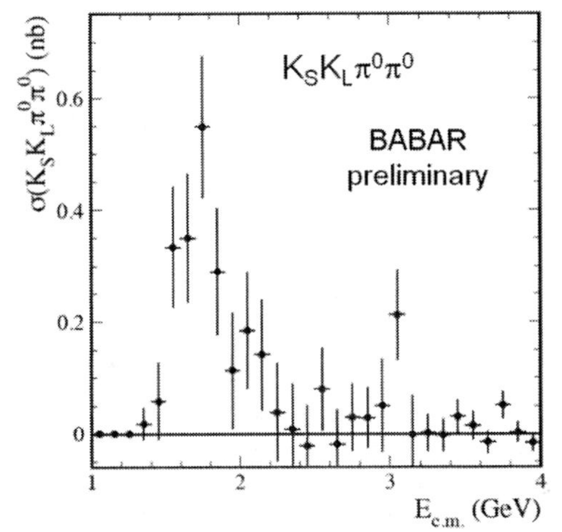

Figure 10. The $e^+e^- \to K_SK_L\pi^0\pi^0$ cross section measured by BABAR.

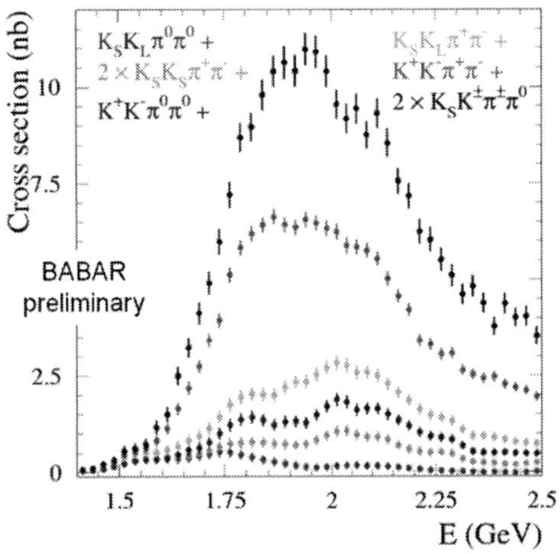

Figure 12. The cross sections for the six charge modes of the $e^+e^- \to K\bar{K}\pi\pi$ process summed cumulatively. The black points represent the total cross section.

$K\bar{K}\pi$ cross section in the maximum is about 12% of the total hadronic cross section.

6 Measurement of the $e^+e^- \to K_SK_L\pi^0\pi^0$ and $e^+e^- \to K_SK^\pm\pi^\mp\pi^0$ cross sections

The process $e^+e^- \to K\bar{K}\pi\pi$ contains six charge combinations. Four of them, $K^+K^-\pi^+\pi^-$, $K^+K^-\pi^0\pi^0$, $K_SK_L\pi^+\pi^-$, and $K_SK_S\pi^+\pi^-$, were measured by BABAR previously [14, 16]. The largest cross section, for $K^+K^-\pi^+\pi^-$ final state, is about 4 nb near 2 GeV.

In Fig. 10 and 11 we present the first measurements of the cross sections for two last charge modes: $K_SK_L\pi^0\pi^0$ and $K_SK^\pm\pi^\mp\pi^0$. Only statistical errors are shown. The $K_SK_L\pi^0\pi^0$ cross section is found to be relatively small and measured with a 25% uncertainty in the peak. The dominant intermediate state for this process is $K^*\bar{K}\pi$. The correlated production of two K^*'s is observed to be small. The $K_SK^+\pi^-\pi^0$ cross section is measured with 6–7% incertainty below 3 GeV. The dominant intermediate states for this reaction are $K^*\bar{K}\pi$ and $K_SK^\pm\rho^\mp$. The correlated $K^*\bar{K}^*$ production is small, less than 15%, and dominated by the charged mode, $K^{*+}\bar{K}^{*-}$.

Now we can calculate the total $e^+e^- \to K\bar{K}\pi\pi$ cross section without any model assumptions. Figure 12 shows the cross sections for the six charge modes summed cumulatively. The black points represent the total cross section. Largest contributions come from the $K^+K^-\pi^+\pi^-$ and $K_SK^+\pi^-\pi^0$ modes. The $e^+e^- \to K\bar{K}\pi\pi$ cross section is about 25% of the total hadronic cross section near 2 GeV.

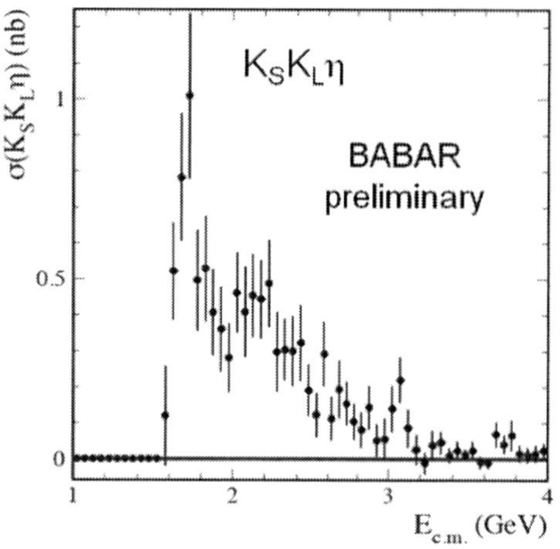

Figure 13. The $e^+e^- \to K_SK_L\pi^0\eta$ cross section measured by BABAR.

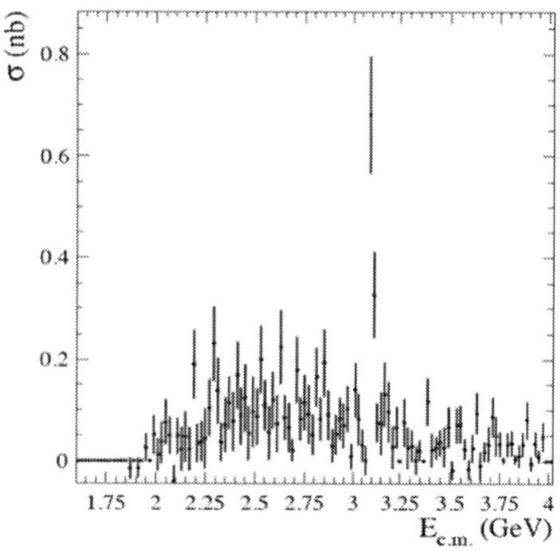

Figure 14. The $e^+e^- \to K_SK^\pm\pi^\mp\eta$ cross section measured by BABAR.

7 The $e^+e^- \to K_SK_L\pi^0\eta$ and $e^+e^- \to K_SK^\pm\pi^\mp\eta$ reactions

The processes $e^+e^- \to K_SK_L\pi^0\eta$ and $e^+e^- \to K_SK^\pm\pi^\mp\eta$ are studied for the first time. Their measured cross sections are shown in Figs. 13 and 14.

The $e^+e^- \to K_SK_L\eta$ reaction is dominated by the $\phi\eta$ intermediate state. The cross section agrees with previous BABAR measurements in the $K^+K^-\eta$ mode [15].

The $e^+e^- \to K_SK^\pm\pi^\mp\eta$ reaction is dominated by the $K^*\bar{K}\eta$ intermediate state. The cross section is relatively small and does not contribute to the total hadronic cross section below 2 GeV.

8 Summary

The precise low-energy e^+e^- cross section data are needed to obtain an accurate SM prediction for the muon anomalous magnetic moment. New results from BABAR reduce the respective uncertainty in the contributions from the processes $e^+e^- \to \pi^+\pi^-\pi^0\pi^0$, $\pi^+\pi^-\eta$, $K\bar{K}\pi$, and $K\bar{K}\pi\pi$. Several previously unmeasured processes with kaons, $e^+e^- \to K_SK_L\pi^0$, $K_SK_L\eta$, $K_SK_L\pi^0\pi^0$, $K_SK_L\pi^0\eta$, $K_SK^\pm\pi^\mp\pi^0$, and $K_SK^\pm\pi^\mp\eta$, have been observed and studied. The BABAR ISR program has provided new information about hadronization at low energies and properties of low-mass resonances.

References

[1] B. Aubert *et al.* (BABAR Collaboration), Nucl. Instrum. Meth. A **479**, 1 (2002).
[2] B. Aubert *et al.* (BABAR Collaboration), Nucl. Instrum. Meth. A **729**, 615 (2013).
[3] C. Patrignani *et al.* (Particle Data Group), Chin. Phys. C, 40, 100001 (2016).
[4] G. W. Bennett *et al.* (Muon g-2) Phys. Rev. D **73**, 072003 (2006).
[5] J. Grange *et al.* (Muon g-2 Collaboration), Muon (g-2) Technical Design Report. arXiv:1501.06858 [physics.ins-det].
[6] V. V. Anashin *et al.* (KEDR Collaboration), arXiv:1610.02827 [hep-ex].
[7] P. A. Baikov *et al.*, Phys. Lett. B **714**, 62 (2012).
[8] M. N. Achasov *et al.* (SND Collaboration), Phys. Rev. D **94**, 032010 (2016).
[9] V. P. Druzhinin *et al.* (SND Collaboration), EPJ Web Conf. **130**, 05004 (2016).
[10] M. Davier, A. Hoecker, B. Malaescu and Z. Zhang, Eur. Phys. J. C **71**, 1515 (2011); Erratum: [Eur. Phys. J. C **72**, 1874 (2012)].
[11] V. P. Druzhinin, arXiv:0710.3455 [hep-ex].
[12] V. M. Aulchenko *et al.* (SND Collaboration), Phys. Rev. D **91**, 052013 (2015).
[13] K. Inami *et al.* (Belle Collaboration), Phys. Lett. B **672**, 209 (2009).
[14] J. P. Lees *et al.* (BABAR Collaboration), Phys. Rev. D **89**, 092002 (2014).
[15] B. Aubert *et al.* (BABAR Collaboration), Phys. Rev. D **77**, 092002 (2008).
[16] J. P. Lees *et al.* (BABAR Collaboration), Phys. Rev. D **86**, 012008 (2012).

Light Meson Decays at BESIII

Shuangshi Fang (for the BESIII Collaboration)[1, a]

[1] *Institute of High Energy Physics, Beijing 100049, China*

Abstract. At present the world's largest sample of 1.3 billion J/ψ events was accumulated at the BESIII detector, which offers a unique place to study light meson decays. The recent results on the light meson decays are reviewed in this talk. An emphasis is put on the significant progresses on the study of η/η' decays, including Dalitz plot analysis of $\eta/\eta' \to \pi\pi\pi$, observation of new decay modes ($\eta' \to \pi^+\pi^-\pi^{+(0)}\pi^{-(0)}$, $\eta' \to \rho^\pm\pi^\mp$, $\eta' \to \gamma e^+e^-$ and $\eta' \to e^+e^-\omega$), study of $\eta' \to \gamma\pi^+\pi^-$ and search for the rare decay of $\eta' \to K\pi$. In addition, a prospect on the Dalitz plot analysis of $\omega \to \pi^+\pi^-\pi^0$ is presented.

1 Introduction

η/η' and ω are the ground states of pseudoscalar and vector nonets, respectively. Since η and η' were discovered half a century ago, both of them have been attracted both theoretical and experimental attentions due to their special role in understanding low energy Quantum Chromodynamics (QCD). In particular for η', it is much heavier than the Goldstone bosons of broken chiral symmetry, and it has a special role in hadron physics because of its interpretation as a singlet state arising from the axial $U(1)$ anomaly. Therefore η' decay dynamics remains a subject of extensive theoretical studies aiming at extensions of chiral perturbation theory (ChPT). Recently a dispersive analysis of $\omega/\phi \to \pi^+\pi^-\pi^0$ was performed in Ref. [1, 2], and it was found that the dispersive relation could provides a good description of the data of $\phi \to \pi^+\pi^-\pi^0$ from KLOE experiment [3]. Since the limited statistics of ω decays, up to date the experimental Dalitz plot investigation on ω has never been performed.

In addition to the many achievements on J/ψ decays and the light meson spectroscopy, the BESIII experiment also offers a unique a place to investigate the light meson decays because of their high production rate in J/ψ decays. At present a sample of 1.31×10^9 J/ψ events [4] were accumulated at the BESIII detector, which corresponds to 1.4×10^6 η (6.8×10^6 η') after taking into account the branching fraction of J/ψ radiative decays into η (η'). Though the ω meson can not be directly produced from J/ψ radiative decays, the production rate for ω in J/ψ hadronic decays is quite high, e.g., $B(J/\psi \to \omega\eta) = (1.74 \pm 0.20) \times 10^{-3}$ [5], which allows to select a clean sample of 8×10^5 ω events to investigate the dynamics of $\omega \to \pi^+\pi^-\pi^0$ even after considering the branching fraction of $\eta \to \gamma\gamma$. In this talk I would like made a review on the recent results on the light mesons, mainly focusing on the significant progresses on η/η' decays, achieved at the BESIII experiment.

[a] e-mail: fangss@ihep.ac.cn

2 η/η' hadronic decays

2.1 Observation of $\eta' \to \pi^+\pi^-\pi^+\pi^-, \pi^+\pi^-\pi^0\pi^0$ [6]

The hadronic decays $\eta' \to \pi^+\pi^-\pi^{+(0)}\pi^{-(0)}$ are not suppressed by approximate symmetries. Recently Guo, Kubis and Wirzba [7], using a combination of chiral perturbation theory (ChPT) and a vector-meson dominance (VMD) model, obtained the following prediction : $\mathcal{B}(\eta' \to \pi^+\pi^-\pi^+\pi^-) = (1.0 \pm 0.3) \times 10^{-4}$ and $\mathcal{B}(\eta' \to \pi^+\pi^-\pi^0\pi^0) = (2.4 \pm 0.7) \times 10^{-4}$. The $\pi^+\pi^-\pi^{+(0)}\pi^{-(0)}$ invariant mass distributions are shown in Figs. 1(a) and (b), respectively, where the η' peak is clearly seen. To ensure that the η' peak is not from background, an extensive MC study was performed. It was found, as displayed by the hatched histograms in Figs. 1(a) and (b), that none of these background sources produces a peak in the $\pi^+\pi^-\pi^+\pi^-$ invariant mass spectrum near the η' mass.

The signal yields are obtained from extended unbinned maximum likelihood fits to the $\pi^+\pi^-\pi^+\pi^-$ and $\pi^+\pi^-\pi^0\pi^0$ invariant mass distributions and the statistical significances for $\eta' \to \pi^+\pi^-\pi^+\pi^-$ and $\eta' \to \pi^+\pi^-\pi^0\pi^0$ are calculated to be 18σ and 5σ, respectively. The branching fractions of $\eta' \to \pi^+\pi^-\pi^{+(0)}\pi^{-(0)}$ are determined to be $\mathcal{B}(\eta' \to \pi^+\pi^-\pi^+\pi^-) = (8.53 \pm 0.69 \pm 0.64) \times 10^{-5}$ and $\mathcal{B}(\eta' \to \pi^+\pi^-\pi^0\pi^0) = (1.82 \pm 0.35 \pm 0.18) \times 10^{-4}$, which are consistent with the theoretical predictions based on a combination of chiral perturbation theory and vector-meson dominance [7], but not with the broken-$SU_6 \times O_3$ quark model [8].

2.2 Dalitz plot analysis of $\eta \to \pi^+\pi^-\pi^0$ and $\eta/\eta' \to \pi^0\pi^0\pi^0$ [9]

Considerable theoretical efforts, see *e.g.* [10–17], have been invested to explain the discrepancy that the predicted

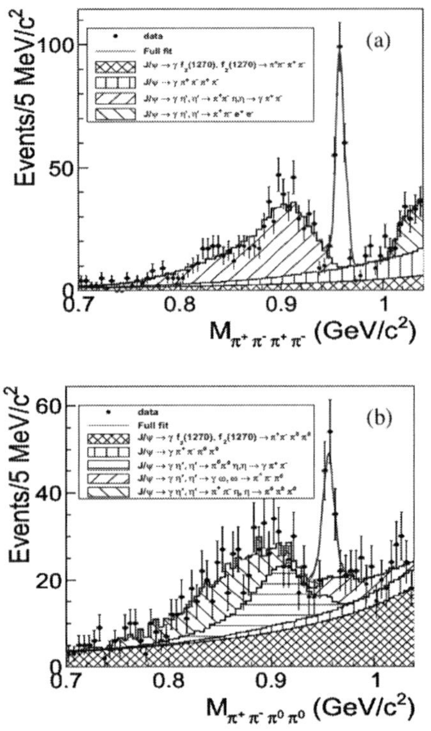

Figure 1. Results of the fits to (a) $M_{\pi^+\pi^-\pi^+\pi^-}$ and (b) $M_{\pi^+\pi^-\pi^0\pi^0}$, where the background contributions are displayed as the hatched histograms.

decay width of $\eta \to \pi^+\pi^-\pi^0$ [18] at the tree level of ChPT is much lower than the experimental value of 300 ± 11 eV [19]. To distinguish between the different theoretical approaches, precise measurements of the matrix elements for $\eta \to \pi^+\pi^-\pi^0$ and the decay width are needed.

For $\eta \to \pi^+\pi^-\pi^0$ decays, a sample of 8×10^4 $\eta \to \pi^+\pi^-\pi^0$ candidate events are selected and the background contamination is estimated to be about 0.1% as shown in Fig. 2 (a), which is neglected in the extraction of the Dalitz plot parameters. The two the Dalitz plot variables are defined as $X = \frac{\sqrt{3}}{Q}(T_{\pi^+} - T_{\pi^-})$ and $Y = \frac{m_\eta + 2m_\pi}{m_\pi}\frac{T_\eta}{Q} - 1$, where where T_π denotes the kinetic energy of a given pion in the η rest frame, $Q = m_\eta - m_{\pi^+} - m_{\pi^-} - m_{\pi^0}$ is the excess energy of the reaction. The distributions of X and Y are shown in Fig. 2 (b) and (c). Using the same parameterization as in Ref. [20], the Dalitz plot parameters for $\eta \to \pi^+\pi^-\pi^0$ are determined to be $a = -1.128 \pm 0.015 \pm 0.008$, $b = 0.153 \pm 0.017 \pm 0.004$, $d = 0.085 \pm 0.016 \pm 0.009$, $f = 0.173 \pm 0.028 \pm 0.021$, which are in reasonable agreement with previous measurements.

Figure 3(a) shows the $\pi^0\pi^0\pi^0$ mass spectrum, where the η peak is quite clean and the background level is less than 1%. The distribution of the variable Z is displayed in Fig. 3(b). Due to the kinematic boundaries, the interval of $0 < Z < 0.7$, corresponding to the region of phase space where the Z distribution is flat, is used to extract the slope parameter α from the data. Analogous to the measurement for $\eta \to \pi^+\pi^-\pi^0$, an unbinned maximum likelihood fit, as displayed in the inset of Fig. 3(b), yields the Dalitz plot slope parameter $\alpha = -0.055 \pm 0.014 \pm 0.004$, which is compatible with the recent results from other experiments and in agreement with the prediction from ChPT at NNLO within two standard deviations of the theoretical uncertainties.

For $\eta' \to \pi^0\pi^0\pi^0$, the η' signal is also clearly observed in $\pi^0\pi^0\pi^0$ mass spectrum [Fig. 3(c)], where the hatched and shaded histogram show the background contributions estimated from the inclusive J/ψ decays and $\eta' \to \pi^0\pi^0\eta$, respectively. The Dalitz plot slope parameter for $\eta' \to \pi^0\pi^0\pi^0$ is measured to be $\alpha = -0.640 \pm 0.046 \pm 0.047$ [Fig. 3(d)], which is consistent with but more precise than previous measurements. The value deviates significantly from zero, which implies that final state interactions play an important role in the decay. Up to now, there are just a few predictions about the slope parameter of $\eta' \to \pi^0\pi^0\pi^0$. In Ref. [21], the slope parameter is predicted to be less than 0.03, which is excluded by our measurement.

2.3 Observation of $\eta' \to \rho^\pm\pi^\mp$ [22]

With the isobar model, a combined amplitude analysis of the isospin violating decays $\eta' \to \pi^+\pi^-\pi^0$ and $\eta' \to \pi^0\pi^0\pi^0$ is performed. The results shown in the invariant mass spectra of $\pi^+\pi^-$, $\pi^+\pi^0$ and $\pi^-\pi^0$ (Fig. 4) indicate that significant P-wave contribution from $\eta' \to \rho^\pm\pi^\mp$ is observed for the first time in $\eta' \to \pi^+\pi^-\pi^0$, which is also clearly indicated as the two clusters in the Dalitz plot of $M^2_{\pi^+\pi^-}$ versus $M^2_{\pi^-\pi^0}$ [Fig. 5 (a)]. The branching fraction $\mathcal{B}(\eta' \to \rho^\pm\pi^\mp)$ is determined to be $[7.44 \pm 0.60(stat) \pm 1.26(syst) \pm 1.84(model)] \times 10^{-4}$. In addition to the non-resonant S-wave, the resonant π-π S-wave with a pole at $(512\pm15)-i(188\pm12)$ MeV, interpreted as the broad σ meson, plays an essential role in the $\eta' \to \pi\pi\pi$ decays. Due to the large interference between non-resonant and resonant S-waves, only the sum is used to describe the S-wave contribution, and the branching fractions are determined to be $\mathcal{B}(\eta' \to \pi^+\pi^-\pi^0)_S = [37.63 \pm 0.77(stat) \pm 2.22(syst) \pm 4.48(model)] \times 10^{-4}$.

For $\eta' \to \pi^0\pi^0\pi^0$, the P-wave contribution in two-body rescattering is forbidden by Bose symmetry. The Dalitz plot for $\eta' \to \pi^0\pi^0\pi^0$ is shown in Fig. 5 (b) and the amplitude result is displayed in Fig. 5 (c). The corresponding branching fraction is measured to be $\mathcal{B}(\eta' \to \pi^0\pi^0\pi^0) = [35.22 \pm 0.82(stat) \pm 2.60(syst)] \times 10^{-4}$.

The branching fractions of $\eta' \to \pi^+\pi^-\pi^0$ and $\eta' \to \pi^0\pi^0\pi^0$ are in good agreement with and supersede the previous BESIII measurements [23]. The value for $\mathcal{B}(\eta' \to \pi^0\pi^0\pi^0)$ is two times larger than that from GAMS [$(16 \pm 3.2) \times 10^{-4}$] [24]. The significant resonant S-wave contribution also provides a reasonable explanation for the negative slope parameter of the $\eta' \to \pi^0\pi^0\pi^0$ Dalitz plot [25]. The ratio of the branching fractions between the S-wave components $\mathcal{B}(\eta' \to \pi^0\pi^0\pi^0)/\mathcal{B}(\eta' \to \pi^+\pi^-\pi^0)S$ is determined as $0.94 \pm 0.029 \pm 0.13$, where the common systematic are cancelled. With the branching fractions of

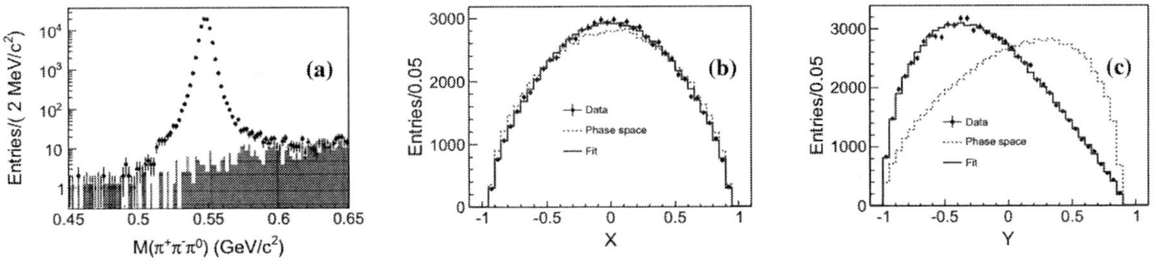

FIGURE 2. (a) Distribution of $\pi^+\pi^-\pi^0$ invariant mass. Projections of the Dalitz plot as a function of (b) X and (c) Y for $\eta \to \pi^+\pi^-\pi^0$ obtained from data (dots with error bars) and phase space distributed MC events (dashed line). The result of the fit described in the text (solid line) is also plotted.

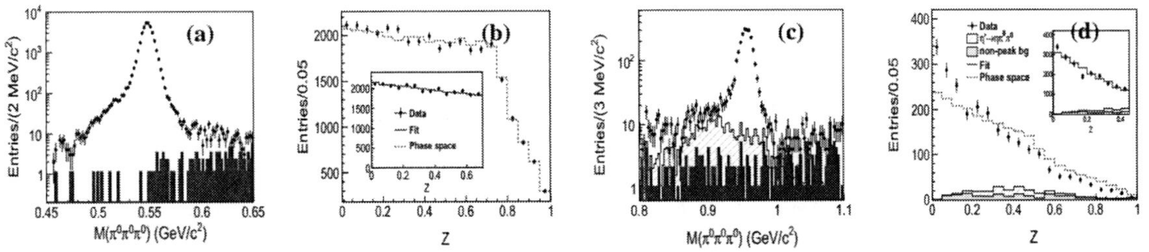

FIGURE 3. (a) Distribution of $\pi^0\pi^0\pi^0$ invariant mass. (b) Distribution of the kinematic variable Z for $\eta \to \pi^0\pi^0\pi^0$. (c) Distribution of $\pi^0\pi^0\pi^0$ invariant mass. (d) Distribution of the kinematic variable Z for $\eta' \to \pi^0\pi^0\pi^0$. Dots with error bars are for data, histograms for background contributions, dashed histograms for phase space distributed MC events and the solid lines in the inset are the results of the fit.

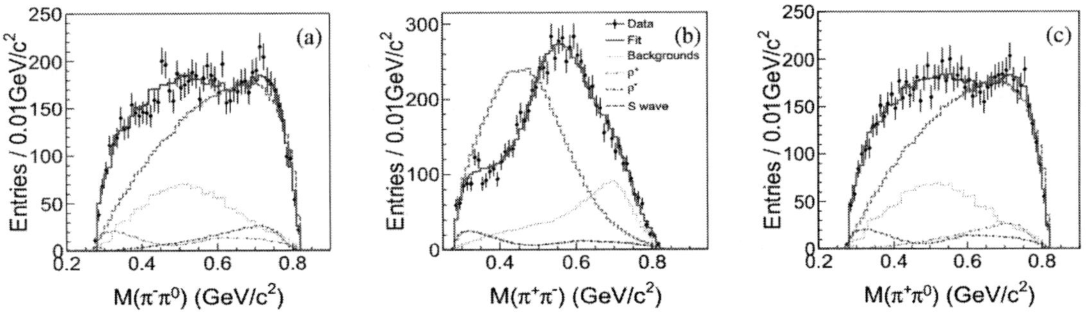

FIGURE 4. Comparison of the invariant mass distributions of (a) $\pi^-\pi^0$, (b) $\pi^+\pi^-$, (c) $\pi^+\pi^0$ (dots with error bars) and the fit result projections (solid histograms). The dotted, dashed, dash-dotted, and dash-dot-dotted histograms show the contributions from background, S wave, ρ^-, and ρ^+, respectively.

$\eta' \to \pi\pi\eta$ taken from PDG [19], $r_\pm = \frac{\mathcal{B}(\eta' \to \pi^+\pi^-\pi^0)}{\mathcal{B}(\eta' \to \pi^+\pi^-\eta)}$ and $r_0 = \frac{\mathcal{B}(\eta' \to \pi^0\pi^0\pi^0)}{\mathcal{B}(\eta' \to \pi^0\pi^0\eta)}$ are now calculated to be $(8.8 \pm 1.2) \times 10^{-3}$ and $(16.9 \pm 1.4) \times 10^{-3}$, respectively.

3 η/η' radiative decays

3.1 Observation of $\eta' \to \gamma e^+e^-$ [26]

Electromagnetic (EM) Dalitz decays of light pseudoscalar mesons, $P \to \gamma l^+l^-$ ($P = \pi^0, \eta, \eta'$; $l = e, \mu$), play an important role in revealing the structure of hadrons and the interaction mechanism between photons and hadrons [27]. In this work, we report the first observation of the Dalitz decay $\eta' \to \gamma e^+e^-$ as shown in Fig. 6 (a) and the branching fraction $\mathcal{B}(\eta' \to \gamma e^+e^-)$ is measured to be $(4.69 \pm 0.20 \pm 0.23) \times 10^{-4}$.

In addition, we present measurements of the transition form factor (TFF) as a function of $M(e^+e^-)$ [Figs. 6 (b) and (c)], which could be described with a single

DOI: 10.1051/epjconf/201714201014

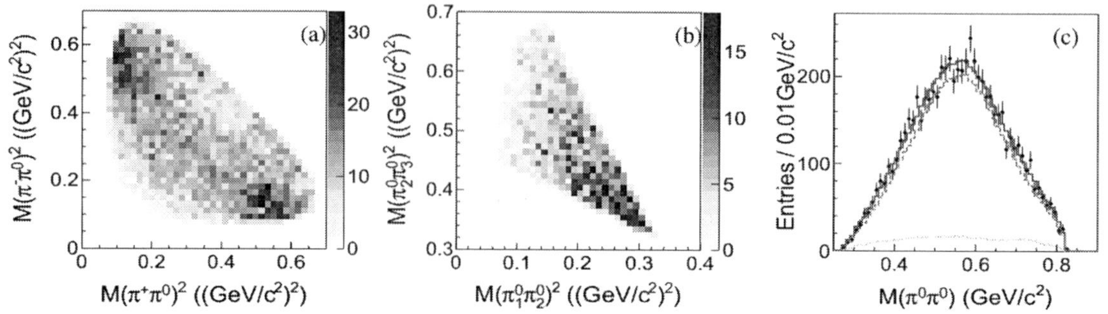

FIGURE 5. (a) Dalitz plot of $M^2_{\pi^+\pi^0}$ versus $M^2_{\pi^-\pi^0}$ for $\eta' \to \pi^+\pi^-\pi^0$. b) $\eta' \to \pi^0\pi^0\pi^0$ Dalitz plot for candidate events selected from data. (c) Comparison of $\pi^0\pi^0$ mass spectrum between data (dots with error bars) and the fit result projections (solid histograms).

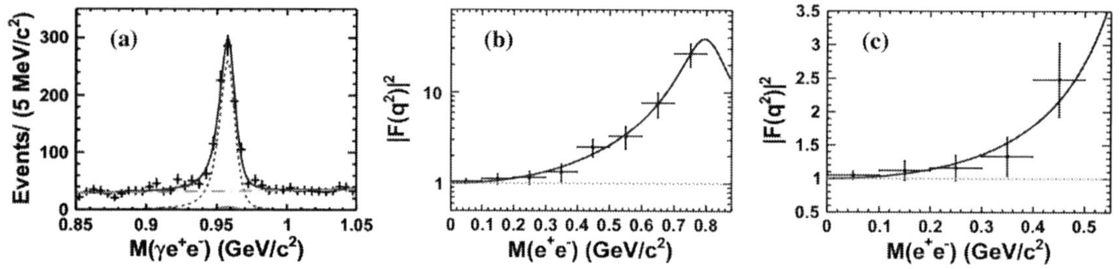

FIGURE 6. (a) Invariant γe^+e^- mass distribution for the selected signal events. The (black) crosses are the data, the (red) dashed line represents the signal, the (green) dot-dashed curve shows the non-peaking background shapes, the (orange) shaded component is the shape of the peaking background events. (b) Fit to the single pole form factor $|F|^2$. (c) Determination of the form factor slope by fitting to $|F|^2$.

pole parameterization of $F(q^2) = \frac{1}{(1-q^2/\Lambda^2)}$. The mass and width parameters of $\Lambda_{\eta'}$ and $\gamma_{\eta'}$ are determined to be (0.79 ± 0.04 ± 0.02) GeV, and $\gamma_{\eta'}$ = (0.13 ± 0.06 ± 0.03) GeV, respectively. The slope of the TFF corresponds to (1.60 ± 0.17 ± 0.08) GeV^{-2} and agrees within errors with the Vector Meson Dominance (VMD) model. predictions. The uncertainty of the η' transition form factor slope is in good agreement the best determination in the space-like region from the CELLO collaboration $b_{\eta'}$ = (1.60 ± 0.16) GeV^{-2} [28], and improves the previous determination of the slope in the time-like region $b_{\eta'}$ = (1.7 ± 0.4) GeV^{-2} [27, 29].

3.2 Observation of $\eta' \to e^+e^-\omega$ [30]

In 2013 BESIII reported the measurement of $\eta' \to \pi^+\pi^-e^+e^-$ [31], which is found to be dominated by $\eta' \to \rho e^+e^-$, in agreement with theoretical predictions [32?]. Based on theoretical models [32, 34], the branching fraction of $\eta' \to \omega e^+e^-$ is predicted to be around 2.0×10^{-4}, but until now there has been no measurement of this decay.

The scatter plot of $M(\pi^0\pi^+\pi^-e^+e^-)$ versus $M(\pi^0\pi^+\pi^-)$ is shown in Fig. 7 (a), where the cluster in the η' and ω region corresponds to the decay $\eta' \to \omega e^+e^-$. The η' peak is clearly seen in the distributions of $M(\pi^0\pi^+\pi^-e^+e^-)$ shown in Fig. 7 (b). To improve the mass resolution, the fit to $M(\pi^0\pi^+\pi^-e^+e^-)$- $M(\pi^0\pi^+\pi^-)$ is performed to extract the signal yields as indicated in Fig. 7 (c). For the first time, the decay of $\eta' \to \omega e^+e^-$ is observed with a statistical significance of 8σ, and its branching fraction is measured to be $\mathcal{B}(\eta' \to \omega e^+e^-)$ = [1.97 ± 0.34(stat)±0.17(syst)]×10^{-4}, which is consistent with theoretical prediction, 2.0×10^{-4} [32].

3.3 Observation of $\eta' \to \gamma\gamma\pi^0$ [35]

Most recently the observation of $\eta' \to \gamma\gamma\pi^0$ was reported by the BESIII Collaboration. Figure 8(a) shows the $\gamma\gamma\pi^0$ invariant mass spectrum, where the clear η' peak is observed. Assuming that the inclusive decay $\eta' \to \gamma\gamma\pi^0$ can be attributed to the vector mesons ρ and ω and the non-resonant contribution, we apply a fit to the $\gamma\pi^0$ invariant mass [Fig. 8 (b)] to determine the branching fraction for the non-resonant $\eta' \to \gamma\gamma\pi^0$ decay using the η' signal events with $|M_{\gamma\gamma\pi^0} - m_{\eta'}| < 25$ MeV/c^2. In the fit, the ρ-ω interference is considered, but possible interference between the ω (ρ) and the non-resonant process is neglected.

The branching fraction of the inclusive decay is measured for the first time to be $\mathcal{B}(\eta' \to \gamma\gamma\pi^0)_{\text{Incl.}}$ = (3.20 ± 0.07 ± 0.23) × 10^{-3}, which is much lower than that of the theoretical predictions[36, 37]. In addition, the branching fraction for the non-resonant decay is determined to be

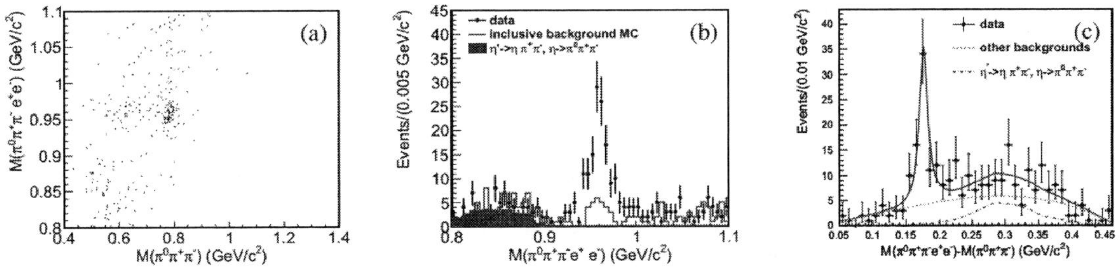

FIGURE 7. (a) Distribution of $M(\pi^0\pi^+\pi^-e^+e^-)$ versus $M(\pi^0\pi^+\pi^-)$; Distribution of $M(\pi^0\pi^+\pi^-e^+e^-) - M(\pi^0\pi^+\pi^-)$ and the fit results. The dots with error bars are for data, the histograms or dash-dotted line are for background contributions and the dotted curve shows the background except for the background events from $\eta' \to \pi^+\pi^-\eta$.

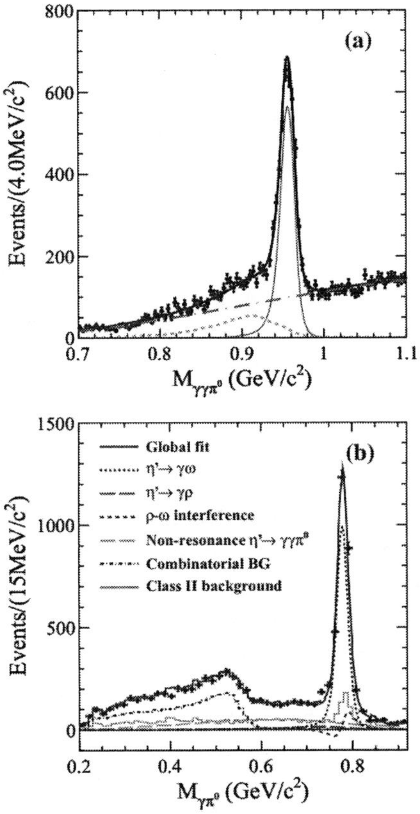

FIGURE 8. (a) Distribution of $M_{\gamma\gamma\pi^0}$ for the selected inclusive $\eta' \to \gamma\gamma\pi^0$ signal events via $J/\psi \to \gamma\eta'$ decay. The dots with error bars are for data, the dotted and dot-dashed curve are the shapes of the background events. The total fit result is shown as the solid line. (b) Distribution of the invariant mass $M_{\gamma\pi^0}$ and fit results. The dots with error bars are data; the dotted-curve is from the ω-contribution; the long dashed-curve is from the ρ-contribution; the short dashed-curve is the contribution of ρ-ω interference; the long dashed curve is the non-resonance; the histogram is from the background events; the short dot-dashed curve is the combinatorial backgrounds of $\eta' \to \gamma\omega, \gamma\rho$. The solid line shows the total fit function.

$\mathcal{B}(\eta' \to \gamma\gamma\pi^0)_{NR} = (6.16 \pm 0.64 \pm 0.67) \times 10^{-4}$, which agrees with the upper limit measured by the GAMS-2000 experiment [38]. As a validation of the fit, the product branching fraction with the omega intermediate state involved is obtained to be $\mathcal{B}(\eta' \to \gamma\omega) \cdot \mathcal{B}(\omega \to \gamma\pi^0) = (2.37 \pm 0.07 \pm 0.18) \times 10^{-3}$, which is consistent with the PDG value [5]. These results are useful to test QCD calculations on the transition form factor, and provide valuable inputs to the theoretical understanding of the light meson decay mechanisms.

3.4 Preliminary results : study of $\eta' \to \gamma\pi^+\pi^-$ decay dynamics

Theoretically the non-resonant part of coupling in $\eta' \to \gamma\pi^+\pi^-$ is accounted for by the higher term of Wess-Zumino-Witten (WZW) ChPT Lagrangian (also known as the box anomaly). However the decay dynamics of $\eta' \to \gamma\pi^+\pi^-$ has been explored with very limited statistics only and new measurements are needed to clarify the scenario. In this work, a sample of 9×10^5 $\eta' \to \gamma\pi^+\pi^-$ events is selected to investigate its decay dynamics and the $\pi^+\pi^-$ mass spectrum is shown in Fig. 9 with a background level of 1%.

The shape of the $M_{\pi^+\pi^-}$ spectrum is analyzed using parameterization relying on model-dependent and model-independent approaches respectively. For the model-dependent approach, the results show that only $\rho(770)$ resonance is insufficient to describe the data even if one takes into account both $\rho(770)$ and omega resonances and the interference between them. The fit performance gets much better after including the box anomaly [Fig. 9 (a)] with a statistical significance larger than 37σ. We also try to replace the box anomaly with $\rho(1450)$ [Fig. 9 (b)] by fixing its mass the width to be the world average values, the fit can also provide a reasonable description of data. Therefore, we conclude that in addition to $\rho(770)$ and ω, the box anomaly is necessary, but the contribution from $\rho(1450)$ could not be ruled out.

With the inclusion of ω into the expansion, $M_{\pi^+\pi^-}$ spectrum is very well described with the model-independent parameterisation [39] as shown in Fig. 9 (c), the values of parameters being $\alpha = 0.992 \pm 0.039 \pm 0.067$

 DOI: 10.1051/epjconf/201714201014

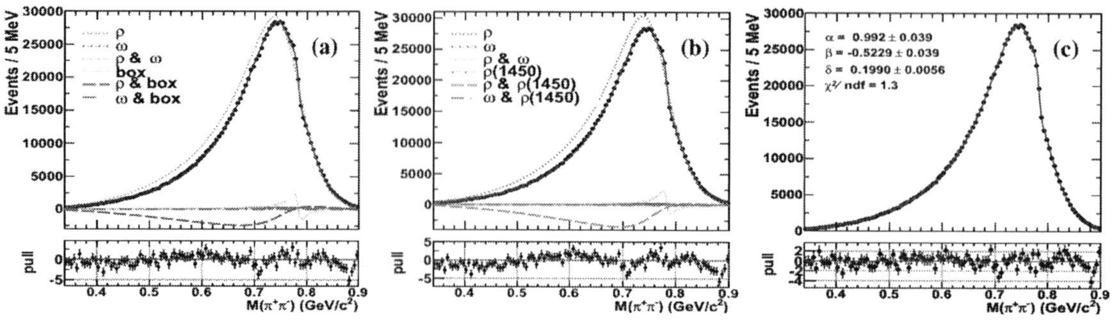

FIGURE 9. The results of the model-dependent fits to $M_{\pi\pi}$ with (a) $\rho - \omega - box\ anomaly$ and (b) $\rho - \omega - \rho(1450)$. (c) The results of model-independent fit with ω interference.

GeV^{-2}, $\beta = -0.523 \pm 0.039 \pm 0.066$ GeV^{-4}, $\delta = 0.199 \pm 0.006 \pm 0.010$. The value of α found in this work is compatible with the work of [39].

4 Search for rare decays of $\eta' \to K^{\pm}\pi^{\mp}$ [40]

Non-leptonic weak decays are valuable tools for exploring physics beyond the Standard Model. Among non-leptonic decays, the decay of the light pseudoscalar meson $\eta' \to K^{\pm}\pi^{\mp}$ is interesting because it is fundamental to understand the long-standing problem of the $\Delta I = 1/2$ rule in weak non-leptonic interactions. The branching fraction of $\eta' \to K^{\pm}\pi^{\mp}$ decay is predicted to be of the order of 10^{-10} or higher [41], with a large long-range hadronic contribution expected.

A search for the non-leptonic weak decay $\eta' \to K^{\pm}\pi^{\mp}$ is performed for the first time through the $J/\psi \to \phi\eta'$ decay, while no evidence for $\eta' \to K^{\pm}\pi^{\mp}$ is seen. Thus the 90% C.L. upper limit on $\mathcal{B}(\eta' \to K^{\pm}\pi^{\mp})$ is calculated to be 3.8×10^{-5}.

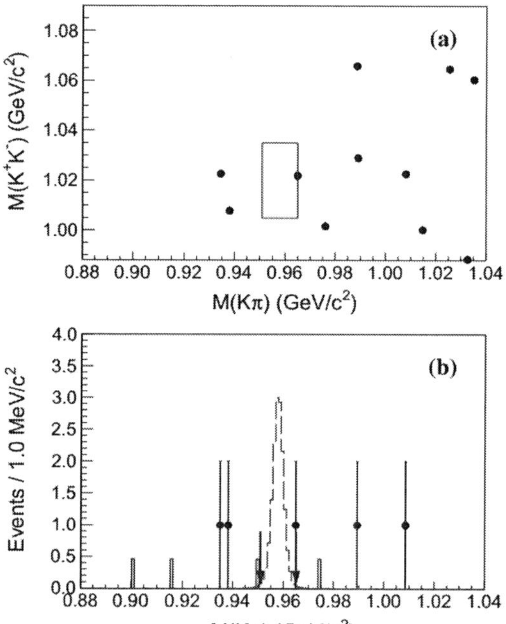

FIGURE 10. (a) Scatter plot of $M(K^+K^-)$ versus $M(K^{\pm}\pi^{\mp})$, where the box indicates the signal region with $|M(K^+K^-) - M(\phi)| <$ 15 MeV/c^2 and $|M(K^{\pm}\pi^{\mp}) - M(\eta')| <$ 7 MeV/c^2. (b) The $K^{\pm}\pi^{\mp}$ invariant mass distribution, where the arrows show the signal region. The dots with error bars are for data, the dashed histogram is for the signal MC with arbitrary normalization, and the solid histogram is the background contamination from a MC simulation of $J/\psi \to \phi\pi^+\pi^-$.

5 Summary

In summary the recent results on η/η' decays at BESIII are presented with a sample of 1.3×10^9 J/ψ events. The hadronic decays of $\eta' \to \pi^+\pi^-\pi^+\pi^-$ and $\eta' \to \pi^+\pi^-\pi^0\pi^0$ are also observed and the measurement branching fractions are consistent with the theoretical predictions ; The Dalitz plots of $\eta \to \pi^+\pi^-\pi^0$ and $\eta/\eta' \to \pi^0\pi^0\pi^0$ are analyzed and the corresponding matrix elements are extracted ; The observation of $\eta' \to \rho^{\mp}\pi^{\pm}$ is reported for the first time ; the new decay mode of $\eta' \to \gamma e^+e-$ is also observed and the TFF is measured with the function of $M_{e^+e^-}$; the decay dynamics of $\eta' \to \gamma\pi^+\pi^-$ is investigated with both model-dependent and model-independent parameterizations. It is found that the extra contribution is necessary to describe data besides the contributions from $\rho(770)$ and ω. The above results indicate that there is a rich field to be explored in η/η' physics, which motivates to search for their rare and forbidden decays.

Based on the roughly estimation, the available $1.3{\times}10^9$ J/ψ events enables to investigate the Dalitz plot of $\omega \to \pi^+\pi^-\pi^0$, which could be used to test the recent theoretical calculations based on the dispersive theory. This analysis is undergoing and the preliminary results will be released soon. However it is still not sufficient to search for ω rare decays. Hopefully a sample of 10 billion J/ψ events will be collected at the BESIII detector, which allows the study of light meson decays, such as η/η' and ω, with unprecedented precision.

References

[1] F. Niecknig, B. Kubis, S. P. Schneider, Eur. Phys. J. C **72**,2014 (2012).
[2] I. V. Danilkin, C. Fernndez-Ramrez *et al.*, Phys. Rev. D **91**,094029 (2015).
[3] A. Aloisio *et al.* (KLOE Collaboration), Phys. Lett. B **561** (2003).
[4] M. Ablikim *et al.* (BESIII Collaboration), Chin. Phys. C 36, 915 (2012) ; M. Ablikim *et al.* (BESIII Collaboration), arXiv :1607.00738 [hep-ex].
[5] C. Patrignani *et al.* (Particle Data Group), Chin. Phys. C **40**, 100001 (2016).
[6] M. Ablikim *et al.* (BESIII Collaboration), Phys. Rev. Lett. **112**, 251801 (2014).
[7] F. K. Guo, B. Kubis and A. Wirzba, Phys. Rev. D **85**, 014014 (2012).
[8] D.Parashar, Phys. Rev. D **19**, 268 (1979).
[9] M. Ablikim *et al.* (BESIII Collaboration), Phys. Rev. D **92**, 012014 (2015).
[10] A. Anisovich and H. Leutwyler, Phys. Lett. B **375**, 335 (1996).
[11] S. P. Schneider, B. Kubis, and C. Ditsche, JHEP **1102**, 028 (2011).
[12] J. Gasser and H. Leutwyler, Nucl. Phys. B **250**, 539 (1985).
[13] J. Kambor, C. Wiesendanger and D. Wyler,Nucl. Phys. B **465**, 215 (1996).
[14] J. Bijnens and K. Ghorbani, JHEP **0711**, 030 (2007).
[15] K. Kampf, M. Knecht, J. Novotny and M. Zdrahal, Phys. Rev. D **84**, 114015 (2011).
[16] G. Colangelo, S. Lanz, H. Leutwyler and E. Passemar, PoS EPS **-HEP2011**, 304 (2011).
[17] P. Guo *et al.*, Phys. Rev. D **92**, 054016 (2015).
[18] H. Osborn and D. J. Wallace, Nucl. Phys. B **20**, 23 (1970) ; J. A. Cronin, Phys. Rev. **161**, 1483 (1967).
[19] K. A. Olive *et al.* (Particle Data Group), Chin. Phys. C **38**, 090001 (2014).
[20] F. Ambrosino *et al.* (KLOE Collaboration), JHEP **0805**, 006 (2008).
[21] C. Roiesnel, T. N. Truong, Paris Ec. Polytech. A **515**, 0982(1982).
[22] M. Ablikim *et al.* (BESIII Collaboration), Phys. Rev. Lett. **118**, 012001 (2017).
[23] M. Ablikim *et al.* (BESIII Collaboration), Phys. Rev. Lett. **108**, 182001 (2012).
[24] D. Alde, F. G. Binon, and C. Bricman, Z. Phys. C **36**, 603 (1987).
[25] M. Ablikim *et al.* (BESIII Collaboration), Phys. Rev. D **92**, 012014 (2015).
[26] M. Ablikim *et al.* (BESIII Collaboration), Phys. Rev. D **83**, 012003 (2011).
[27] L. G. Landsberg, Phys. Rept. **128**, 301 (1985).
[28] H. J. Behrend *et al.* (CELLO Collaboration), Z. Phys. C **49**, 401 (1991).
[29] R. I. Dzhelyadin *et al.*, Phys. Lett. B **88**, 379 (1979).
[30] M. Ablikim *et al.*, (BESIII Collaboration), Phys. Rev. D **92**, 051101(R) (2015).
[31] M. Ablikim *et al.*, (BESIII Collaboration), Phys. Rev. D **87**, 092011 (2013).
[32] A. Faessler, C. Fuchs, M. Krivoruchenko, Phys. Rev. C **61**, 035206 (2000).
[33] B. Borasoy, R. Nissler, Eur. Phys. J. A **33**, 95 (2007).
[34] C. Terschlüsen, S. Leupold, M. Lutz, Eur. Phys. J. A **48**, 190 (2012).
[35] M. Ablikim *et al.*, (BESIII Collaboration), arXiv :1612.0572 [hep-ex].
[36] R. Jora, Nucl. Phys. Proc. Suppl. 207-208, 224 (2010).
[37] R. Escribano, PoS QNP 2012, 079 (2012).
[38] D. Alde *et al.* (GAMS-2000 Collaboration), Z. Phys. **C 36**, 603 (1987).
[39] F. Stollenwerk, C. Hanhart, A. Kupsc, U.-G. Meissner, A. Wirzba, Phys. Lett. B**707**, 184 (2012).
[40] M. Ablikim *et al.*, (BESIII Collaboration), Phys. Rev. D **93**, 072008 (2016).
[41] L. Bergström and H. R. Rubinstein, Phys. Lett. B **203**, 183 (1988).
[42] D. M. Asner *et al.*, Int. J. Mod. Phys. A **24**, 1 (2009).

The APEX experiment at JLab

Searching for the vector boson A' decaying to e^+e^-

Gregg B. Franklin[1,⋆] for the APEX collaboration at JLab

[1] *Dept. of Physics; Carnegie Mellon University; Pittsburgh, PA 15213, USA*

Abstract. Jefferson Lab's A' Experiment (APEX) will search for a new vector boson, the A', in the mass range 65 MeV< $m_{A'}$ < 550 MeV, with sensitivity for the A' coupling to electrons of $\alpha' > 6 \times 10^{-8}\alpha$, where $\alpha = e^2/4\pi$. New vector bosons with such small couplings arise naturally from a small kinetic mixing of the "dark photon", A', with the photon — one of the very few ways in which new forces can couple to the Standard Model — and have received considerable attention as an explanation of various dark-matter related anomalies. In this experiment, A' bosons produced by radiation off an electron beam could appear as narrow resonances with small production cross-sections in the e^+e^- invariant mass distribution. The two Jefferson Lab HRS spectrometers will provide a reconstructed invariant-mass resolution for the A' of $\delta M/M < 0.5\%$. With a 33-day run, the experiment will achieve high sensitivity by taking advantage of this mass resolution and high statistics of the e^+e^- pairs, which will be orders of magnitude larger than in previous searches for the A' boson in this mass range. This paper will review the key concepts of the experiment and the status of the preparations for running APEX. The results of a completed pilot run will be presented.

1 Introduction

The APEX experiment [JLab Exp. E12-10-009][1] will search for a dark photon — a new massive vector boson that couples weakly to ordinary matter— by looking for a signature of the A' produced through a bremsstrahlung-like process as JLab's electron beam passes through a high-Z target, illustrated in Fig. 1. The experiment is based on the concepts presented in J.D. Bjorken *et al.* [2]. The APEX collaboration consists of upwards of 100 researchers from about 26 institutions.

The A' production is forward peaked, with a production angle $\sim (m_{A'}/E)^{3/2}$. The experiment will search a kinematic regime in which the A' can decay to an electron-positron pair; electron-positron pairs will be detected in the twin High Resolution Spectrometers positioned at small angles relative to the Hall A incident electron beam. APEX is designed to reach the required invariant mass resolution, 0.5%, for forward-peaked lepton pairs by using three key ingredients: 1) JLab's twin high-resolution spectrometers (HRS left and HRS right) with improved rate capability trackers, 2) a septum magnet, shown schematically in Fig. 1, that will bend the electron-positron pairs produced around ±7.5° outwards and into the acceptance of the spectrometers, and 3) a novel multi-foil tungsten target.

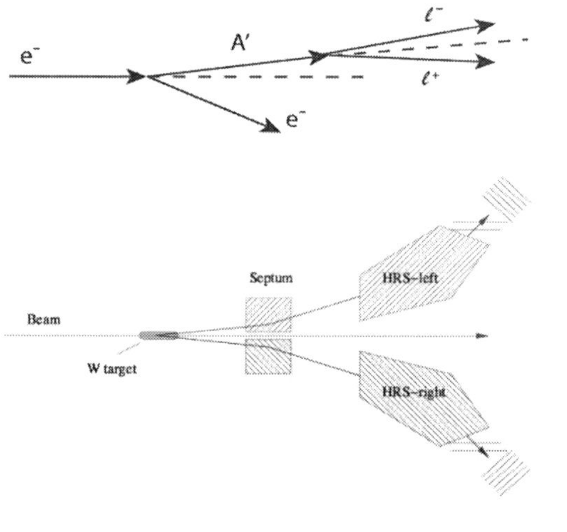

Figure 1. The APEX experiment will attempt to produce the A' through a bremsstrahlung-like process (top diagram) and search for a subsequent decay into a electron-positron pair using JLab's High Resolution Spectrometers. (bottom diagram).

2 The APEX reach

The production cross-section can be calculated using a QED-like Feynman diagram shown in Fig. 2, in which we assume the coupling constant e' can be written as an unknown dimensionless parameter ϵ times the electron charge e. This gives a production cross section that is proportional to $\epsilon^2/m_{A'}^2$. Thus, the production rate is parame-

⋆e-mail: gbfranklin@cmu.edu

DOI: 10.1051/epjconf/201714201015

terized by $m_{A'}$ and ϵ^2, or, equivalently, it is customary to write ϵ^2 as α'/α where $\alpha = e^2/4\pi$.

The background processes include trident diagrams of a very similar nature and are larger by order $1/\epsilon^2$, but this background is broadly distributed in mass so APEX can take advantage of the fact that the signal will be a very narrow invariant-mass peak sitting on top of this broad background.

Since the background is known and the signal is a function of α' and m, we have designed an experiment that has a region of sensitivity down to $\alpha'/\alpha = 9 \times 10^{-8}$ over an A' mass region of 65 MeV to 525 MeV. Reaching this level of sensitivity requires an invariant-mass resolution of 0.5% or better.

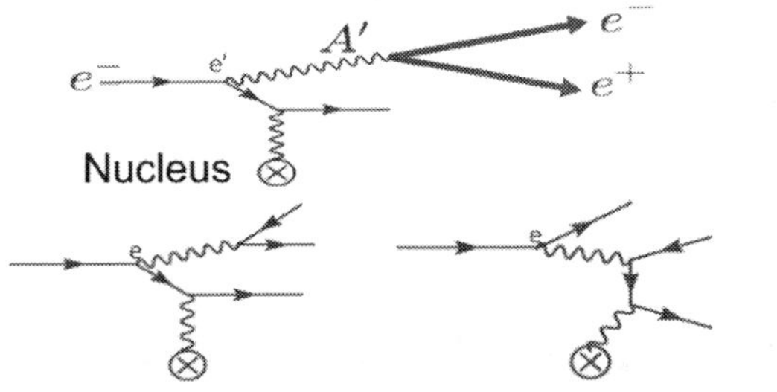

Figure 2. The Feynman diagram for production of the A' involves the coupling constant $e' = \epsilon e$ (top diagram) while the background Bethe Heitler pair production processes, shown in the bottom diagrams, are larger by a factor of ϵ.

The High Resolution Spectrometers have excellent resolution but relatively small kinematic acceptances compared to an apparatus based on large open geometries. However, the proposed search region of 65 MeV to 525 MeV can be covered with only four kinematic settings, shown in Fig. 3, due to the strong forward peaking of both the A' and the subsequent lepton decay pairs. Figure 3 shows that APEX can reach a sensitivity of ϵ^2 down to about 10^{-7} with just 30 days of production running.

The multi-foil target is shown in Fig. 4. The incident electron-beam passes through a series of tungsten foils. These foils are arranged so that the outgoing positron-electron pairs produced in one foil do not pass through any of the downstream foils on their way to the septum magnet. This minimizes the multiple scattering and is required to reach the desired mass resolution. In addition, the spatial distribution of the foils actually increases the experiment's angular acceptance and mass coverage.

3 Achieving $\delta m/m = 0.5\%$

The experiment is based on reconstructing the invariant mass of lepton pairs using the two four-momenta measured in the High Resolution Spectrometers (HRSs).

$$M^2 = (p_1^\mu + p_2^\mu)^2 \approx 2|\vec{p}_1||\vec{p}_2| - 2\vec{p}_1 \cdot \vec{p}_2 = 4 p_1 p_2 sin^2\frac{\theta}{2} \approx p_1 p_2 \theta^2.$$

Uncertainty in the energy of the incident beam does not affect the experiment. The errors in the pair of lepton

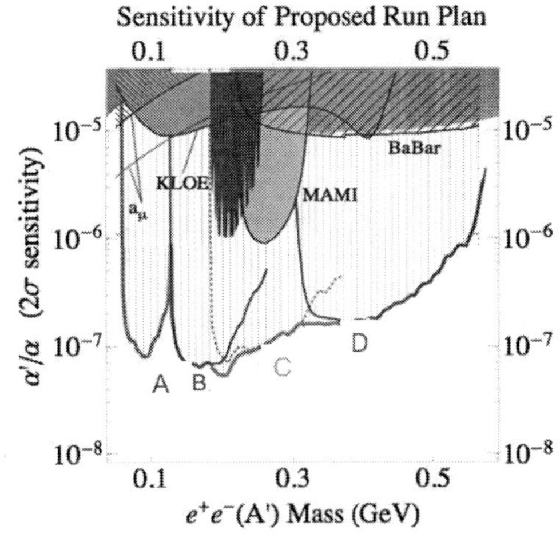

Figure 3. The predicted region of sensitivity and reach of the APEX experiment assuming 6 days at 1.1 GeV electron-beam energy (region A), 6 days at 2.3 GeV (region B), 6 days at 3.3 GeV (region C), and 12 days at 4.5 GeV (region D). The exclusion region of the APEX test run is shown in blue along with the exclusion regions from KLOE, MAMI and BaBar.

Figure 4. Left: The foil production target ladder design with three foil production targets at bottom and auxiliary targets used for alignment and optics calibrations at top. Right: Photo of a production target showing line of vertical tungsten target foils.

momenta and opening angle would introduce an error in the extracted invariant mass,

$$\frac{\delta M}{M} = \frac{\delta p_1}{2p_1} + \frac{\delta p_2}{2p_2} + \frac{\delta\theta}{\theta}.$$

The trajectories of the out-going decay leptons are measured by the instrumentation packages located at the back of each of the two HRS spectrometers. Determination of the lepton-pair opening angle θ for each event is complicated by the need to track back through the 23 meters of the QQDQ magnetic optics and then further back through the septum to determine the lepton trajectories as they exit the target foils. Our ability to reconstruct the lepton trajectories back to the target region was demonstrated during the APEX test run. As part of the HRS calibration procedures, two tungsten sieves (tungsten plates containing a pattern of holes) were used. The sieves were inserted in each spectrometer arm in front of the septum magnet during low intensity running with both spectrometer arms tuned for electrons. The analysis magnetic-optics parameters were then tuned to accurately reconstruct the hole pattern using the trajectories of the electrons that had passed through the holes. This calibration data demonstrated that the opening angle between two outgoing particles (one in each arm) was determined to 0.3 mrad. A systematic error in the decay-lepton opening angle of this size would correspond to a reconstructed mass shift of only 0.17%.

This uncertainty in the determination of the absolute mass is not important for APEX, but the resolution of the mass measurement is important since we are taking advantage of the HRS's ability to provide a signature of a narrow peak sitting on a broad background. It is the projected APEX resolution that will give the experiment its sensitivity compared with the competition. Assuming both spectrometers have the same fractional rms resolution, σ_p/p, and using σ_θ/θ for the accuracy of the reconstructed opening angle, the expected width of the peak is

$$\left(\frac{\sigma_M}{M}\right)^2 = \frac{1}{2}\left(\frac{\sigma_p}{p}\right)^2 + \left(\frac{\sigma_\theta}{\theta}\right)^2.$$

The APEX experiment is designed to achieve an invariant mass resolution of $\delta m/m = 0.5\%$. At better than 10^{-3}, the HRS momentum resolution does not contribute significantly to our error budget so the missing mass resolution is determined by our ability to reconstruct the opening angle of the outgoing lepton pair. The multiple scattering of the leptons as they exit a single target foil contributes an uncertainty of around 0.5 mrad for the worst-case kinematics setting, so we need to measure the outgoing angles to about 0.7 mrad. The reconstruction of the sieve holes during the APEX test run demonstrated that we can achieve a resolution of 0.54 mrad for the opening decay angle and, in turn, this indicates that APEX can achieve a resolution of $\delta M/M < 0.5\%$.

4 Preparation for the APEX run

As discussed above, the APEX reach in ϵ^2 is achieved through its excellent invariant mass resolution and this, in turn, is determined by our ability to reconstruct the lepton-pair opening angle at the target. The tungsten-sieve calibration data are used for tuning the magnetic optics parameters and provide a direct measurement of our ability to determine the lepton trajectories as they leave the target. The APEX test run analysis demonstrated that we achieved a resolution of $\delta\theta < 0.54$ mrad. However, this angular resolution measurement is extracted from data taken with both spectrometers tuned for electrons. Additional uncertainties arise from the need to reverse the field in one spectrometer arm to accept positrons after the sieve-based optics calibration is completed; aberrations that may arise when one half of the septum magnetic is reversed may reduce the experiment's resolution. In principle, it would be best to calibrate with one arm tuned for positrons, but the analysis of the test run data taken with this configuration was not able to clearly identify positrons coming from the sieve holes. This observation agrees with Monte Carlo simulations showing that positrons produced in the target and passing through the sieve holes cannot be separated from positrons produced in the sieve itself. For the test run, the limit on the degradation of the angular resolution caused by reversing the fields had to be estimated.

For the actual APEX production runs, we plan to calibrate the positron-arm with positrons by implementing active sieves based on arrays of scintillating fibers, shown in Fig. 5. This will avoid uncertainties associated with reversing the fields. There will be one active sieve for each of the two arms. Each active sieve will consist of two planes of 1 mm diameter scintillating fibers with 32 fibers per plane. The scintillator fibers will be coupled to multi-anode PMTs whose signals will be processed by JLab's F250 16-channel 250 MHz flash ADCs after amplification by x10 signal amplifiers. The active sieves will be located inside the target vacuum housing and can be inserted into the spectrometer acceptance regions using remotely controlled linear actuators.

Figure 5. Top: An active sieve attached to a linear actuator and vacuum bellows. Bottom: The APEX septum magnet.

The APEX production run will take advantage of the well-understood Hall A equipment, data acquisition, and analysis software associated with the High Resolution Spectrometers. The active-sieve and the septum magnet (shown in Fig 5) are both complete and ready for installation on short notice. In addition, we plan to make small modifications to the high voltage electronics used for the spectrometer's vertical drift chambers. These modifications will allow us to run with higher luminosity during the APEX production running.

5 APEX test run and upcoming production running

When reviewing the preparations for the APEX run, the most important point to make is that the collaboration already ran a test run in July of 2010 that not only verified all key aspects of the experiment, but also produced respectable results around $m_{A'} \sim 200$ MeV and these results have already been published as a Physical Review Letter[3]. The test run was performed using an existing, but non-optimal, septum magnet and a 22 mg/cm^2 tantalum foil. The two spectrometer arms were positioned for central acceptance of $\pm4.5\%$ and the incident electron beam was delivered at 2.26 GeV. The data collected contained 770,000 coincidence events. The analysis showed that 7.4% of these events were accidental coincidences between the two arms and 0.9% were due to meson contamination, leaving 700,000 good trident events.

The e^+e^- invariant mass spectrum obtained from the test run data, shown in Fig. 6, shows no visible sign of a narrow peak that would correspond to A' production. A quantitative peak search algorithm, described in detail in Ref. [3], was used to produce the 90% confidence upper limit shown in Fig. 7. It can be seen that the test run data saw no evidence for $A' \to e^+e^-$ and set an upper limit of $\alpha'/\alpha \simeq 10^{-6}$.

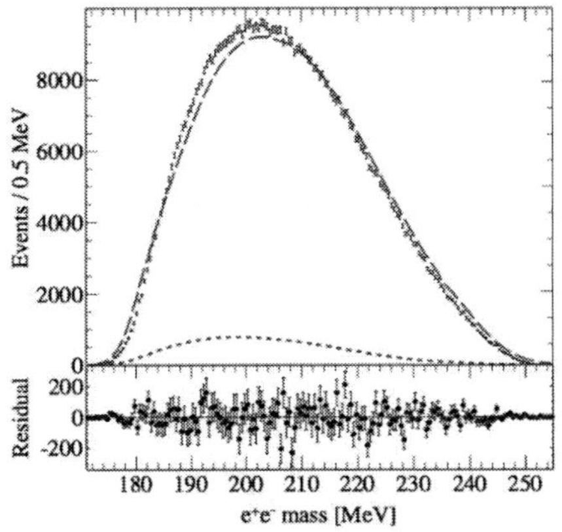

Figure 6. Invariant mass spectrum obtained from the APEX test run data. The upper panel shows the data (black points, with error bars), the contributions from e^+e^- coincidence events (blue short-dash line), and the calculated QED trident background added to the accidental event sample (red long-dash line). The lower panel shows the bin-by-bin residuals with respect to a 10-parameter fit to the global distribution.

The APEX 90% confidence level reach for 30 days of production running with the improved equipment is shown in Fig. 8. The landscape of exclusion zones based on the results of other experiments has been filled in considerably since the APEX experiment was first proposed. However, the APEX calculated reach, down to $\alpha' \sim 6 \times 10^{-8}\alpha$, remains an order of magnitude better than our current state of knowledge in the mass region 65 MeV$< m_{A'} <$ 550 MeV.

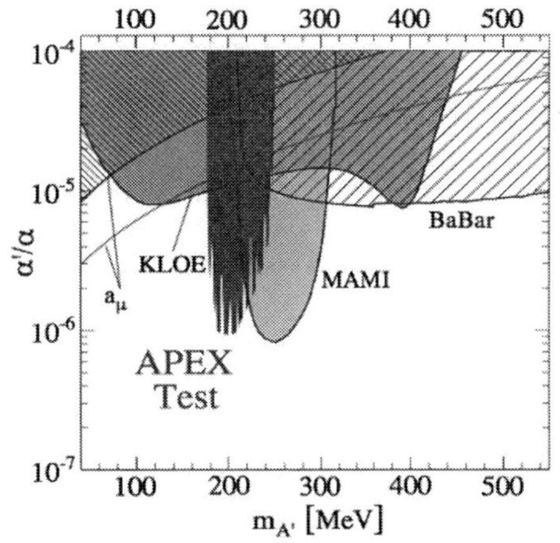

Figure 7. The 90% confidence upper limit on α'/α vs the A' mass for the APEX test run (solid blue). Also shown are the exclusion regions known at the time of the test-data publication: from the muon anomalous magnetic moment a_μ (fine hatched) [4], KLOE (solid gray) [5], Mainz (solid green) [6], and BABAR (wide hatched) [2, 7] as presented in Ref. [3]. Calculations indicate that the A' could explain the observed discrepancy between the calculated and measured muon anomalous magnetic moment if (α'/α) lies below the fine-hatched region and above the red line.

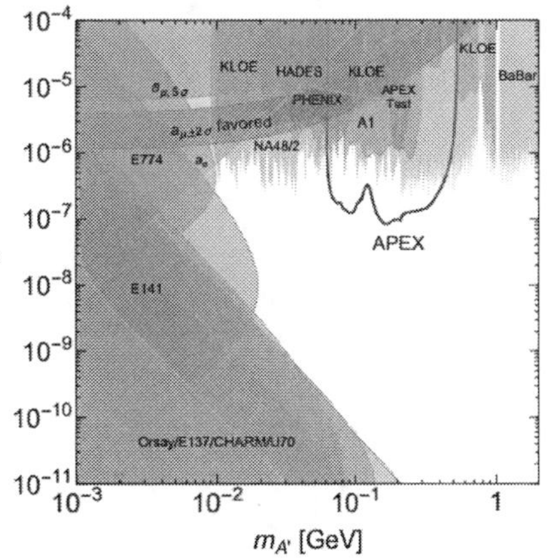

Figure 8. The predicted reach of the APEX experiment assuming 30 days of production running compared to existing results.

Although the experiment was prepared to run in the spring of 2016, the run did not take place due to complications with the Jefferson Lab 12 GeV-era accelerator turn on. It now appears that the earliest likely running of APEX would be in the fall of 2018.

References

[1] P. Essig *et al.*, JLab Experiment E12-10-009 (2009); R. Essig, P. Schuster, N. Toro, and B. Wojtsekhowski, J. High Energy Phys. **02**, 009, (2011).

[2] J.D. Bjorken, R. Essig, P. Schuster, and N. Toro, Phys. Rev. D **80**, 075018 (2009).

[3] S. Abrahamyan *et al.*, Phys. Rev. Let. **107**, 191804 (2011).

[4] M. Pospelov, Phys. Rev. D **80** 095002 (2009).

[5] C. Cheung, *et al.* Phys. Rev. D **80** 035008 (2009); N. Arkani-Hamad and N. Weiner, J. High Energy Phys. **12** 104 (2008); D.E. Morrisay, D. Poland, and K.M. Zurek, J. High Energy Phys. **07** 050 (2010).

[6] H. Merkel *et al.* (A1 Collaboration), Phys. Rev. Lett. **106**, 251802 (2011).

[7] M. Reece and L.-T. Wang, J. High Energy Phys. **07** 051 (2009); B. Aubert *et al.* (BABAR Collaboration), arXiv:0902.2176.

EPJ Web of Conferences **142**, 01016 (2017)
Advances in Dark Matter and Particle Physics
DOI: 10.1051/epjconf/201714201016

Geant4 Standard and Low Energy electromagnetic libraries

Vladimir Ivanchenko[1,2,a] and Sebastien Incerti[3] on behalf of the Geant4 Collaboration

[1]CERN, CH1211Geneva 23, Switzerland
[2]Geant4 Associates International Ltd, United Kingdom
[3]Université of Bordeaux 1, CNRS/IN2P3, CENBG, 33175 Gradignan, France

Abstract. Geant4 is a Monte Carlo toolkit well established for large scale simulations for high energy physics, medicine, space science and other applications. Electromagnetic (EM) physics libraries are key components of these simulations. In this work we describe the Standard and Low Energy EM libraries available in Geant4 version 10.3 including descriptions of new models and interfaces relevant for dark matter searches and limits of applicability for these models.

1 Introduction

The Geant4 Monte Carlo toolkit [1-3] is used for simulation of many particle physics experiments including ATLAS and CMS. It is a natural choice to use Geant4 for the design of new experiments to search for dark matter particles. Light dark matter particle may have a coupling with leptons and gammas. A possible dark matter signal in this case should be seen in processes of production of electron, positron, gamma, or in missing energy experiments. Such a fix target experiment - NA64 already started at CERN SPS [4].

In Geant4, such processes are simulated with electromagnetic (EM) physics models, which are organised in standard and low-energy sub-libraries. In this work, a status of Geant4 EM sub-libraries will be described with focus on aspects important for dark matter search. We will discuss the recent Geant4 version 10.3 (December, 2016).

2 GEANT4 electromagnetic physics

Geant4 EM processes and models are implemented using a common interface for all sub-libraries [3], [5], [6]. This allows use of a combination of low-energy and high-energy models in a transparent way. In user applications, it is possible additionally define specific models for particle types and detector regions. For sampling of final states, it is also possible to exchange angular generators between standard and low-energy models.

In this work, we will discuss EM physics models which are important for search of the dark matter particles having EM coupling: conversion to e+e- pair, bremsstrahlung, and positron annihilation.

2.1 Gamma models

The full list of gamma models and their applicability area are discussed in our recent review [3]. The most important process for dark matter search is the gamma conversion process in atomic field. Models from the standard and low-energy EM sub-libraries in recent Geant4 versions have practically the same conversion cross section (Fig.1). They are based on the first Born approximation. For high energies (E > 80 GeV) a new relativistic model was developed which takes into account LPM suppression [7].

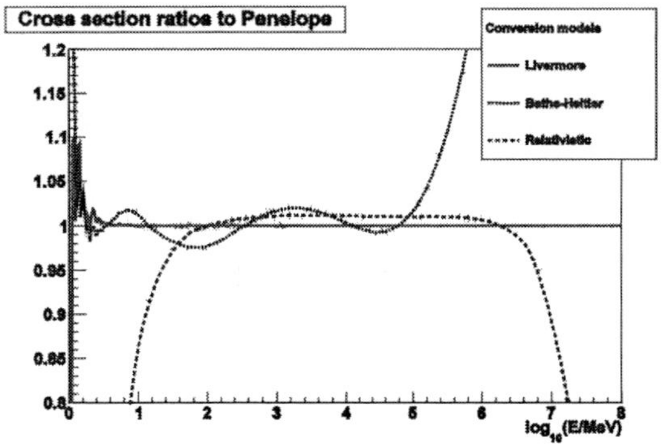

Figure 1. Ratios of cross section of gamma conversion in tungsten for various Geant4 models to the Penelope cross section. The relativistic model does not take into account atomic effects at threshold, but is the only model implemented the LPM suppression at high energy. Some difference at reaction threshold between Livermore and Penelope is due to different binning of cross section data.

[a] Corresponding author: Vladimir.Ivantchenko@cern.ch

In the Bethe-Heilter model, the Geant3 parameterisation of the cross section is implemented (Fig.1). The relative accuracy of this parameterisation is estimated as 2% for all atoms, and the upper limit of applicability is clearly seen in the plot. The relativistic model uses analytical expressions for cross sections. Penelope and Livermore models are more accurate at the reaction threshold and for high energy they use the same theoretical formulas as the relativistic one but without LPM suppression.

All these cross sections agree within 2% in the energy range 100 MeV – 100 GeV for all atoms. These models do not include high order corrections, so their absolute accuracy may be estimated on the level of a few percent. Sampling of final state in the Livermore model is different from others, because it includes a possibility of combining two models: one for sampling of the reaction with a nucleus, and another for reaction with atomic electrons. In the last case, atomic electron recoil is produced as a secondary particle [3], [8], which is a possible background for the dark matter signal search in fix target experiments.

2.2 Energy loss models

For electron and positron the main energy loss processes are ionisation and bremsstrahlung [3], [5-7]. In the case of ionisation, one secondary delta-electron is produced. The process of bremsstrahlung has two components: interaction with atomic nucleus or with atomic electrons. In current EM models electron recoil is not yet simulated but this is foreseen in the future. For Geant4 10.3, a new process was implemented of e+e- pair production by electron or positron. For that, the approach developed for the simulation for the e+e- pair production by muon is used [9]. In Fig.2 two partial cross sections for energy transfer above 10 MeV are shown. The process of e+e- pair production has much less cross section compared with bremsstrahlung but may contribute to background to dark matter signal if an energy transfer is large enough. In Geant4 10.3, this new process is included into Opt3 and Opt4 physics lists.

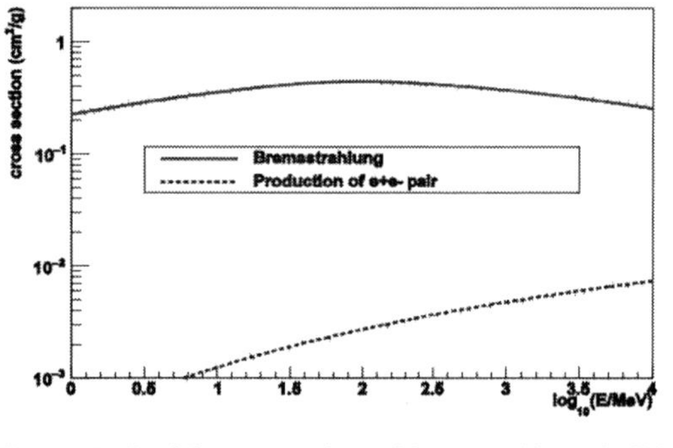

Figure 2. Partial cross section of bremsstrahlung (solid line) and e+e- pair production (dotted line) in lead by electrons with energy transfer above 10 MeV as a function of the projectile electron energy.

2.3 Positron annihilation

Two gamma annihilation of a positron is the main positron annihilation Geant4 process. For version 10.3, a correlation of linear polarisation of produced photons was implemented [10]. Additionally a new UI command and a C++ interface were added to enable rare high energy processes of positron annihilation to muons and hadrons [9], [11]. These UI commands are available for all reference physics lists.

3 Configuration of EM physics

We recommend [3] using reference physics lists or building custom physics list using physics constructors from the Geant4 distribution. These approaches guarantees usage of a well validated combination of physics models. However, a set of reference physics lists cannot cover all use cases. In particular, for simulation of direct interaction of dark matter particles with media, additional models should be added into application. For that, a model may be easily prepared using inheritance from the existing models for gamma, electron, or positron interactions. For estimation of background for dark matter search, some of the necessary EM models are already available with Geant4 10.3 but not all. A user can add a new custom model on top of any physics list using existing interfaces to EM physics.

4 Summary

The existing set of EM physics sub-libraries of Geant4 provides a possibility for designing experiments for the search of the dark matter particles having coupling with gamma, electrons, and positrons. Additional development for the next leading order corrections to EM physics models is required within Geant4 but also may be implemented at the level of user applications.

References

1. S. Agostinelli, et al, Nucl. Instrum. Methods Phys. Res. A. **506**, 250 (2003)
2. J. Allison, et al., IEEE Trans. Nucl. Sci. 53, **270** (2006)
3. J. Allison, et al., Nucl. Instrum. Methods Phys. Res. A **835**, 186 (2016)
4. NA64 Collaboration, arXiV:1610.02988 [hep-ph]
5. J. Apostolakis, et al., Radiat. Phys. Chem., **78**, 859 (2009)
6. V. Ivanchenko, et al., Prog. Nucl., Sci., Technol. **2**, 898 (2011)
7. V. N. Ivanchenko, et al., J. Phys : Conf. Ser. **513**, 022015 (2014)
8. M. L. Iparraquirre, G. O. Depaola, Eur. Phys. J. C **71**, 1778 (2011)
9. A. G. Bogdanov, et al., IEEE Trans. Nucl. Sci. **53**, 513 (2006).
10. M. H. L. Pryce, J. C. Ward, Nature **4065**, 435 (1947)
11. J. Apostolakis, et al., J. Phys : Conf. Ser. **664**, 072021 (2015)

EPJ Web of Conferences **142**, 01017 (2017) DOI: 10.1051/epjconf/201714201017
Advances in Dark Matter and Particle Physics

Recent results from the SND detector

A. A. Korol[1,2,a], M. N. Achasov[1,2], A. Yu. Barnyakov[1,2], K. I. Beloborodov[1,2], A. V. Berdyugin[1,2], A. G. Bogdanchikov[1], A. A. Botov[1], T. V. Dimova[1,2], V. P. Druzhinin[1,2], V. B. Golubev[1,2], L. V. Kardapoltsev[1,2], A. G. Kharlamov[1,2], S. V. Koshuba[1], D. P. Kovrizhin[1,2], A. S. Kupich[1], K. A. Martin[1,2], A. E. Obrazovsky[1], E. V. Pakhtusova[1], S. I. Serednyakov[1,2], D. A. Shtol[1,2], Z. K. Silagadze[1,2], I. K. Surin[1,2], Yu. V. Usov[1,2], and A. V. Vasiljev[1,2]
The SND Collaboration

[1]*Budker Institute of Nuclear Physics, Novosibirsk, 630090, Russia*
[2]*Novosibirsk State University, Novosibirsk, 630090, Russia*

Abstract. Recent results of the SND detector experiments are presented. Data were taken at the VEPP-2000 and the VEPP-2M e^+e^- colliders in the Novosibirsk in the c.m. energy ranges from 0.3 to 2.00 GeV. Integrated luminosity used are 70pb-1 and 25pb-1 respectively. The cross sections of the $e^+e^- \to \pi^0\gamma$, K^+K^-, $\omega\pi^0$, $\pi^+\pi^-\pi^0\eta$, $\omega\eta$, $\omega\eta\pi^0$ processes has been measured.

1 Detector and experiment

We present the results of study the data collected on the SND detector [1–3] at the electron-positron colliders VEPP-2M [4] and VEPP-2000 [5] in the center of mass energy range 0.3 ÷ 2.0 GeV. The SND detector consists of several subsystems (Fig.1) including a spherical electromagnetic calorimeter, a cylindrical tracking system, a threshold Cherenkov counters and a muon detector. Electromagnetic calorimeter covers 0.95 of the full solid angle, has thickness 13.4 X0, energy resolution $\delta E/E = 0.042/\sqrt[4]{E(GeV)}$, and angle resulution $\delta\varphi = 1.5°$. Tracking system covers 0.94 of the full solid angle, has angle resolution $\delta\varphi = 0.45°$, $\delta\theta = 0.8°$.

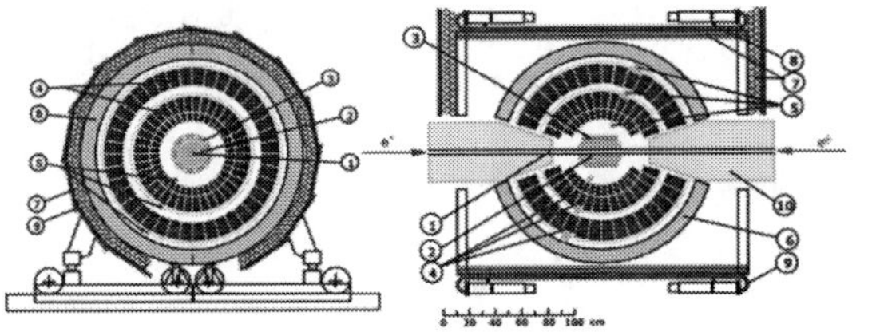

Figure 1. SND detector layout: 1 – beam pipe, 2 – tracking system, 3 – aerogel Cherenkov counter , 4 – NaI(Tl) crystals, 5 – phototriodes, 6 – iron muon absorber, 7–9 – muon detector, 10 – focusing solenoids .

The integrated luminosity for experiments with SND detector at both the VEPP-2M and VEPP2000 e^+e^- colliders is summarized in the Table.1.

During 2010-2013 the luminosity was limited by the shortage of positrons. This problem is to be solved in the ongoing upgrade of the VEPP-2000 complex with electrons and positrons transported from the VEPP-5 injection complex. Experiments at upgraded VEPP-2000 are expected to start by the end of 2016.

Table 1. Used data

VEPP-2M			
	Below ϕ	Around ϕ	Above ϕ
$IL,\ pb^{-1}$	9.1	13.2	8.8
$\sqrt{s},\ GeV$	0.36 – 0.97	0.98 – 1.06	1.06 – 1.38
VEPP-2000			
$IL,\ pb^{-1}$	15.5	6.9	47.0
$\sqrt{s},\ GeV$	0.30 – 0.97	0.96 – 1.05	1.05 – 2.00

2 Known processes cross section measurement

The $e^+e^- \to \pi^0\gamma$ process cross section was measured [6] using full statistics collected at VEPP-2M and VEPP-2000 colliders. This process has the third largest cross section below 1 GeV, its study allows to obtain parameters of electromagnetic radiative decays of the vector mesons (ρ, ω, ϕ), and allows to extract information on the transition form factor $\pi^0\gamma\gamma^*$. The process $e^+e^- \to 2\gamma$ is used for normalization. Common selection criteria for 2γ and 3γ final states are trigger, charged tracks absense, large total energy deposition, low momentum estimated from calorimeter, and muon system veto. Final selection is based on 4C kinematic fit: $\chi^2_{3\gamma} < 30$, $36° < \theta_\gamma < 144°$, $80 < M_{rc} < 190$ MeV. Here M_{rc} is the mass recoiling against largest energy photon. The number of signal events is determined from the fit of π^0 in the M_{rc} spectrum.

[a]e-mail: a.a.korol@inp.nsk.su

This is the most precise measurement of the $e^+e^- \to \pi^0\gamma$ process cross section (Fig.2). Systematic uncertainty at the ω peak is 1.4%, main contributions from luminosity (1.2%) and selection criteria (0.6%).

Using PDG [7] value for $B(\omega \to \pi^+\pi^-\pi^0) \cdot B(\omega \to e^+e^-)$ we have obtained $\Gamma(\omega \to \pi^0\gamma)/\Gamma(\omega \to \pi^+\pi^-\pi^0) = 0.0992 \pm 0.0023$, which is higher than the KLOE [8] value 0.0897 ± 0.0016 by 3.4σ. Measured $B(\rho \to \pi^0\gamma) = (4.20 \pm 0.47 \pm 0.22) \cdot 10^{-4}$ is by 1.8σ lower than the current PDG value $(6.0 \pm 0.8) \cdot 10^{-4}$, but agrees with the branching fraction for the charged mode $B(\rho^\pm \to \pi^\pm\gamma) = (4.5 \pm 0.5) \cdot 10^{-4}$. For ϕ-meson decay we get $B(\phi \to \pi^0\gamma) \cdot B(\phi \to e^+e^-) = (3.92^{+0.71}_{-0.40} \pm 0.51) \cdot 10^{-7}$.

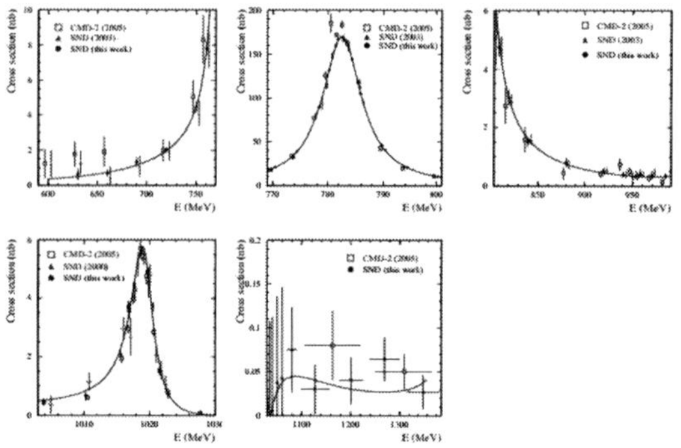

Figure 2. $e^+e^- \to \pi^0\gamma$ cross section

We studied the $e^+e^- \to K^+K^-$ process [9]. Preliminary events selection is trigger and 2 back-to-back central tracks in drift chamber. Then for events with absense of the signal from threshold Cherenkov counter for at least one track the number of events is fitted from normalized energy deposition distribution. Our cross section measurement agrees with the BABAR data and has comparable or better accuracy (Fig.3).

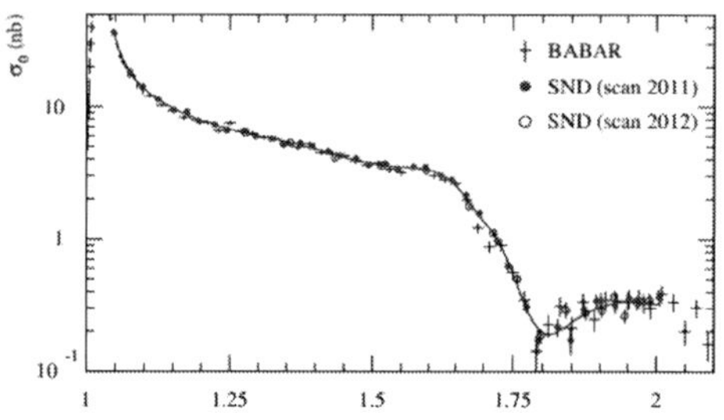

Figure 3. $e^+e^- \to K^+K^-$ cross section

The process $e^+e^- \to \omega \to \pi^0\pi^0\gamma$ was earlier studied in [10] but that study has a mistake in the radiative correction calculation. This is fixed now ([11]) and also additional data are taken into account (37 pb^{-1}instead of 25 pb^{-1}). The result shown at Fig.4.

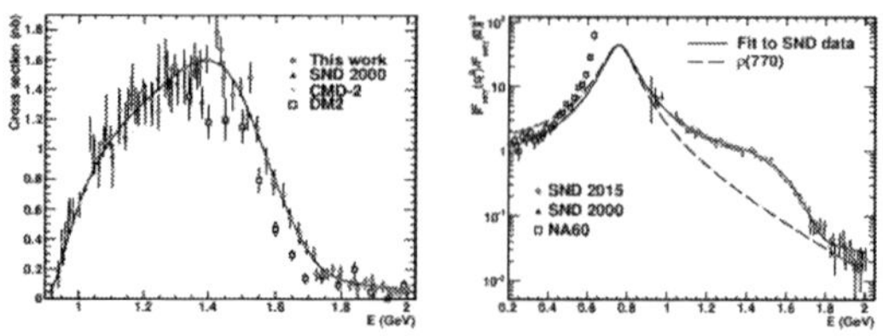

Figure 4. $e^+e^- \to \omega \to \pi^0\pi^0\gamma$ cross section and $\omega\pi^0\gamma$ form factor

3 New processes study

Below 2 GeV the total hadronic cross section is calculated as a sum of exclusive cross sections. Currenly the exclusive and inclusive data below 2 GeV are in reasonable agreement Fig.[12]. But in the energy region 1.5-2.0 GeV exclusive data are incomplete. There are no experimental data on the final states $\pi^+\pi^-\pi^0\eta$, $\pi^+\pi^-\eta\eta$, $\pi^+\pi^-\pi^0\pi^0\pi^0$, $\pi^+\pi^-\pi^0\pi^0\eta$, ..., so their study is very important.

We studied the $e^+e^- \to \pi^+\pi^-\pi^0\eta$ process [13]. This is the first measurement of its cross-section (Fig.5) and intermediate states. The intermediate states found are

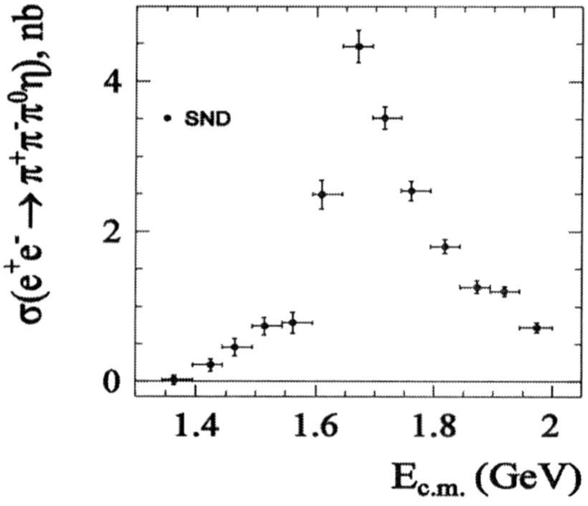

Figure 5. Process $e^+e^- \to \pi^+\pi^-\pi^0\eta$ cross section

$\omega\eta$, $\phi\eta$, structureless $e^+e^- \to \pi^+\pi^-\pi^0\eta$, and $a_0(980)\rho$. Above1.8 GeV the last one is the dominant reaction mechanism (Fig.6). The process $e^+e^- \to \omega\eta$ has been measured separately (Fig.7). There is a significant difference between SND result and the previous BABAR measurement [14].

There is also the first study of the process $e^+e^- \to \omega\pi^0\eta \to 7\gamma$ [15]. Events of the $e^+e^- \to \pi^0\pi^0\eta\gamma \to 7\gamma$ final state are selected . The dominant intermediate state is $\omega\pi^0\eta$. No noticeable $\eta'\gamma$ signal is observed. The $\pi^0\eta$ mass spectrum for selected $\omega\pi^0\eta$ events is well described by the model of the $a^0(980)\omega$ intermediate state (Fig.8).

The cross section energy dependence is described by a single-resonance model. The resonance mass and width

DOI: 10.1051/epjconf/201714201017

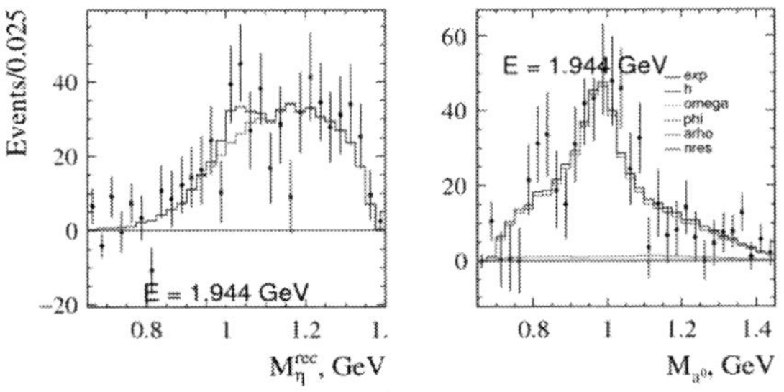

Figure 6. Mass recoling against η and $\eta\pi^0$ invariant mass.

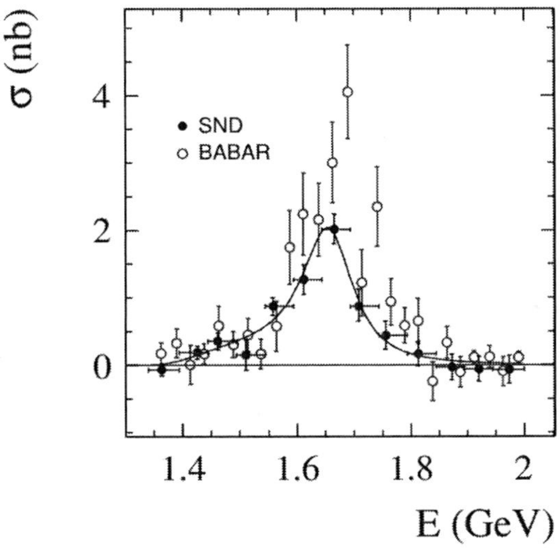

Figure 7. Process $e^+e^- \to \omega\eta$ cross section

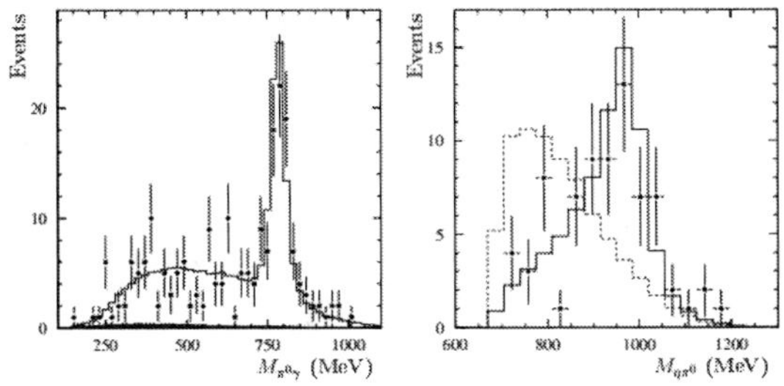

Figure 8. $\pi^0\gamma$ mass, $\eta\pi^0$ mass

are consistent with those for $\rho(1700)$ better than non-resonant at 1.2σ level (Fig.9).

4 Conclusions

During 2010 – 2013 the SND detector accumulated $\sim 70pb^{-1}$ of integrated luminosity at the VEPP-2000 electron-positron collider in the c.m. energy range 0.3 – 2 GeV. Data analysis on hadron production is in progress. The obtained results have comparable or better accuracy than previous measurements ($\pi^0\gamma$, $K^+K^-, \omega\pi^0$). For some

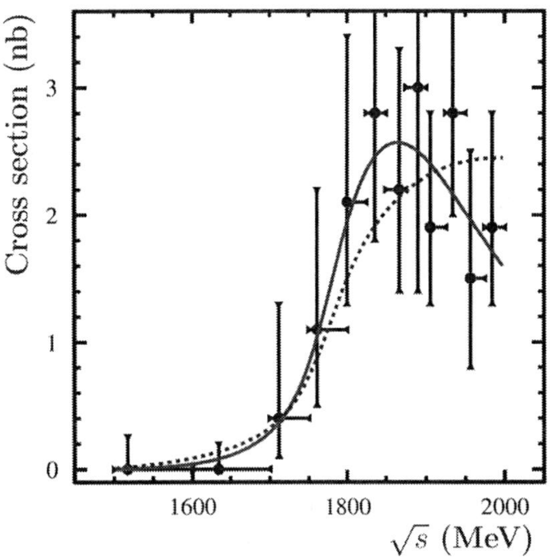

Figure 9. Process $e^+e^- \to \omega\pi^0\eta$ cross section

processes the cross sections have been measured for the first time ($\pi^+\pi^-\pi^0\eta, \omega\pi^0\eta$). $a^0(980)\omega$ and $a^0(980)\rho$ intermediate states found to be significant in these processes. After VEPP-2000 upgrade the data taking runs will be continued with a goal of $\sim 1\, fb^{-1}$ of integrated luminosity.

References

[1] M.N. Achasov et al., Nucl. Instrum. Meth. **A449**, 125 (2000), hep-ex/9909015
[2] G.N. Abramov et al., eConf **C010430**, T10 (2001), [,122(2001)], hep-ex/0105093
[3] V.M. Aulchenko et al., Nucl. Instrum. Meth. **A598**, 102 (2009)
[4] I.A. Koop et al., *Status of VEPP-2M collider, its upgrade plans and Novosibirsk Phi-factory project*, in *Physics and detectors for DAPHNE. Proceedings, 3rd International Workshop, DAPHNE'99, Frascati, Italy, November 16-19, 1999* (1999), pp. 393–404
[5] B.I. Khazin (CMD-3, SND), Nucl. Instrum. Meth. **A623**, 353 (2010)
[6] M.N. Achasov et al. (SND), Phys. Rev. **D93**, 092001 (2016), 1601.08061
[7] K.A. Olive et al. (Particle Data Group), Chin. Phys. **C38**, 090001 (2014)
[8] F. Ambrosino et al. (KLOE), Phys. Lett. **B669**, 223 (2008), 0807.4909
[9] M.N. Achasov et al., Phys. Rev. **D94**, 112006 (2016), 1608.08757
[10] M.N. Achasov et al., Phys. Rev. **D88**, 054013 (2013), 1303.5198
[11] M.N. Achasov et al., Phys. Rev. **D94**, 112001 (2016), 1610.00235
[12] F. Jegerlehner, EPJ Web Conf. **118**, 01016 (2016), 1511.04473

[13] M.N. Achasov et al., Phys. Rev. **D94**, 092002 (2016), 1607.00371
[14] B. Aubert et al. (BaBar), Phys. Rev. **D73**, 052003 (2006), hep-ex/0602006
[15] M.N. Achasov et al., Phys. Rev. **D94**, 032010 (2016), 1606.06481

 DOI: 10.1051/epjconf/201714201018

Background in the search for dark photon in e^+e^- annihilation

Venelin Kozhuharov[1,2,a]

[1] University of Sofia "St. Kl. Ohridski", 5 J. Bourchier Blvd., 1164 Sofia, Bulgaria
[2] Laboratori Nazionali di Frascati, 00044 Frascati RM, Italy

Abstract. An important ingredient in the search for new particles and rare phenomena is the correct estimation of the expected contribution from processes originating from the Standard Model. The search for a new vector degree of freedom, the Dark photon, is affected by the precise knowledge of the electromagnetic interactions. In the case of invisibly decaying Dark photon produced in e^+e^- annihilation the evaluation of the background should be performed with less than 1% uncertainty since the sensitivity of the proposed experiments depend lineary on it. The present status of the background estimation and the considered processes are reviewed and discussed, with a special attention devoted to the possible unresolved issues.

1 Introduction

The nature of Dark Matter (DM) and its possible interaction with the Standard Model (SM) particles is still one of the unresolved puzzles in nowadays physics. A framework in which a new plethora of particles almost completely decoupled from the known degrees of freedom could exist gained attention with the possibility to address few of the existing "smoking guns" in particle physics simultaneously - the muon anomalous magnetic moment [1], the excess of antimatter in cosmic rays [2–5], the annual modulation observed in the DAMA/Libra data [6], and the ^{8}Be anomaly [7, 8]. This hidden sector may be connected to the SM particles through a portal - a single messenger that exhibits non-vanishing interactions with the particles both in the visible and in the hidden sector. The most exploited scenario is the case of a new vector field, a so called Dark photon (DP) A', which could be the mediator of an additional gauge symmetry, U(1) in the simplest case. Some of the SM particles could be charged under the new interaction or the coupling between them and A' could be generated dynamically through the kinetic mixing of A' with any of the neutral electroweak gauge fields. In the most simple case the interaction strength α' could be normalized to the electromagnetic interaction strength α_{EM}, thus leaving only two new parameters - $\epsilon^2 = \alpha'/\alpha_{EM}$ and the mass of the dark photon $m_{A'}$.

The dark sector may contain extra degrees of freedom, χs, with mass m_χ which couple to A' without suppression, with a relative strength α_D. Depending on the mass hierarchy in the invisible sector, the phenomenology of the Dark photon events could be divided into two well distinctive scenarios:

- Visibly decaying A': If $m_\chi > m_{A'}$ then the dark photon decays to SM particles with a rate proportional to ϵ^2. For $m_{A'} < 2m_\mu$ the only allowed final state is the e^+e^-. The experiments addressing this scenario contain high resolution spectrometers to measure the electron/positron momenta and to reconstruct the A' decay vertex.
- Invisible A': If $m_{A'} > 2m_\chi$ then the dominant decay process will be $A' \to \chi\chi$ since it is not suppressed by the small factor ϵ. Since in most of the cases χs can not be detected the A' experimental signature will be events with missing mass or missing energy in the final state.

No DP has been observed so far and the present data put constrain in the DP parameter space in the range $\epsilon^2 < 10^{-5} - 10^{-6}$ for $m_{A'} > 5MeV$. However, most of the experiments have addressed only the visibly decaying A', thus leaving the parameter space for invisible A' scenario largely unexplored.

A huge variety of Dark photon models exist [9], including leptophylic, leptophobic, protophobic [10, 11] and many others may appear. The extremely rich possible phenomenology points towards the necessity of execution of a diverse experimental program with novel techniques, allowing to probe different regions in the parameter space that might even seem currently excluded in some of the simplest cases by a particular process.

2 Invisible dark photon searches

When the DP is quasi-stable (for small ϵ^2) or decays into $\chi\chi$ the full reconstruction of the event by measuring only the final state particles is impossible. To overcome this problem precise knowledge of the initial state is necessary. One of the most suitable processes to search for an associate A' production in annihilation, $e^+e^- \to A'\gamma$, where both the electron and the positron are with known parameters. This can be realized with the positron-on-target technique where the positrons from an accelerator impinge on

[a] e-mail: Venelin.Kozhuharov@cern.ch

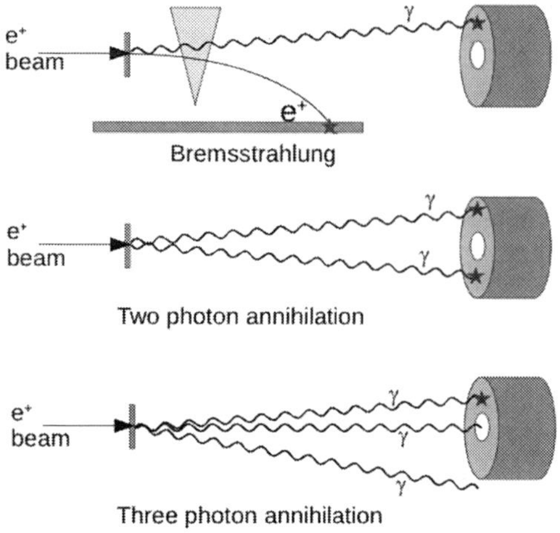

Figure 1. Dominant processes contributing to the search for invisible dark photon production in e^+e^- annihilation

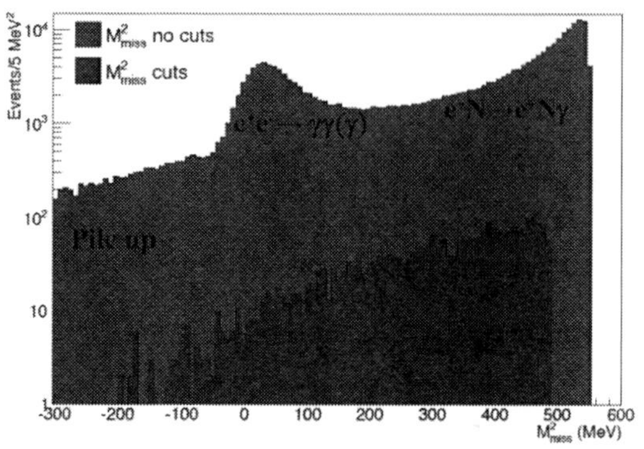

Figure 2. Missing mass squared distribution of the dominant background events, obtained with the simulation of the PADME experiment, based on GEANT4

a thin target and have the possibility to annihilate in flight with the matter electrons. A key ingredient for such an experiment is an electromagnetic calorimeter which should measure the properties of the recoil photon.

The single kinematics quantity, the missing mass, is reconstructed according to

$$M^2_{miss} = (P_{e+} + P_{e-} + P_\gamma)^2, \tag{1}$$

where P_{e+} is taken from the known beam momentum, P_{e-} is assumed as $(m_e, 0, 0, 0)$ (electrons are taken to be at rest, compared to the much higher beam energy) and P_γ is calculated using the measured energy E_γ and the impact point of the accompanying photon in the electromagnetic calorimeter.

Currently three experiments, PADME [12, 13], VEPP3 [14, 15], and MMAPS [16] are planning to exploit this technique to search for possible Dark Photon production in the annihilation process. Their parameters are listed in table 1. Two of them, PADME and VEPP3, got approved and are in construction phase.

The dominant background contributions together with the possible approach to suppress them are shown in figure 1 and are:

- Radiative emission of photons. The average angle of photon emission with respect to the positron beam direction falls as $1/E_{beam}$. Most of the high energy photons are focused in the very forward direction. Since the occupancy of the central part of the electromagnetic calorimeter may reach intolerable values, usually it is made as a ring-shaped structure without any active material close to the beam axis. The only way to suppress the bremsstrahlung background is the addition of a detector able to measure the emitting positron and perform matching with the detected in the calorimeter photon. This could be done by implementing a magnetic field and a charged particle detector, which may also provide momentum measurement.
- Two photon annihilation. The missing mass distribution for $e^+e^- \to \gamma\gamma$ events, when only a single photon is detected, peaks at zero. However, due to symmetric realization of the calorimeter and its central hole, the acceptance for single photon events from two photon annihilation is almost negligible.
- Multiphoton annihilation. The effect of the presence of additional photons in the annihilation process has to be considered when the energy of the extra photons is higher than the beam energy spread or the energy resolution of the calorimeter. These extra photons break the symmetry of the event and decrease the vetoing capability of the cluster multiplicity in the calorimeter. In addition, the missing mass distribution for such events does not peak at zero which further decreases their suppression.

The cross sections for each of these processes are listed in table 2

A typical missing mass distribution spectrum, obtained with a full GEANT4 [17] simulation of the PADME setup, is shown in figure 2.

The signal region is defined as

$$(m^2_{A'} - x\sigma_{m^2_{A'}}) < M^2_{miss} < (m^2_{A'} + x\sigma_{m^2_{A'}}), \tag{2}$$

where $\sigma_{m^2_{A'}}$ depends on $m_{A'}$ and is obtained from Monte Carlo simulation of the expected signal while x usually takes values between 1 and 2. As can be seen from figure 2, the search for DP in annihilation process is not a zero background search. For most of the $m_{A'}$ intervals the expected background contribution from SM processes in the defined signal region is of the order $O(10^4 - 10^5)$ events.

The minimal accessible relative interaction strength of the DP ϵ^2_{min} of the experiment is proportional to the the uncertainty of the estimation of the observed potential signal, and is inversely proportional to the product of the detection

Table 1. Proposed experiments to search for the production of Dark photons in e^+e^- annihilation.

	PADME	MMAPS	VEPP3
Place	LNF	Cornell	Novosibirsk
Beam energy	550 MeV	6 GeV	500 MeV
$m_{A'}$ limit [MeV]	23.7	74	22.6
Target thickness	$2\times10^{22}\ e^-/cm^2$	$O(10^{23})\ e^-/cm^2$	$5\times10^{15}\ e^-/cm^2$
Beam intensity [mA]	8×10^{-11}	2.3×10^{-6}	30
$e^+e^-\to\gamma\gamma$ rate	15	2.2×10^6	1.5×10^6
ϵ^2 limit (plateau)	10^{-6} (10^{-7} SES)	$10^{-6}-10^{-7}$	10^{-7}
Time scale	2017-2018	?	2020
Status	Approved	Funds identification	Approved

Table 2. Dominant background contributions to the search for the production of Dark photons in e^+e^- annihilation. Cross sections are calculated for E_{beam} = 550 MeV and carbon target.

Process	σ [mb]	Comment
$e^+e^-\to\gamma\gamma$	1.55	
$e^+N\to e^+N\gamma$	4000	$E_\gamma > 1MeV$
$e^+e^-\to\gamma\gamma\gamma$	0.16	$E_\gamma > 1MeV$
$e^+e^-\to e^+e^-\gamma$	188	$E_\gamma > 1MeV$

efficiency and the two photon annihilation statistics, which is used to determine the total particle flux.

$$\epsilon^2_{min} \sim \sigma_{tot}/(Acc(A')\times N(\gamma\gamma)), \quad (3)$$

$Acc(A')$ depends on the design of the experiment and $(\gamma\gamma)$ is determined from the target and beam time.

Since the number of the signal events is just the difference between the expected and the observed number of events, $N_{signal} = N_{obs} - N_{exp}$, the total uncertainty of the potential signal is given by

$$\sigma_{tot} \sim (\sqrt{N_{obs}}) \oplus (\delta_{model}\times N_{exp}), \quad (4)$$

assuming that the number of the expected events can be determined with high accuracy through Monte Carlo generation of large samples so that the statistics factor in $\sigma(N_{exp})$ vanishes. The model uncertainty, δ_{model}, represents the accuracy of the simulation of a given background process. For large statistics experiments the term $\delta_{model}\times N_{exp}$ becomes dominant in the total uncertainty determination.

In the particular case of PADME, the relative statistical uncertainty in the number of observed events is expected to be below 1%. A short review of the current status of the precision of the simulation of the different background sources follows.

3 Bremsstrahlung background

In the positron-on-target Dark photon search the positrons do not impinge the target equally spaced in time but are rather bunched in groups with a length of picoseconds to nanoseconds. Only in the long bunch limit or infinite time resolution the experiments turn into single particle measurement. An event is defined by the time resolution of the detectors and usually incorporates $O(10^3-10^4)$ positrons interacting in the target. Since the total cross section for bremsstrahlung emission is high (see table 2) each event contains positrons which have undergone significant energy loss in the target. These positrons could be detected and vetoed to suppress the background from bremsstrahlung events, when a single cluster in the calorimeter with E_γ is identified. However, when applying this technique, in order to keep the acceptance for A' detection at the level of O(20%), it is necessary to match the sum of the positron energy E_{e+} and the photon energy E_γ and veto only events for which

$$|E_{e+} + E_\gamma - E_{beam}| < E_{cut}, \quad (5)$$

where E_{cut} is determined by the resolution.

The total cross section for the emission of a photon with energy k by an electron or positron when passing through matter can be written as [17]

$$\frac{d\sigma}{dk} = \frac{d\sigma_n}{dk} + Z\frac{d\sigma_e}{dk}, \quad (6)$$

where $d\sigma_n/dk$ is the differential cross section for bremsstrahlung emission in the field of the nucleus and $Zd\sigma_e/dk$ accounts for bremsstrahlung produced in the field of Z atomic electrons.

A common approach for the estimation of the background contribution due to bremsstrahlung is the usage of GEANT4 simulation package. Below 1 GeV electron energy, the cross section for bremsstrahlung from neutral atoms in the GEANT4 versions 9.5 and above [18] is obtained from the interpolation of the Seltzer and Berger tabulated data of $d\sigma_n/dk$ and $d\sigma_e/dk$ [19]. This model shows good agreement with the experimental data. However, the uncertainty for $d\sigma/dk$ for electrons with energy 50 MeV $< E_{e^+} <$ 1 GeV is in the range (3-5)%. This uncertainty is few times higher than the expected statistical uncertainty in the determination of the observed number of events N_{obs} and additional tuning of the model parameters might be necessary to reproduce the observed bremsstrahlung spectrum. For positrons with energy higher than 1 GeV, an analytic expression for $d\sigma/dk$, including also the Landau-Pomeranchuk-Migdal effect, is used in GEANT4.

The experiments searching for Dark photon in annihilation employ targets with thickness of the order and below 100 μm. Usually at such distance the energy loss of the traversing particles is sampled in a single step leading to the physical generation of at most a single emit-

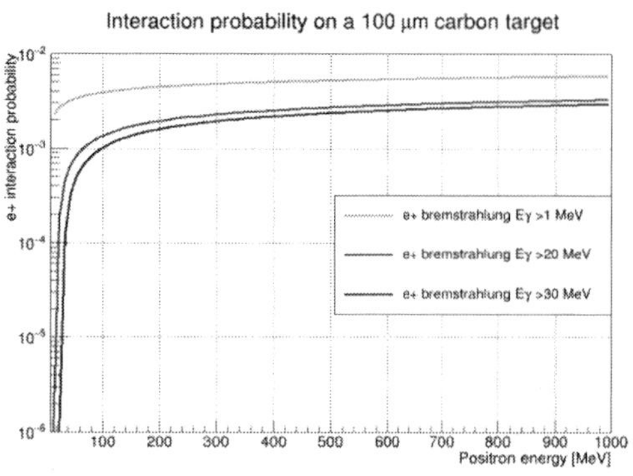

Figure 3. Probability for emission of a bremsstrahlung photon with energy $E > E_{cut}$ as a function of the positron energy. The cross sections are calculated using the GEANT4, version 9.4 parametrized model for bremsstrahlung emission.

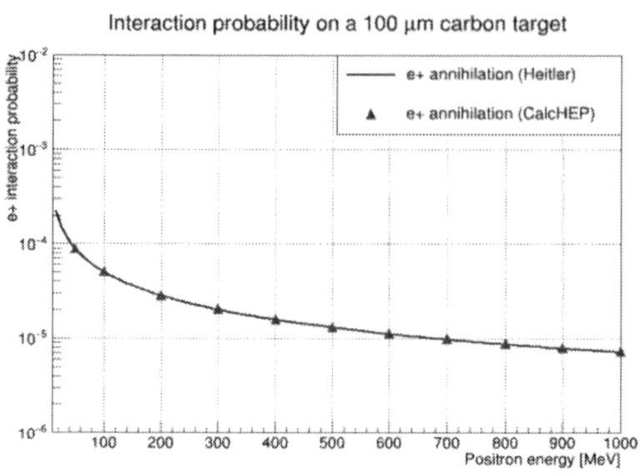

Figure 4. Comparison of the two photon annihilation probability for a positron traversing 100 μm thich carbon target calculated with CalcHEP and according to the Heitler formula.

ted photon. Although the targets are thin, the probability for the emission of a photon with energy $E_\gamma > 20MeV$ for 100 μm carbon target traversed by a positron with energy 550 MeV is 0.28%, as can be seen from figure 3. Since the dependence on the beam energy for 300 MeV < E_{beam}<600 MeV is weak, naively, the probability for the emission of two photons with $E_\gamma > 20MeV$ would be $\sim 10^{-5}$. Although quite low, this probability is not negligible since the $e^+N \to e^+N\gamma\gamma$ events may not satisfy the relation in equation (5), when a single photon is detected. Such second order effects might be treated in a dedicated way, by forcing their generation explicitly and merging them with the full sample used for background estimation, applying an appropriate normalization factor. In any case, the proper background estimation due to multiple bremsstrahlung emission in the target is important.

In the positron-on-target experiments the bremsstrahlung background can be estimated with electron beam with the same geometrics and kinematics characteristics as the nominal positron beam. Ideally, the acquired statistics should be an order of magnitude higher than in the nominal running mode with positrons. In this way only the uncertainty due to the additional factor, which accounts for the difference in the interactions of positron and electrons, remains. The electron data can also be exploited to measure the bremsstrahlung cross section $d\sigma/dk$ for the target materials and to tune the GEANT4 parameters.

4 Annihilation background

The cross section for in flight annihilation of the positron is described by the Heitler formula [20]

$$\sigma(E,Z) = \frac{Z\pi r_e^2}{\gamma+1}\times \left[\frac{\gamma^2+4\gamma+1}{\gamma^2-1} ln\left(\gamma+\sqrt{\gamma^2-1}\right) - \frac{\gamma+3}{\sqrt{\gamma^2-1}}\right], \quad (7)$$

where E is the positron energy, r_e is the classical electron radius and γ is the Lorentz factor. The electron is considered to be at rest. This process is included in the default GEANT4 electromagnetic library and is taken into account when propagating positrons through matter. However, this is not the case with the e^+e^- annihilation into three or more photons, which are completely ignored. While the symmetric design of the calorimeters of PADME, VEPP3, and MMAPS efficiently suppresses the two photon annihilation, the production of extra photons in the final state might result in the detection only of a single particle in the calorimeter. Then the invariant mass of the undetected particles will be reconstructed as missing mass for the event, $M^2_{miss} = (\sum P^\gamma_{miss})^2$. Such multiphoton events with just a single detected photon represent an irreducible background to the search for Dark photon, leaving the only possibility to estimate their contribution precisely.

The cross section for three photon annihilation was calculated with CalcHEP - a tool allowing to calculate Feynman diagrams and to perform integration over a multiparticle phase space [21]. The CalcHEP model was updated to account for the non-zero electron mass and the values of the interaction constants at low energy were introduced. The difference between the the calculated with CalcHEP cross section for $e^+e^- \to \gamma\gamma$ and the analytic expression in equation (7) is shown in figure 4. It was less than 1% for positron energy in the interval 100 MeV < E_{e^+} < 1 GeV and proved the applicability of CalcHEP for this low energy regime.

The divergency in $\sigma(e^+e^- \to 3\gamma)$ when $E_\gamma \to 0$ is cancelled by the virtual photon corrections to the $\sigma(e^+e^- \to 2\gamma)$. However, experimentally it is easier to introduce a cut on E_γ below which any sensitivity to extra photons is lost - this cut is determined by the beam energy spread. For $E_{e^+} = 550MeV$ and $E_\gamma > 5MeV$, the relative contribution of the 3γ annihilation to the 2γ cross section was estimated to be of the order of 4.5%, as can be obtained from figure 5. This value increases gradually with the positron beam energy. In the case of a single ring shaped calorimeter, the three photon annihilation turns to be the dominant

DOI: 10.1051/epjconf/201714201018

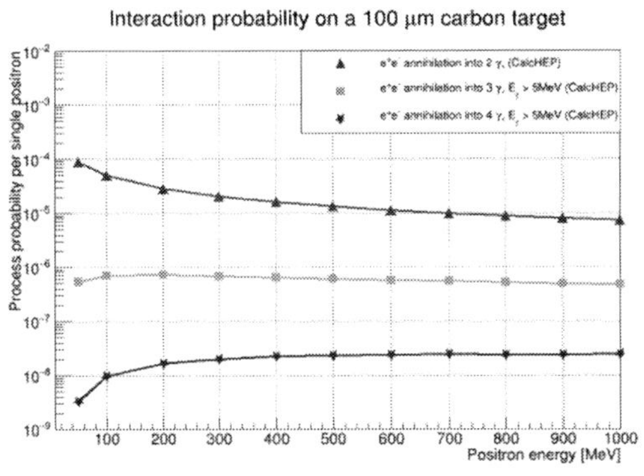

Figure 5. Probability for multiphoton production through annihilation of positrons on 100 μm carbon target into photons with energy $E > 5MeV$, as a function of the positron energy. The cross sections are calculated with CalcHEP.

background contribution to the dark photon search in the intermediate $m_{A'}$ region.

Proceeding further, the total annihilation width $\Gamma(e^+e^- \to N\gamma)$ is given by the sum of the different possible final states

$$\begin{aligned}\Gamma_{annih} =&\Gamma(e^+e^- \to \gamma\gamma) + \Gamma(e^+e^- \to \gamma\gamma\gamma)+ \\ &\Gamma(e^+e^- \to 4\gamma) + ... \approx 1.05 \times \Gamma(e^+e^- \to \gamma\gamma).\end{aligned} \tag{8}$$

For $E_{e^+} \approx$ 500MeV the value differs from the Born approximation by as much as 5 %.

To estimate the background from three photon annihilation events a dedicated sample generation might be attempted. As in the multi bremsstrahlung case, this sample has to be merged with the full background sample, after applying a factor to account for the three photon annihilation cross section. However, since the four photon contribution to the three photon sample is of the same order as of the three photon to the two photon, currently the uncertainty of this scale factor is of the order of 5%, as seen from figure 5. This uncertainty is again few times higher than the expected $\sigma(N_{obs})$.

A different approach is to use dedicated Monte Carlo event generators, which take into account the radiative corrections in the proper way. Two of them are currently available, BABAYGA [22, 23] and MCGPJ [24], but at least one of them has to be modified to include the mass of the electron due to the small invariant mass regime, O(10-100 MeV), of the positron-on-target experiments searching for Dark photons.

5 Conclusions

The correct and precise estimation of the background contribution is crucial for the interpretation of the results of the experiments aiming at a search for Dark photon production in positron-on-target annihilation. The required precision is of the order of 1% and below. Since the uncertainty in the models used to simulate the background may extend up to 5% a more detailed investigation is necessary.

A dedicated measurement of the differential cross section for radiative emission of photons by positrons and electrons with low energy (O(10 - 100 MeV)) in the field of the nuclei and the atomic electrons is highly desirable and will allow to perform the tuning of the existing Monte Carlo simulation tools. A possible alternative for the positron-on-target annihilation experiments is to exploit also data driven methods for background estimation, which may include acquiring data with electron beam, keeping the conditions unchanged. In addition, the multiphoton annihilation cross section, even if pure electromagnetic process, has to be known to the percent level and may also require experimental measurements for the tuning of the Monte Carlo generators.

Acknowledgements

This work has been partially supported by LNF-INFN under grant LNF-SU 70-06-497/07-10-2014 and NSF-Bulgaria under grant MON-FNI DN08-14/14.12.2016.

References

[1] G. W. Benett *et al.* (The g-2 Collaboration), Phys. Rev. **D73**, 072003 (2006).

[2] O. Adriani *et al.* [PAMELA Collaboration], Nature **458**, 607 (2009).

[3] M. Ackermann *et al.* [Fermi LAT Collaboration], Phys. Rev. Lett. **108**, 011103 (2012).

[4] M. Aguilar *et al.* [AMS Collaboration], Phys. Rev. Lett. **110**, 141102 (2013).

[5] M. Aguilar *et al.* [AMS Collaboration], Phys. Rev. Lett. **117**, no. 9, 091103 (2016).

[6] R. Bernabei *et al.*, Eur. Phys. J. **C67**, 39 (2010).

[7] A.J. Krasznahorkay *et al.*, Phys. Rev. Lett. **116**, 042501 (2016).

[8] A.J. Krasznahorkay, talk at ADMPP, Messina, 24-27 October, Italy (2016).

[9] M. Raggi and V. Kozhuharov, Riv. Nuovo Cim. 38, no. 10, 449 (2015).

[10] J. L. Feng, B. Fornal, I. Galon, S. Gardner, J. Smolinsky, T. M. P. Tait and P. Tanedo, Phys. Rev. Lett. **117**, no. 7, 071803 (2016)

[11] I. Galon, talk at ADMPP, Messina, 24-27 October, Italy (2016).

[12] M. Raggi and V. Kozhuharov, Adv. High Energy Phys. **2014**, 959802 (2014)

[13] M. Raggi, talk at ADMPP, Messina, 24-27 October, Italy (2016).

[14] B. Wojtsekhowski, D. Nikolenko and I. Rachek, arXiv:1207.5089 [hep-ex].

[15] I. Rachek *et. al*, talk at ADMPP, Messina, 24-27 October, Italy (2016).

[16] J. Alexander, talk at ADMPP, Messina, 24-27 October, Italy (2016).

[17] S. Agostinelli *et al.* [GEANT4 Collaboration], Nucl. Instrum. Meth. A **506**, 250 (2003).

[18] V. Ivantchenko, talk at ADMPP, Messina, 24-27 October, Italy (2016).

[19] S.M.Seltzer and M.J.Berger. Nucl.Inst.Meth. **B12**, 95 (1985).

[20] W. Heitler., The Quantum Theory of Radiation, Clarendon Press, Oxford (1954).

[21] A.Pukhov *et al.*, Preprint INP MSU 98-41/542,arXiv:hep-ph/9908288 A.Pukhov, e-Print Archive: hep-ph/0412191.

[22] C. M. Carloni Calame, Phys. Lett. B **520**, 16 (2001).

[23] C. M. Carloni Calame, talk at ADMPP, Messina, 24-27 October, Italy (2016).

[24] A. B. Arbuzov, G. V. Fedotovich, F. V. Ignatov, E. A. Kuraev and A. L. Sibidanov, Eur. Phys. J. C **46**, 689 (2006).

DOI: 10.1051/epjconf/201714201019

On the creation of the 17 MeV X boson in the 17.6 MeV M1 transition of ^{8}Be

A.J. Krasznahorkay[1,a], M. Csatlós[1], L. Csige[1], J. Gulyás[1], T.J. Ketel[2], A. Krasznahorkay[3], I. Kuti[1], Á. Nagy[1], B.M. Nyakó[1], N. Sas[1], and J. Timár[1]

[1] *Institute for Nuclear Research, Hungarian Academy of Sciences, MTA Atomki*
[2] *Nikhef National Institute for Subatomic Physics, Science Park 105, 1098 XG Amsterdam, The Netherlands*
[3] *CERN,European Organization for Nuclear Reasearch, Geneva, Switzerland*

Abstract. Electron-positron angular correlations were remeasured for the 17.6 MeV ($J^\pi = 1^+ \to 0^+$) ground state transition in ^{8}Be using an improved setup compared to one we used previously. Significant deviations from the internal pair creation was observed at large angles in the angular correlations, which supports that, in an intermediate step, a neutral isoscalar particle with a mass of 17.0±0.5 (stat)±0.5 (sys) MeV/c^2 and $J^\pi = 1^+$ was created.

1 Introduction

Dark matter is currently one of the greatest unsolved mysteries in physics. It is accepted already that all the ordinary matter we can find accounts for only about 4 percent of the universe. The quest to find the hidden universe is one of the key efforts that has brought cosmologists and particle physicists together. In spite of the great success of the standard model of particle physics (SM), which was crowned by the discovery of the Higgs boson, it is not completely satisfactory. It does not explain dark matter (DM), neutrino masses, the cosmological baryon asymmetry, etc.

Although dark matter has been gravitationally confirmed by astrophysical observations in various ways, one has no information on its properties. Whether it is made up of single, yet undiscovered particle. If yes, what would be its mass and couplings to other particles, and so on.

The leading dark-matter candidates are Weakly Interacting Massive Particles (WIMPs), axions, dark photons, etc. predicted by some physics theories but never detected. WIMPs are kind of like heavy neutrons, with a mass of 10 to 100 times heavier than a neutron. The axion was proposed to ensure that the strong nuclear force treated matter and antimatter on equal footing (so as to agree with observations). There have been dozens of big experiments that have looked for dark matter and found nothing yet.

In various dark matter theories, a kind of popular model includes a light new boson X which mediates between the Standard Model and a dark sector, e.g., Ref. [1, 2].

Very recently, we have observed an anomaly in the nuclear decay of ^{8}Be. This could be a first hint for a 17 MeV X-boson signal [3]. The ^{7}Li(p,γ)^{8}Be reaction was used at the E_p = 0.441 MeV and E_p = 1.03 MeV resonances to populate the excited states in ^{8}Be selectively and the differential internal pair conversion coefficients were studied for the 17.6-MeV, and 18.15-MeV, $J^\pi = 1^+ \to 0^+$, M1 transitions in ^{8}Be. Significant peak-like enhancement of the internal pair creation was observed at large angles in the angular correlation of the 18.15 MeV transition, but not in the 17.6 MeV one [3]. This observation was interpreted as the creation and the subsequent decay of a neutral boson with mass m_0c^2=16.70±0.35(stat)±0.5(sys) MeV. The branching ratio of the e^+e^- decay of such a boson to the γ decay of the 18.15 MeV level of ^{8}Be is found to be 5.8×10^{-6} for the best fit [3].

A theoretical group lead by J. Feng [4] studied our data as well as all other previous experiments in this area and showed that the evidence strongly disfavors dark photons. They proposed a new theory, however, that synthesizes all existing data and determined that the discovery could indicate a fifth fundamental force [4]. If confirmed by further experiments, this discovery of a possible fifth force would completely change our understanding of the universe, with consequences for the unification of forces and dark matter.

The possible relation of the X boson to the dark matter problem as well as the fact that it might explain the $(g\text{-}2)_\mu$ puzzle, triggered an enormous theoretical and experimental interest in the particle and hadron physics community.

The protophobic 17 MeV gauge boson can mediate isovector transitions, so there is no dynamical suppression of this decay. However, its mass is near the 17.64 MeV threshold, so that the decay is kinematically suppressed noted Feng et al. [4]. They calculated the suppression factor for particle mass of 17 MeV and 17.4 MeV and obtained values of 2.3 and 5.2, respectively. In spite of that suppression, it would be important to see the anomaly also in the 17.64 MeV transition, since it is a much cleaner case without having any interference effect, so we deceided to

[a] e-mail: kraszna@atomki.hu

DOI: 10.1051/epjconf/201714201019

repeat that experiment with better conditions than before [3].

In the present work we reinvestigated the anomaly observed previously by using a new and more stable Tandetron accelerator of our Institute. The multi wire proportional counters were replaced with silicon DSSD detectors as well as the complete electronics and data acquisition system from CAMAC to VME.

The expected signature of the new particle is a very characteristic angular correlation of the e^+e^- pairs from its two-particle decay, as shown in Fig.1. The predicted [5, 6] angular correlation between the e^+e^- pairs emitted in the internal pair creation (IPC) process however drops rapidly with the separation angle θ, as shown in Fig.2.

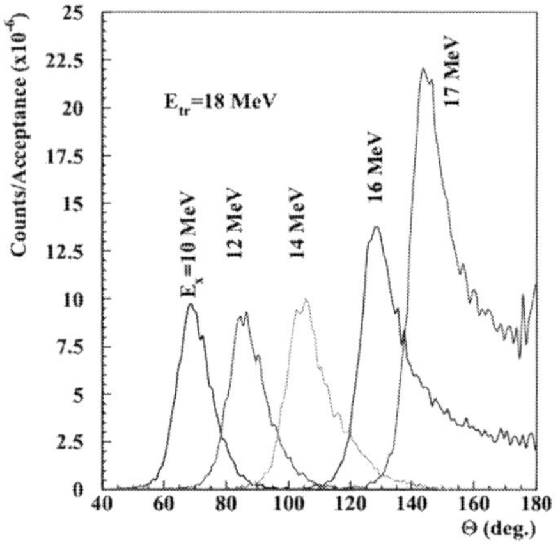

Figure 1. *Simulated angular correlations of IPC and of 1% boson decay e^+e^- pairs for boson masses indicated with the different curves.*

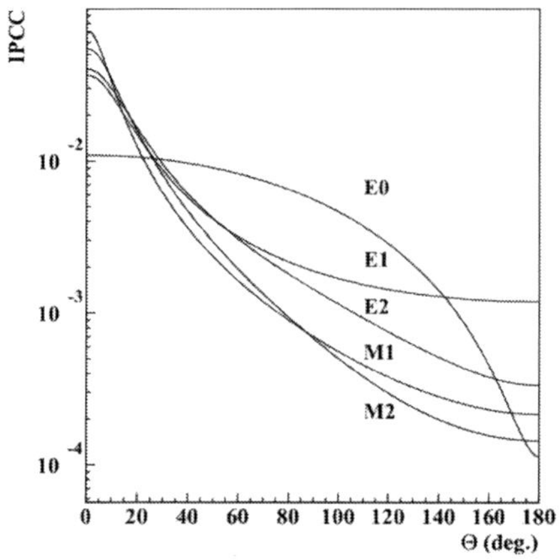

Figure 2. *Simulated angular correlations of IPC for different multipolarity nuclear transitions*

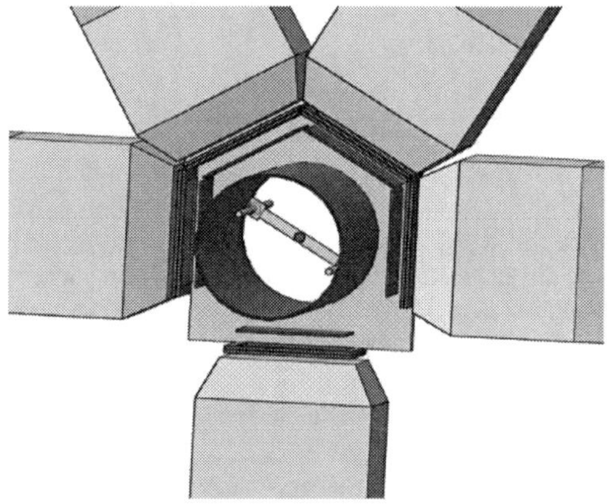

Figure 3. *CAD drawing of the e^+e^- spectrometer with five DSSD+ΔE – E detector telescopes. The target (blue spot in the center of the figure) is evaporated onto 10 μm Al strip foil spanned between 3 mm thick perspex rods to minimize the scattering and external pair creation in the vicinity of the target. The beam pipe is shown in black around which the ΔE and the DSSD detectors are arranged. The 1 mm thick ΔE detectors are shown in violet and red, while the E scinillators in yellow and their light guides are in blue.*

2 Experiments

To populate the 17.6 MeV 1^+ states in ^{8}Be selectively, we used the ^{7}Li(p,γ)^{8}Be reaction at the E_p=441 keV resonance [7]. The experiments were performed at the 2 MV Tandetron accelerator in Debrecen. A proton beam with a typical current of 1.0 μA impinged on 15 μg/cm^2 thick LiF target evaporated on 10 μm thick Al backing.

The e^+e^- pairs were detected by five plastic scintillator ΔE–E detector telescopes placed perpendicularly to the beam direction at azimuthal angles of 0°, 60°, 120°, 180° and 270° [8]. The positions of the hits were registered by double sided silicon strip detectors having a strip widths of 3 mm. The target strip foil was perpendicular to the beam direction. The telescope detectors were placed around the vacuum chamber made of a carbon fiber tube with a wall thickness of 1 mm. The arrangement of the detectors is shown in Fig.4.

γ rays were also detected for monitoring purposes. A ϵ_{rel}=100% HPGe detector (measured at 1.33 MeV relative to that of a standard 3"-diameter, 3"-long NaI(Tl) scintillator) was used at 25 cm from the target to detect the 17.6 MeV γ rays in the ^{7}Li(p,γ)^{8}Be reaction. Typical γ-ray spectra measured at E_p=441 keV is shown in Fig.4.

The 17.6 ($1^+ \rightarrow 0^+$) photopeak and their single and double escape peaks are clearly visible. The broad peaks at 14-15 MeV correspond to transitions to the first excited 2^+ level at E_x = 3.0 MeV, which has a width of Γ = 1.5 MeV [7]. The branching ratios of γ-transition to the ground state

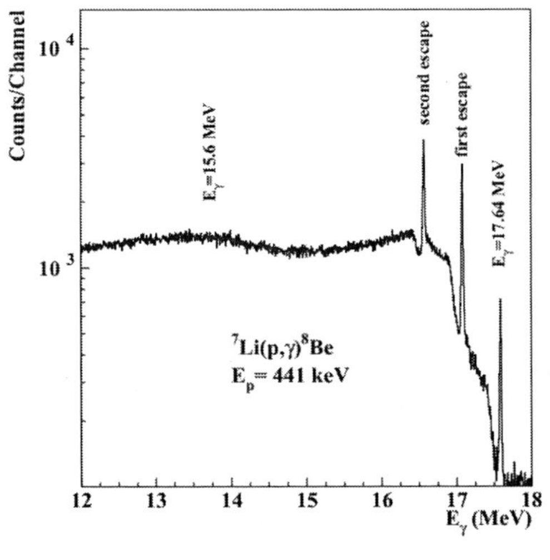

Figure 4. *A typical γ-ray spectrum measured at E_p= 441 keV.*

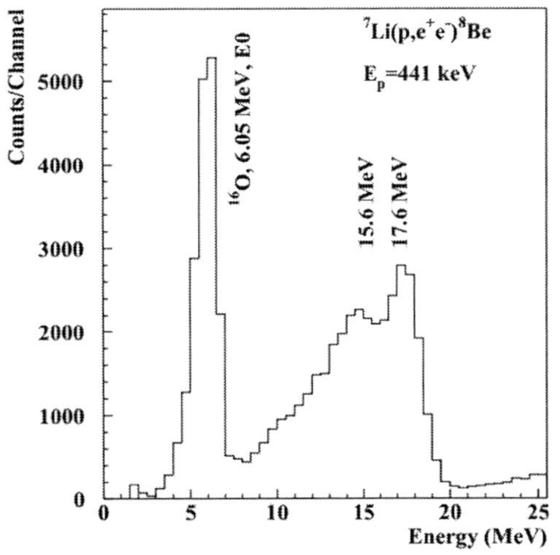

Figure 5. *Total energy spectra of e^+e^- pairs measured at E_p= 441 keV.*

and to the 2^+ are about 70% and 30% for the 17.6 MeV 1^+ state [7].

The excitation function of the reaction was also measured around the 441 keV resonance in order to check the target thickness. The measured width of the resonance was found to be 15 keV compared to the real width of Γ=12.2 keV taken from the literature [7]. From this we concluded that the energy loss of the protons in the target was sufficienly small (8.7 keV). In this way we can be sure that the multipolarity of the transition is dominated by M1. The contribution of the direct capture process gives a small background with multipolarity of E1, but according to the excitation function measurements their contribution is less than 1% [10].

Figure 5 shows the total energy spectrum of e^+e^- pairs measured at the proton absorption resonance at E_p=441 keV.

The strong 6.05-MeV peak comes from the ^{19}F(p, α)^{16}O reaction followed by the 100% IPC transition ($0^+ \rightarrow 0^+$, E0). This transition was used for energy calibration of the spectrometer and also for checking their efficiency calibration, since the angular correlation of the e^+e^- pairs coming from this transition is well known.

The efficiency calibration of the telescopes was made by using the same dataset but with uncorrelated e^+e^- pairs coming from different events.

The energy dependence of the calibration was simulated by the GEANT3 simulation code and taken also into account.

Fig.6 shows our experimental results for the angular correlation of e^+e^- pairs measured at the proton absorption resonance at E_p= 441 keV. In order to check the efficiency of the experimental setup we calculated the angular correlation also for the 6.05 MeV E0 transition coming from ^{16}O. It is shown in the upper curve of Fig.6a together with the simulated results for an E0 transition.

For the 17.6 MeV transition we observed a slight deviation from the simulated pure M1 internal pair conversion correlation (IPCC) curve at large angles. A smoothly increasing difference could be originated from the direct (non-resonant) proton capture, the multipolarity of which dominates E1 [9], and it adds to the M1 decay of the resonance. The contribution of the direct capture depends on the target thickness if the energy loss of the beam in the target is comparable to the width of the resonance. The full simulated curve in Fig. 6b is obtained by adding a small E1 contribution (2.0%) to the dominant M1 one, which describes the experimental data reasonably well, but not a peak-like deviation observed at about 150 degree.

The e^+e^- decay of a hypothetical boson with mass of 17 MeV/c^2 emitted isotropically from the target has been simulated together also with the normal IPC emission of e^+e^- pairs. The results of the angular correlation measurements together with such simulations in a magnified angular range is illustrated in Fig.6b.

3 Fitting the measured angular correlations

The final e^+e^- angle correlation distribution is described by an exponentially falling distribution modeled after the IPC simulation, and the signal distribution modeled from the simulation of a boson decaying to e^+e^- pairs.

The fit was performed with RooFit [11]. Describing the e^+e^- angular correlation distribution with the following probability density function (PDF):

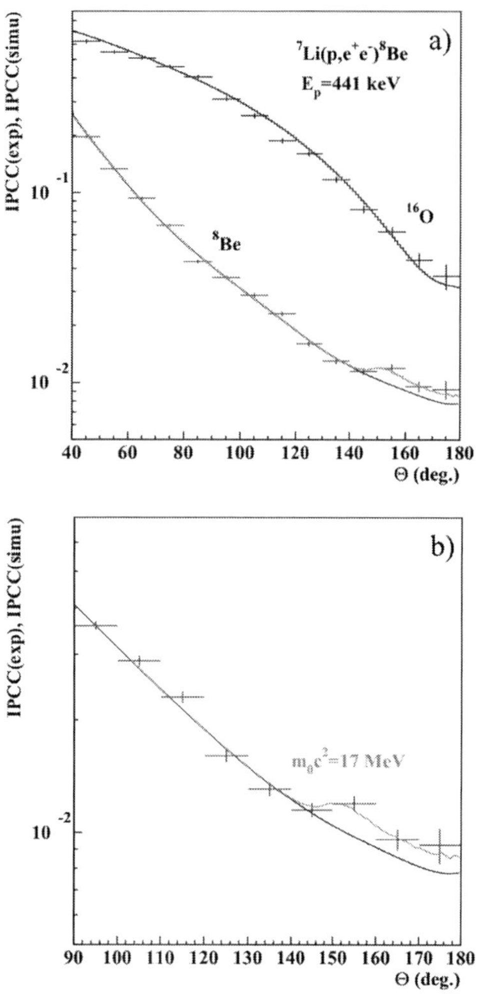

Figure 6. *Measured angular correlation of the e^+e^- pairs originated from the decay of the 17.6 MeV resonance compared with the simulated angular correlations [8] assuming M1+1.4%E1 mixed transitions (full (blue) curve).*

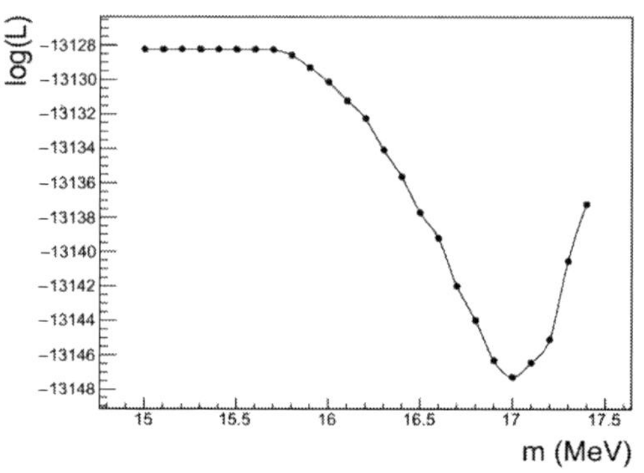

Figure 7. Fit results as a function of the hypothetical new particle's mass.

RooFit estimate the best values for N_{Sig} and N_{Bkgd}. The best fitted values of the likelihood used to minimise the fit is shown in Figure 7, and has a clear minimum close to 17 MeV/c^2 simulated particle mass.

Letting the particle mass lose in the fit the best fitted mass is calculated as $m = 17.0 \pm 0.2\ MeV/c^2$. The result of this fit is shown in Figure 8. The branching ratio of the e^+e^- decay of such a boson to the γ decay of the 17.64 MeV level of ^{8}Be is found to be 4.0×10^{-6} for the best fit.

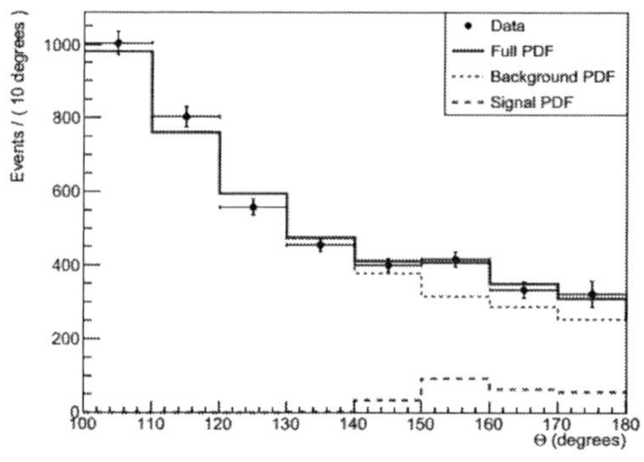

Figure 8. Fit of the e^+e^- correlation angle, letting the fit find the best value of m, N_{Sig} and N_{Bkgd} at the same time.

$$PDF(e^+e^-) = N_{Bkgd} * PDF(IPC) + N_{Sig} * PDF(signal)\,, \tag{1}$$

where N_{Bkgd} and N_{Sig} are the fitted number of background and signal events, respectively.

The signal PDF was constructed as a 2-dimensional model as a function of the e^+e^- opening angle and the mass of the simulated particle. To construct the mass dependence, the PDF linearly interpolates the e^+e^- opening angle distributions simulated for discrete particle masses.

Using the composite PDF described in Equation 1 we first performed a list of fits, by fixing the simulated particle mass in the signal PDF to a certain value, and letting

4 Repeating the experiment for the 18.15 MeV transition

Using the new setup described above we repeated our previus experiment showing the presence of the X boson as well. The 18.15 MeV 1$^+$ state in ^{8}Be was populated by the ^{7}Li(p,γ)^{8}Be reaction at bombarding proton energy of E_p=1100 keV. The typical beam current was 1.0 μA, and 700 μg/cm^2 thick Li_2O target evaporated on 10 μm thick Al backing was used. The target thickness was monitored continously during the experiment by detecting the 18.15 MeV γ-ray originated from the target. A

 DOI: 10.1051/epjconf/201714201019

typical γ-spectrum measured at E_p= 1100 keV is shown in Fig.9. The 18.15 MeV ($1^+ \rightarrow 0^+$ gs.) photopeak and its single and double escape peaks are clearly visible. The broad peak at 15.15 MeV correaponds to the 15.15MeV ($1^+ \rightarrow 2^+$ 3.03MeV) transition.

The energy resolution of the peaks reflects both the widths of the resonance (Γ=168 keV) and the energy loss in the target. The branching ratio of the γ-transition from the 18.15 MeV 1^+ state to the ground state and to the 2^+ is 30% and 70%, respectively [7]. The population of the ground state from this state is much less favored, then from the 17.6 MeV state.

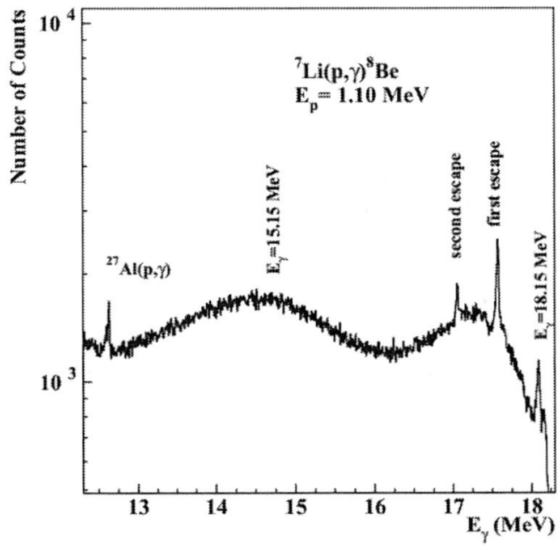

Figure 9. *A typical γ-ray spectrum (left panel) and total energy spectra of e^+e^- pairs (right panel) measured at E_p= 1100 keV.*

Figure 10 shows the e^+e^- sum energy spectrum.

Energy calibration was performed for the electron and positron energies correcting for the different energy losses in the wall of the chamber, and in the ΔE and DSSD detectors.

The 6.05 MeV ^{16}O transition was much stronger in this case, as the cross setction of the ^{19}F(p, α)^{16}O reaction is much larger at this bombarding energy than at 441 keV. We have used an energy threshold of 4 MeV for both the e^+ and e^- energies to reduce the intensity of the 6.05 MeV transition.

Another contaminant line marked by ^{27}Al is coming from the ^{27}Al(p,γ)^{28}Si reaction induced on the backing of the target.

As the branching ratio for the decay of the 18.15 MeV state was very much unfavored, for calculating the angular correlations we set a wide gate from 13 MeV to 20 MeV, covereing both the ground state transition and the transition going to the first excited state. The result is shown in Fig. 11.

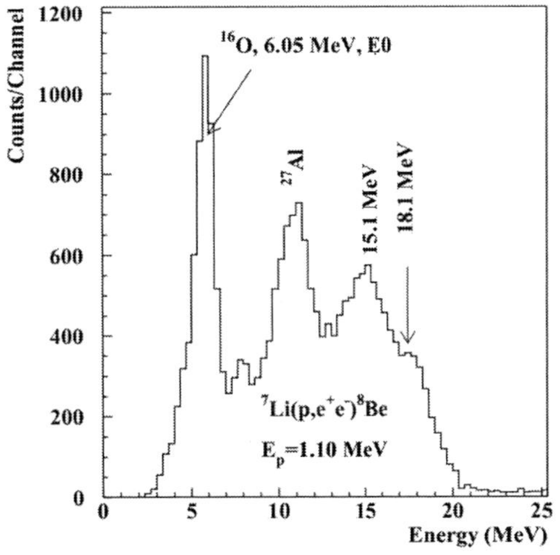

Figure 10. *Total energy spectra of e^+e^- pairs measured at E_p= 1030 keV.*

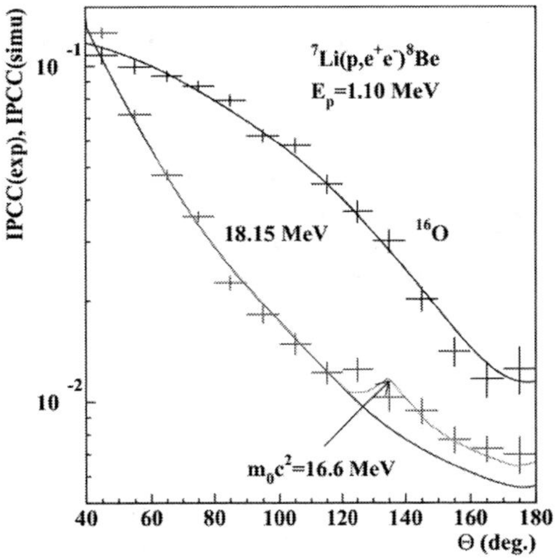

Figure 11. *Measured angular correlation of the e^+e^- pairs originated from the decay of the 18.15 MeV resonance compared with the simulated angular correlations [8] assuming M1+1.4%E1 mixed transitions (full (blue) curve).*

5 Conclusion

We have measured the e^+e^- angular correlations for the M1 transitions depopulating the 17.64 MeV and 18.15 MeV states in ^{8}Be, and observed peak-like deviations from the predicted IPC in both cases.

The peaks were located at 155° and 135° for the 17.64 MeV and 18.15 MeV transitions, respectively. The deviation between the experimental and theoretical angular correlations are significant and can be described by assuming the creation and subsequent decay of a m_0c^2=17.0 MeV boson.

To the best of our knowledge, no nuclear physics related description of such deviations can be made. The branching ratio of the e^+e^- decay of such a boson to the γ decay is less for the 17.6 MeV transition (4.0×10^{-6}), then for the 18.15 MeV one (6.8×10^{-6}), which agrees nicely with the prediction of Feng et al. [4].

6 Future plans

As a next step of the project, we plan to check the creation of the X-boson in the $0^- \to 0^+$ 21.01 MeV transition in ^{4}He. The $J^\pi = 1^+$ X boson can be emitted with L=1 angular momentum in the above transition. As the energy of the transition is considerably larger in this case, the created 17 MeV X boson would move with much larger speed, so the expected maximum of the correlation angle is much smaller ($\Theta \approx 105°$). The expected background hovewer, is considerably smaller, since the internal pair cration is forbidden for that transition.

The wide (Γ=0.84 MeV) 0^- state will be excited in the ^{3}H(p,γ)^{4}He reaction. The reaction has a very large and positive Q value (Q=19.814 MeV), so the low energy tail of the state (40% of the strength) can be excited with 1.00 MeV protons wihout creating any background as the threshold energy for the ^{3}H(p,n)^{3}He reaction is: E_{th}= 1.019 MeV. However, at this bombarding energy we excite also the first excited state of ^{4}He (E_x= 20.21 MeV, Γ=0.50 MeV), which has a J^π of 0^+ so we are expecting e^+e^- pairs from its E0-decay to the ground state by internal pair creation.

More recently, Ellwanger and Moretti made another possible explanation of the experimental results through a light pseudoscalar particle [12]. Given the quantum numbers of the ^{8}Be* and ^{8}Be states, the X boson can indeed be a $J^\pi = 0^-$ pseudoscalar particle, if it was emitted with $L = 1$ orbital momentum. We plan to study the $\gamma\gamma$-decay of the 17-MeV particle, as well in ^{4}He, in order to distinguish between the vector boson and the pseudoscalar boson scenarios. According to the Landau-Yang theorem, the decay of a vector boson by double γ emission is forbidden, however the decay of a pseudoscalar one is allowed. The angular correlation of the γ-rays will be measured by using 15 large (3"x3") $LaBr_3$ detectors. If the X boson with a mass of 17 MeV is created in the decay of the 0^- state, and also decays to two γ rays, their angular correlation should peak at an angle of:

$$\cos(\theta) = 1 - \frac{m_x^2}{2E_{\gamma_1}E_{\gamma_2}}, \qquad (2)$$

where m_x is the mass of the X boson (17 MeV/c^2) and $E_{\gamma 1}$ and $E_{\gamma 2}$ are the energies of the γ-rays. In our case this expression gives $\theta = 105°$ for γ rays with equal energies.

7 Outlook

Current and near future experiments all over the world are looking for the X boson as well. The projected sensitivities of a few relevant experiments will be briefly discussed below.

A recent study has explored the possibility of using BES III and BaBar to probe the 17 MeV X boson [13]. They are already analyzing the data.

The DarkLight experiment use electrons scattering off a gas hydrogen target to produce dark photons, which later decay to e^+e^- pairs [14]. It is sensitive to masses in the range of 10 -100 MeV and coupling constants down to $\epsilon \geq 4x10^{-4}$, covering the majority of the allowed X boson parameter space. Phase I of the experiment is already taking data for the next 18 months, whereas phase II could run within two years after phase I.

The Mu3e experiment will look at the muon decay channel and will be sensitive to masses in the 10 MeV - 80 MeV range [15]. The first phase (2015 – 2016) probed the region $\epsilon \geq 10^{-3}$, while phase II (2018 and beyond) will extend this reach almost down to $\epsilon \geq 10^{-4}$, which will include the whole region of interest for the X boson.

A search for dark photons at LHCb experiment during Run 3 (scheduled for the years 2021 – 2023) has been proposed [16] using the charm meson decay. For dark photon masses below about 100 MeV, the experiment can explore nearly all of the remaining parameter space. In particular, it can probe the entire region relevant for the X gauge boson explaining the ^{8}Be anomaly.

An experiment for a new gauge boson search at the VEPP-3 facility at Novosibirsk was also proposed [17]. The experiment will consist of a positron beam incident on a gas hydrogen target and will look for missing mass spectra. The search will be independent of the X particle decay modes and lifetime. Its region of sensitivity extends down to $\epsilon \geq 2 \times 10^{-4}$, and includes the entire region relevant for X. Once accepted, the experiment will take 3 – 4 years.

The MESA experiment will use an electron beam incident on a gaseous target to produce dark photons of masses between 10 - 40 MeV with electron coupling as low as $\epsilon \geq 3 \times 10^{-4}$. The commissioning is scheduled for 2020.

The Heavy Photon Search experiment (HPS) is using a high-luminosity electron beam incident on a tungsten target to produce dark photons and search for their e^+e^- and $\mu^+\mu^-$ decays [19]. HPS is expected to complete its dataset by 2020.

The PADME experiment will look for new light gauge bosons resonantly produced in collisions of a positron beam with a diamond target. The collaboration aims to complete the detector assembly by the end of 2017. The expected sensitivity after one year of running is $\epsilon \geq 3x10^{-3}$, with plans to get as low as $\epsilon \geq 10^{-4}$ [20].

Acknowledgements

This work has been supported by the Hungarian OTKA Foundation No. K106035, and by the European Community FP7 - Contract ENSAR n° 262010.

References

[1] M. Pospelov, A. Ritz, and M. B. Voloshin, Secluded WIMP dark matter, Phys. Lett. B 662, 53 (2008).
[2] T. Kitahara and Y. Yamamoto, Protophobic light vector boson as a mediator to the dark sector, Phys. Rev. D 95, 015008 (2017).
[3] A.J. Krasznahorkay et al., Phys. Rev. Lett. **1**16 042501 (2016)
[4] J. Feng et al., Phys. Rev. Lett. **1**17, 071803 (2016)
[5] M.E. Rose, Phys. Rev. 76, 678 (1949)
[6] P. Schlüter, G. Soff, W. Greiner, Phys. Rep. 75, 327 (1981)
[7] D.R. Tilley et al., Nucl. Phys. **A**745, 155 (2004)
[8] J. Gulyás et al., Nucl. Instr. and Meth. in Phys. Res. A **8**08, 21 (2016)
[9] G.A.Fisher et al., Phys.Rev. **C**14, 28 (1976)
[10] D. Zahnow et al., Z. Phys. **A** 351, 229 (1995)
[11] W. Verkerke and D. P. Kirkby, "The RooFit toolkit for data modeling," eConf C **0303241** (2003) MOLT007 [physics/0306116].
[12] U. Ellwanger and S. Moretti, arXiv:1609.01669v2
[13] L.-B. Chen, Y. Liang, and C.-F. Qiao, arXiv:1607.03970 [hep-ph].
[14] J. Balewski et al., arXiv:1412.4717 [physics.ins-det].
[15] B. Echenard, R. Essig, and Y.-M. Zhong, JHEP 01 (2015) 113, arXiv:1411.1770 [hep-ph]. J. Heeck and W. Rodejohann, Phys. Lett. B705 (2011) 369
[16] P. Ilten, J. Thaler, M. Williams, and W. Xue, Phys. Rev. D92 no. 11, (2015) 115017, arXiv:1509.06765 [hep-ph].
[17] B. Wojtsekhowski, D. Nikolenko, and I. Rachek, arXiv:1207.5089 [hep-ex].
[18] T. Beranek et al., Phys. Rev. D88 (2013) 015032, arXiv:1303.2540 [hep-ph].
[19] O. Moreno, arXiv:1310.2060 [physics.ins-det].
[20] M. Raggi, V. Kozhuharov, and P. Valente, EPJ Web Conf. 96 (2015) 01025, arXiv:1501.01867 [hep-ex].

EPJ Web of Conferences **142**, 01020 (2017)
Advances in Dark Matter and Particle Physics
DOI: 10.1051/epjconf/201714201020

The LDMX Experiment

Jeremiah Mans[1,*] for the LDMX collaboration
[1] *University of Minnesota, Minneapolis, MN, USA*

Abstract. The thermal-relic mechanism for the production of dark-matter in the early universe, combined with the precision measurements of the cosmic microwave background and the lack of observation of dark matter at the LHC, motivates the search for dark matter in the mass range between 1 MeV and 1 GeV. We describe the concept for the Light Dark Matter Experiment (LDMX), which could carry out such a search using the missing-momentum technique. The experiment would utilize a unique high-rate, low-current electron beam with an energy between 4 GeV and 10 GeV, and would have sensitivity to the thermal-relic hypothesis over a wide range of masses below 1 GeV.

1 Introduction

While the existence of dark matter in the universe is well-established through astrophysical observations including galaxy rotation curves[1], gravitational-lensing studies[2], and measurements of the cosmic microwave background[3], the discovery of the particle which makes up that dark matter has remained stubbornly difficult. Many possible theoretical frameworks have been developed to explain the nature of dark matter and how it was produced in the early universe[4]. Among the simplest possibilities is that cosmological dark matter arose as a thermal relic from the hot early Universe, a concept which only requires small non-gravitational interactions between dark and familiar matter, and is robustly viable over the MeV to TeV mass range.

To achieve the dark matter abundance established by cosmic microwave background measurements, the thermal relic process generically requires a weaker coupling between the dark matter and normal matter when the mass of the dark matter particle is lower as number density of relic dark matter particles is proportional to their annihilation cross-section into normal matter. To date, most experimental results have focused on dark matter particle masses in a range from a few GeV to approximately one TeV, the mass range consistent with the weak coupling scale of the Standard Model. These results include both direct-detection of collisions of dark matter with nuclear targets and searches for direct production of dark matter in colliders. To date, these searches have excluded the thermal relic hypothesis for a wide range of masses [5].

However, these measurements are difficult to extend to lower masses. For colliders, the decrease in coupling predicted by the thermal relic hypothesis reduces the production rate and the low mass hinders the clear identification of the process. For direct-detection experiments, the recoil of a nucleus due to a sub-GeV mass particle impact is very difficult to measure. However, the sub-GeV mass scale is very natural from analogy with normal matter – indeed most of the electrically-charged fermions have masses in this range. A natural minimal model of dark-matter interactions contains a vector mediator A' boson and a neutral dark matter particle χ which may be either a scalar or fermion [6]. The A' will be assumed to mix kinetically with the photon, but there are a wide array of models where the vector coupling arises from a new interaction. For the purposes of dark matter detection, these different models have broadly similar predictions, however, so we will consider the kinetically-mixed A' as the example.

Figure 1 shows the limits achieved by colliders, direct-detection experiments, and fixed-target experiments, as compared to the targets for either scalar or fermion relic dark matter particles. As can be seen, collider detectors can produce limits over a wide range of masses, but are limited in cross-section sensitivity compared to the thermal-relic target low masses. Direct-detection experiments typically have a low-mass cutoff due to the small amplitude of the recoil compared with the noise. The strongest limits in the low mass region have been established using fixed-target techniques, which allow for very high luminosity to combat the very small predicted cross-section in the thermal relic hypothesis.

The dark matter community has identified this mass region as one of the primary targets for new experiments [8]. Due to the large backgrounds from neutrinos in energetic hadron fixed-target configurations, many of the experiments propose to use electron beams. The production of dark matter particles in the collisions of an electron beam with a dense target is analogous to the usual brehmsstrahlung process in normal matter where an energetic photon is produced in the interaction between a charged particle and the electric field of the nucleus. In the "dark-brehmsstrahlung" process, an A' is produced in-

*e-mail: jmmans@physics.umn.edu

DOI: 10.1051/epjconf/201714201020

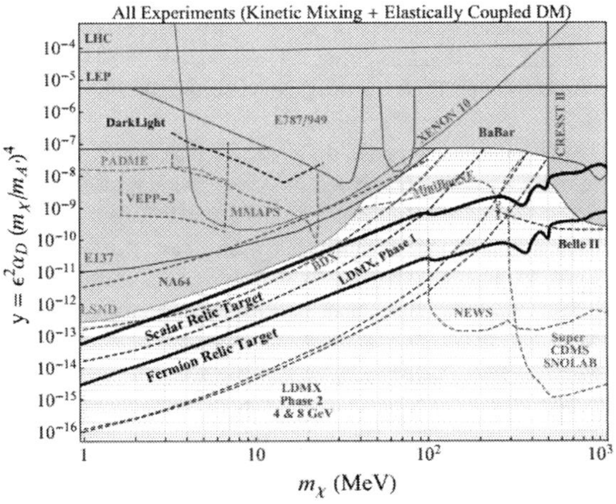

Figure 1. The parameter space for current results (solid areas) and future experimental projections (dashed lines) in the y vs. m_χ plane plotted against the thermal relic targets for scalar and fermion dark matter. The results and projects assume that the mediator couples to either scalar or fermion dark matter which scatters elastically off standard model particles. Results are evaluated for $\frac{m_\chi}{m_{A'}} = \frac{1}{3}$ and $\alpha_D = 0.5$ which is a conservative assumption as discussed in [7].

stead, typically through a mixing process with the normal photon. That dark-field mediator then produces a pair of χ which escape the detector. The kinematics of the "dark-brehmsstrahlung" process are different from the usual process, however. While the normal photon is mass-less, the A' boson must have a mass of a similar scale to the χ. As as result, most of the momentum after the collision is typically carried by the A' boson and the electron will typically have a moderate transverse momentum after the interaction. This contrasts strongly with normal brehmsstrahlung where the photon energy spans a wide range and is often nearly co-linear with the electron.

The dark-brehmsstrahlung process provides several signatures which can be used for identification of the process[7]. As the χ particles will carry away most of the energy of the interaction, the experiment can be performed as a missing energy search where the signature is a full-energy incoming beam electron which leaves an anomalously-small energy in the detector. The NA64 collaboration has recently reported results from an initial run of a detector applying this concept using secondary electron beam at CERN [9]. This technique neglects the information which can be gained from the outgoing electron, however. The production of the dark matter will not only cause the electron to lose a great deal of energy, it will also typically result in a significant momentum transverse to the original electron direction. The LDMX experiment proposes to combine the energy-loss of the electron, its transverse momentum after the interaction, and the lack of other activity in the detector to achieve definitive results on the thermal relic hypothesis over a wide range of masses.

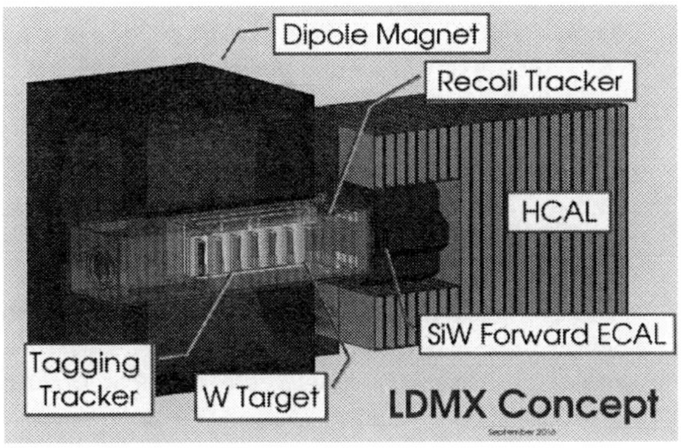

Figure 2. Conceptual layout of the LDMX Experiment.

2 Experiment Concept

The LDMX experiment aims to identify events where an incoming beam-energy electron has interacted with a target, with the result of a low-energy, moderate transverse-momentum electron and no other observable particles in the final state. This section describes the conceptual design of the detector and initial projections of the detector performance based on simulation studies with simulations based on GEANT4 [10]. The detector design is not final, and adjustments are continuing based on the results of studies, and new performance estimates will be developed. However, the general concept of the detector is established and indicative performance information is available for many of the elements. The projections shown in Fig. 1 are given for two phases of detector operation, phase one with an exposure of $\approx 10^{14}$ electrons-on-target (EOT) and a second phase with $\approx 10^{16}$ EOT.

The design of the detector, shown in Figure 2, is driven by the requirement to fix the kinematics of the interaction precisely and reject an array of background processes which could fake the signal. The experiment consists of a tungsten target in a 1.5 T dipole magnetic field, preceded by a low-mass tagging tracker and followed by a recoil tracker in the dipole magnet fringe field. The recoil tracker is followed by a high-granularity silicon/tungsten electromagnetic calorimeter and a plastic-scintillator/iron hadron veto calorimeter.

2.1 Backgrounds

The primary backgrounds of concern are:

- *Beam-related backgrounds* – electrons with less than the full-momentum expected in the beam could impact the target, resulting naturally in a low energy observed in the detector. Suppressing this background requires good beam quality and a incoming beam "tagging tracker" to confirm that each incoming electron has the expected momentum.
- *Brehmsstrahlung followed by photo-nuclear interaction* – the most-common energetic interaction between a beam electron and the target will be a normal

brehmsstrahlung interaction. For some fraction of these events, the energy transferred to the forward-going photon will be large so that electron energy is low. A calorimeter is needed to identify these photons, but if the photon undergoes a photo-nuclear interaction in the calorimeter, it may leave a relatively-small energy deposit. A sensitive and high-granularity calorimeter is required to identify these photo-nuclear interactions and their showers from the recoil electrons. A hadron veto calorimeter is also needed to contain and identify neutrons and other hadrons produced in the photo-nuclear interactions.

- *Neutrino production in the target* – in contrast to fixed-target experiments utilizing hadron beams, neutrino production in electron fixed-target experiments is very low, particularly at energies below 20 GeV. For the luminosities required for LDMX, neutrino processes should be unobservable.

2.2 Beam and Magnet

To achieve the necessary cross-section sensitivity, the LDMX experiment requires a high-charge, low-current electron beam with an energy in the range of 4 – 10 GeV. The beam would ideally deliver equally-spaced single electrons at a rate of $10^7 - 10^8$ electrons per second. Such a beam could be produced using the LCLS II beam-line at SLAC by putting a small number of electrons into the unused accelerator buckets between the primary high-current pulses used for X-ray science. A beamline could be constructed to transport these particles to SLAC's End Station A hall for an experiment. To smooth the radiation load of the electromagnetic calorimeter and allow the spatial separation of multiple electrons in a single accelerator bunch, the beam will be spread to an area of approximately 10 cm^2.

The experimental analyzing magnet is a common 18D36 dipole magnet with a 35 cm vertical gap and operated at a central field of 1.5 T. The magnet is rotated by approximately 100 mrad about the vertical axis with respect to the upstream beamline so that the incoming 4 GeV beam follows the desired trajectory to the target, arriving at normal incidence and centered on the target, which is laterally centered in the vacuum chamber at 40 cm downstream of the center of the magnet.

2.3 Tagging Tracker

The role of the tagging tracker is to ensure that each incoming beam electron has the full beam momentum and arrives at the target at normal incidence. The tagger tracker also measures the impact position of the incoming beam electron, for comparison with tracks in the recoil tracker and clusters in the calorimeters. The design for the tagging tracker is heavily based on the successful design of the HPS tracker [11]. It consists of seven double-sided modules of silicon microstrips arranged at 10 cm intervals. The modules contain a pair of 4 cm × 10 cm sensors, with one sensor oriented vertically and the other at ±100 mrad stereo angle, allowing improved pattern recognition and three-dimensional tracking. These standard $p^+ - \text{in} - n$ sensors have 30 (60) μm sense (readout) pitch, allowing for excellent spatial resolution.

The readout is based on CMS APV25 ASICs operated in multi-peak mode, which provides reconstruction of hit times with a resolution of approximately 2 ns. At very high occupancies, six-sample readout can be used to distinguish hits that overlap in time down to 50 ns, which limits the readout rate, and therefore the trigger rate, to approximately 50 kHz. However, at the low hit occupancies anticipated in LDMX, three-sample readout may suffice, enabling a maximum trigger rate approaching 100 kHz. The mechanical and thermal design of the tracker is based on the HPS tracker, but with some more-relaxed requirements due to the lower beam currents and radiation expected in the LDMX experiment.

The tagging tracker design is expected to provide 1% resolution for 4 GeV incoming electrons. Full simulation of the detector design indicates transverse momentum resolution less than 1.5 MeV along both the bending and non-bending planes, which is small compared with the 4 MeV effect expected from multiple scattering in the target. The impact parameter resolution is expected to be approximately 7 μm (48 μm) in horizontal (vertical) direction. Finally, simulated 1.2 GeV electrons which are injected such that they can contact all seven layers of the tagging tracker are misidentified as full-energy electrons at a rate of less then 10^{-6}.

2.4 Target

The separation point between the tagging tracker and the recoil tracker is the target. The baseline target design is a 10% X_o tungsten foil which measures 3 cm by 8 cm. The target is followed by a thin (2 mm) plastic scintillator plane which is read out directly (without wavelength-shifting) by an array of SiPM phototransducers. This scintillator pad provides information about the energy of fragments emerging from the target and and is used in the trigger as described below. An alternative to the tungsten foil of using an active target based on LYSO is also being considered.

2.5 Recoil Tracker

The recoil tracker is designed to identify low-momentum (50 MeV to 1.2 GeV) recoil electrons and precisely determine their momentum, direction, and impact position at the target. In addition, it must work together with the calorimeters to correctly distinguish low-momentum signal recoils from scattered beam electrons and multi-particle backgrounds. The key elements of the design are determined by these goals. First, the recoil tracker is placed at the end of the magnet in the beginning of the fringe field to optimize tracking for particles up to two orders of magnitude softer than the beam-energy electrons measured by the tagging tracker. Second, the recoil tracker is short and wide for good acceptance in angle and momentum and to minimize the distance from the target to the

DOI: 10.1051/epjconf/201714201020

calorimeters to improve their angular coverage. Finally, the recoil tracker provides 3-d tracking near the target for both direction and impact parameter resolution but emphasizes low mass-density over the longest possible lever-arm further downstream to deliver the best possible momentum resolution. The recoil tracker consists of four stereo layers located immediately downstream of the target and two axial layers at larger intervals in front of the ECAL. The area of the axial layers is significantly larger than that of the stereo layers to maintain angular acceptance. This design delivers transverse-momentum measurements at the 4 GeV resolution limit from multiple-scattering in the target. It maintains $> 99\%$ track reconstruction efficiency for recoil tracks with momentum greater than 1 GeV and falls below 80% at $\approx$ 100 MeV.

2.6 Electromagnetic Calorimeter

In the LDMX design, every beam electron which does not interact in the target can be expected to produce an energetic electromagnetic shower in the calorimeter. As result, the electromagnetic calorimeter must be capable of handling high rates of particle arrival, highly-granular to separate the showers of multiple beam electrons in a single integration interval, and sufficiently radiation-tolerant to absorb the load induced by serving as an active beam dump. To meet these requirements, a high-granularity, Si-W sampling calorimeter will be used for the ECAL of LDMX. The ECAL will draw heavily on the endcap calorimeter upgrade of the CMS experiment for the HL-LHC [12]. In particular, the hexagonal sensors, front-end readout electronics and front-end trigger architecture will be the same. This is facilitated by the fact that the CMS experiment has an instantaneous event rate of 40 MHz while that of LDMX will be 46 MHz.

The choice of silicon is important for handling the impact of fairly substantial radiation doses expected in LDMX, particularly in the central sensors of the calorimeter. Simulations of the radiation environment performed with FLUKA [13] indicate that, while electromagnetic interactions are the dominant source of irradiation in the LDMX environment, a substantial hadronic component is also present. For 10^{16} EOT, the effective 1 MeV neutron-equivalent fluence at shower max peaks at around $3 \times 10^{15}\,\mathrm{n/cm^2}$, which is roughly an order of magnitude less than the peak fluence expected for the HL-LHC CMS endcap calorimeter. Such a fluence requires the operation of the silicon at reduced temperature, but appropriate sensor characteristics have been established by the LHC experiments.

The calorimeter design requires 30 layers of 300 μm-thick silicon sensors with absorber layers of tungsten and copper. Each layer would contain seven hexagonal modules and would be about 51 cm wide. Because photo-nuclear backgrounds are most dangerous when the interactions occur as the first or second interaction of an electromagnetic shower, the absorbers will be thinner in the first layers of the calorimeter and thicker after $\approx 3X_0$. This will increase the probability that highly-ionizing fragments from an early photo-nuclear interaction are measured by the silicon rather than being lost in the absorber. The full depth of the electromagnetic calorimeter would be $40X_0$ and approximately 30 cm.

The electromagnetic calorimeter is expected to have a resolution of approximately $17\%/\sqrt{E} \oplus 1.4\%$, which will allow good separation of 4 GeV beam particles from recoil electrons, which are expected to have less than 1.2 GeV of energy. Higher precision is not critical for the experiment, as the recoil tracker will provide excellent momentum measurements for low-momentum recoil electrons. Instead, the calorimeter must have the ability to identify low-activity photo-nuclear events. The excellent MIP-tracking of the calorimeter is important for this effort, as photo-nuclear interactions may result in the production of a small number of charged pions or muons, which can be difficult to identify in some calorimeters.

2.7 Hadron Veto Calorimeter

The hadron veto calorimeter is responsible for identifying penetrating hadrons produced from photo-nuclear reactions in the target or the ECAL, particularly energetic neutrons. The occupancy of the calorimeter will be quite low, as relatively few events ($< 0.5\%$) will produce penetrating hadrons. As a result, the calorimeter can be constructed with limited transverse segmentation to save cost. The active material of the calorimeter is plastic scintillator read out with wavelength-shifter fibers while the absorber material is expected to be iron or steel. The absorber thickness and number of layers are under optimization, however the full depth of the calorimeter is expected to be at least 5λ, and will need good angular coverage for hadrons initiated either in the target or the electromagnetic calorimeter. The hadron calorimeter will be read out using SiPM-based readout developed for the CMS Phase-1 Hadron Calorimeter Upgrade [14]. Single-neutron inefficiencies of less than 10^{-3} are expected for neutrons with kinetic energy above 1 GeV.

2.8 Trigger and DAQ Systems

The primary trigger for LDMX is based on the electromagnetic calorimeter. The CMS-developed front-end ASIC will produce energy measurements for every 46 MHz accelerator bunch. These energies will be summed over a set of the front layers of the calorimeter, including the shower-max region. The trigger will require that the observed energy be significantly lower than what is expected for a 4 GeV beam electron. Such a trigger can provide a rejection factor of 2×10^{-5} with no loss of efficiency for signal. To avoid the trigger firing continuously on empty accelerator buckets, the trigger will also utilize the scintillator pad mounted after the target to identify that a particle passed through the target.

The rejection factor provided by the primary physics trigger is sufficient to allow a total trigger budget of 5 kHz, including multiple background-measurement triggers and detector monitoring triggers. The DAQ is designed based on the Reconfigurable Cluster Element (RCE) developed at SLAC [15], and is scaled to be capable of a readout bandwidth of 25 kHz. The design provides a substantial safety factor in trigger rate and event size, and allows the potential for additional physics triggers to cover other topics.

3 Conclusion

The thermal-relic mechanism for the production of dark-matter in the early universe, combined with the precision measurements of the cosmic microwave background and the lack of observation of dark matter at the LHC, motivates the search for dark matter in the mass range between 1 MeV and 1 GeV. The Light Dark Matter Experiment (LDMX), which could carry out such a search using the missing-momentum technique. The experiment would utilize a unique high-rate, low-current electron beam with an energy between 4 GeV and 10 GeV. It is based on a highly-granular and radiation-tolerant silicon-tungsten calorimeter and a precision tracking system including measurements of both the incoming and recoil electrons. These elements are assisted by a hadronic veto system and a trigger and DAQ system, to complete a design which would have sensitivity to the thermal-relic hypothesis over a wide range of masses below 1 GeV. The conceptual design is currently in an advanced state, with design optimization ongoing and performance studies underway using simulated data samples.

References

[1] V.C. Rubin, N. Thonnard, W.K. Ford, Jr., Astrophys. J. **238**, 471 (1980)

[2] D. Clowe, A. Gonzalez, M. Markevitch, Astrophys. J. **604**, 596 (2004), `astro-ph/0312273`

[3] P.A.R. Ade et al. (Planck), Astron. Astrophys. **594**, A13 (2016), `1502.01589`

[4] H. Baer, K.Y. Choi, J.E. Kim, L. Roszkowski, Phys. Rept. **555**, 1 (2015), `1407.0017`

[5] T. Marrodán Undagoitia, L. Rauch, J. Phys. **G43**, 013001 (2016), `1509.08767`

[6] M. Pospelov, Phys. Rev. **D80**, 095002 (2009), `0811.1030`

[7] E. Izaguirre, G. Krnjaic, P. Schuster, N. Toro, Phys. Rev. Lett. **115**, 251301 (2015), `1505.00011`

[8] J. Alexander et al., *Dark Sectors 2016 Workshop: Community Report* (2016), `1608.08632`, `http://inspirehep.net/record/1484628/files/arXiv:1608.08632.pdf`

[9] D. Banerjee et al. (NA64), Phys. Rev. Lett. **118**, 011802 (2017), `1610.02988`

[10] S. Agostinelli et al. (GEANT4), Nucl. Instrum. Meth. **A506**, 250 (2003)

[11] P. Hansson Adrian (HPS), *The Silicon Vertex Tracker for the Heavy Photon Search Experiment*, in *Proceedings, 2015 IEEE Nuclear Science Symposium and Medical Imaging Conference (NSS/MIC 2015): San Diego, California, United States* (2016)

[12] CMS Collaboration, Tech. Rep. CERN-LHCC-2015-010. LHCC-P-008. CMS-TDR-15-02 (2015), `https://cds.cern.ch/record/2020886`

[13] G. Battistoni, S. Muraro, P.R. Sala, F. Cerutti, A. Ferrari, S. Roesler, A. Fasso, J. Ranft, AIP Conf. Proc. **896**, 31 (2007), [,31(2007)]

[14] N.J. Pastika, JINST **11**, C03020 (2016)

[15] R. Herbst et al., *Design of the SLAC RCE Platform: A general purpose ATCA based data acquisition system*, in *Proceedings, 21st Symposium on Room-Temperature Semiconductor X-ray and Gamma-ray Detectors (RTSD 2014): Seattle, WA, USA, November 8-15, 2014* (2016), p. 7431254

EPJ Web of Conferences **142**, 01021 (2017) DOI: 10.1051/epjconf/201714201021
Advances in Dark Matter and Particle Physics

Compton Scattering Program Studying Nucleon Polarizabilities

Philippe Martel[1,2,⋆], *Maik* Biroth[1], *Cristina* Collicott[1,3,4,5], *Dilli* Paudyal[4], and *Ali* Rajabi[5]
on behalf of the A2 collaboration at MAMI

[1] *Institut für Kernphysik, Universität Mainz, D-55099 Mainz, Germany*
[2] *Department of Physics, Mount Allison University, Sackville, New Brunswick E4L 1E6, Canada*
[3] *Department of Physics, The George Washington University, Washington, D.C. 20052, USA*
[4] *Department of Physics and Atmospheric Science, Dalhousie University, Halifax, Nova Scotia B3H 4R2, Canada*
[5] *Department of Astronomy and Physics, Saint Mary's University, Halifax, Nova Scotia B3H 3C3, Canada*
[6] *Department of Physics, University of Regina, Regina, Saskatchewan S4S 0A2, Canada*
[7] *Department of Physics, University of Massachusetts Amherst, Amherst, Massachusetts 01003, USA*

Abstract. The A2 collaboration at the Mainz Microtron has undertaken a program of Compton scattering experiments with the goal of extracting the proton and, in the future, neutron polarizabilities. These are fundamental parameters that describe the response of the nucleon to an electromagnetic field. An extraction of these parameters will provide a crucial test of QCD theories. This talk presented the concept and status of this program, including some preliminary results.

1 Introduction

Nuclear Compton scattering, the interaction of an electromagnetic wave with a nucleon, can be described by a set of effective Hamiltonians. At lowest order this interaction is solely dependent on the mass, m, and electric charge, e, of the nucleon through

$$H_{\text{eff}}^{(0)} = \frac{\vec{\pi}^2}{2m} + e\phi, \tag{1}$$

where $\vec{\pi} = \vec{p} - e\vec{A}$. At the next order, the anomalous magnetic moment, κ, becomes relevant in

$$H_{\text{eff}}^{(1)} = -\frac{e(1+\kappa)}{2m}\vec{\sigma}\cdot\vec{H} - \frac{e(1+2\kappa)}{8m^2}\vec{\sigma}\cdot\left[\vec{E}\times\vec{\pi} - \vec{\pi}\times\vec{E}\right] \tag{2}$$

These first two orders encompass what are called the 'Born' terms, describing a point-like charged particle without internal structure.

1.1 Scalar Polarizabilities

As one goes to higher energies, the internal structure of the nucleon becomes evident by

$$H_{\text{eff}}^{(2)} = -4\pi\left[\frac{1}{2}\alpha_{E1}\vec{E}^2 + \frac{1}{2}\beta_{M1}\vec{H}^2\right], \tag{3}$$

where α_{E1} and β_{M1}, so called scalar polarizabilities, represent the response of the nucleon to an applied electric or magnetic field, respectively. These effects can be visualized by imagining the nucleon being placed in a static field, as shown in Fig. 1. The application of an electric field causes a current within the pion cloud, effectively 'stretching' the nucleon. The application of a magnetic field causes both a diamagnetic moment within the pion cloud and a paramagnetic moment in the constituent quarks. Though the scalar polarizabilities for the proton have been determined through various Compton scattering experiments [1], some recent studies [2][3] have caused the Particle Data Group (PDG) to adjust the pre-

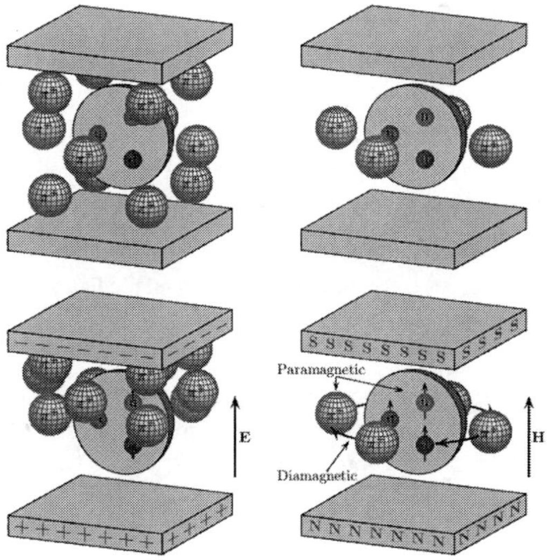

Figure 1. Simplistic visual representation of the response of a nucleon to an applied electric (left) or magnetic (right) field.

⋆Speaker - e-mail: martel@uni-mainz.de

viously quoted numbers [4]. Additionally, there is still interest in reducing the errors on these two factors, with the current values of $\alpha_{E1} = (11.2 \pm 0.4) \times 10^{-4}\,\mathrm{fm}^3$ and $\beta_{M1} = (2.5 \pm 0.4) \times 10^{-4}\,\mathrm{fm}^3$ [5].

1.2 Spin Polarizabilities

For interactions at increasingly higher energies, the response of the nucleon *spin* to an electric or magnetic field begins to contribute through

$$H^{(3)}_{\mathrm{eff}} = -4\pi\Big[\frac{1}{2}\gamma_{E1E1}\vec{\sigma}\cdot(\vec{E}\times\dot{\vec{E}}) + \frac{1}{2}\gamma_{M1M1}\vec{\sigma}\cdot(\vec{H}\times\dot{\vec{H}}) - \gamma_{M1E2}E_{ij}\sigma_i H_j + \gamma_{E1M2}H_{ij}\sigma_i E_j\Big]. \quad (4)$$

The four γ factors, called the spin polarizabilities, depict the response to an excitation by an electric or magnetic dipole (E1 or M1) or quadrupole (E2 or M2) followed by a de-excitation by an electric or magnetic dipole. Unlike the scalar polarizabilities, there had been no experimental determination of the values, other than several linear combinations of them such as the forward spin polarizability

$$\gamma_0 = -\gamma_{E1E1} - \gamma_{E1M2} - \gamma_{M1E2} - \gamma_{M1M1}, \quad (5)$$

for which $\gamma_0 = (-1.0 \pm 0.08) \times 10^{-4}\,\mathrm{fm}^4$ [6][7], and the backward spin polarizability

$$\gamma_\pi = -\gamma_{E1E1} - \gamma_{E1M2} + \gamma_{M1E2} + \gamma_{M1M1}, \quad (6)$$

for which $\gamma_\pi = (8.0 \pm 1.8) \times 10^{-4}\,\mathrm{fm}^4$ [8]. With the use of these two relations, one can change to a basis of γ_{E1E1}, γ_{M1M1}, γ_0, and γ_π, the use of which will be demonstrated in Sec. 4.

Though limited in experimental numbers, there have been various theoretical calculations and predictions of the spin polarizabilities, as given in Table 1. To address this need for experimentally extracted spin polarizabilities, as well as improved scalar polarizabilities, the A2 collaboration at the Mainz Microtron (MAMI) facility, located in Mainz, Germany, began a program of Compton scattering experiments to investigate the proton polarizabilities. This program utilized polarized photon beams with both polarized and unpolarized proton targets to measure the unpolarized cross section as well as three asymmetries

Table 1. Spin polarizabilities, in units of $10^{-4}\,\mathrm{fm}^4$, K-matrix calculation [9], dispersion relation calculations HDPV [10] and DPV [11][12], chiral Lagrangian calculation L_χ [13], heavy baryon and covariant chiral perturbation theory calculations HBχPT [3] and BχPT [14].

	K-mat.	HDPV	DPV	L_χ	HBχPT	BχPT
γ_{E1E1}	−4.8	−4.3	−3.8	−3.7	−1.1±1.8	−3.3
γ_{M1M1}	3.5	2.9	2.9	2.5	2.2±0.7[1]	3.0
γ_{E1M2}	−1.8	−0.02	0.5	1.2	−0.4±0.4	0.2
γ_{M1E2}	1.1	2.2	1.6	1.2	1.9±0.4	1.1
γ_0	2.0	−0.8	−1.1	−1.2	−2.6	−1.0
γ_π	11.2	9.4	7.8	6.1	5.6	7.2

[1]Note in additional to the theory error given, there's an additional error on γ_{M1M1} of 0.5, as described in the paper [3].

$$\Sigma_{2x/z} = \frac{N^R_{x/z} - N^L_{x/z}}{P_\gamma P_T\left(N^R_{x/z} + N^L_{x/z}\right)}, \quad \Sigma_3 = \frac{N_\parallel - N_\perp}{P_\gamma(N_\parallel + N_\perp)}. \quad (7)$$

The first two of these asymmetries rely on a circularly polarized photon beam and a transversely or longitudinally polarized proton target, respectively, where $N^{R/L}_{x/z}$ represents the number of events with x/z target polarization and R/L beam helicity, P_γ represents the magnitude of the beam polarization, and P_T represents the magnitude of the target polarization. The third relies on a linearly polarized photon beam on an unpolarized proton target, where $N_{\parallel/\perp}$ represents the number of events with $\parallel$ / $\perp$ beam polarization, and P_γ represents the magnitude of the beam polarization.

2 Experiment

2.1 Beam

The MAMI facility is a cascade of racetrack microtrons which can accelerate electrons up to energies of 1.6 GeV [15], with the additional capability of polarizing the beam longitudinally to a high degree [16]. The A2 hall is a real photon facility in which the electron beam strikes a thin radiator, producing Bremsstrahlung photons. The choice of a diamond radiator produces linearly polarized photons through coherent Bremsstrahlung, whereas an amorphous radiator combined with a longitudinally polarized electron beam produces circularly polarized photons through a helicity transfer process. The residual electron is bent in a large spectrometer magnet into an array of 352 scintillators, permitting the tagging of the photon and determination of its energy with a resolution dependent upon the electron beam energy [17]. The photon beam is then collimated before intersecting with the target.

2.2 Targets

For measuring $d\sigma_0/d\omega$ and Σ_3, an unpolarized liquid hydrogen target is sufficient. For Σ_{2x} and Σ_{2z}, however, a target of polarized protons is required. This is accomplished with a Frozen Spin Target (FST) [18] through the process of Dynamic Nuclear Polarization [19]. The process begins by cooling the target to approximately 0.2 K and using a large dipole magnet to align the electron spins. The polarization is then transferred to the proton by injecting microwaves at the appropriate frequency to cause spin-flip transitions. Once the desired polarization is achieved, the target is further cooled to about 25 mK in order to 'freeze' the spins in place. Even at this temperature, the target would still depolarize in a few hours without a field. To make room for the detectors, however, the polarizing magnet needs to be removed, at which point a weaker holding coil is energized to maintain the polarization. The Mainz FST has been online since 2009, achieving polarizations up to 90%, with relaxation times of over 1000 h.

DOI: 10.1051/epjconf/201714201021

Free radicals are needed in the FST in order for it to polarize, which necessitates the target material to be composed of more than just hydrogen. While several options are available, the Mainz group has opted to work with butanol, C_4H_9OH. This does complicate the analysis due to the inclusion of carbon and oxygen in the target, as well as the helium bath used to cool it down. A special carbon target was manufactured such that it could be inserted into the same cryostat, and run as a separate beamtime, to subtract out the contribution of the non-hydrogen nuclei in the butanol target.

2.3 Detectors

To analyze the final state of the interaction, a nearly 4π detector system is put in place over the target. This detector system is composed of the Crystal Ball (CB) [20] and Two Arms Photon Spectrometer (TAPS) [21] systems. The CB is an array of 672 NaI crystals, covering $20° < \theta < 160°$ and $-180° < \phi < 180°$. This is coupled with a barrel of 24 plastic scintillators called the Particle Identification Detector (PID), as well as a pair of Multi-Wire Proportional Chambers (MWPC), for charged particle identification and tracking [22]. TAPS is a vertical wall of 384 BaF_2 and 72 $PbWO_4$ crystals, which covers the downstream hole of the CB. It has an additional wall of plastic scintillators for charged particle identification. This system is shown in Fig. 2.

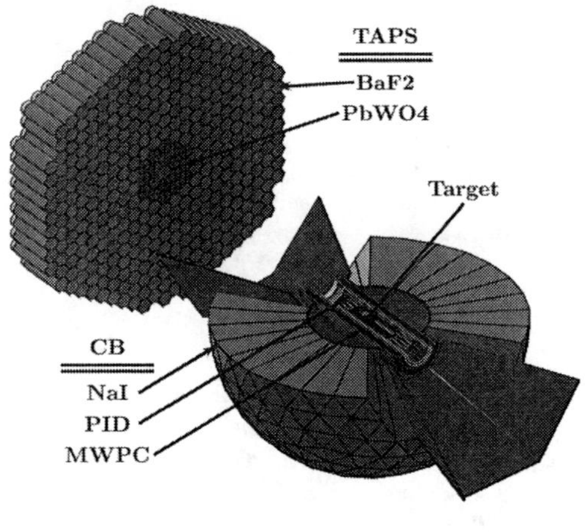

Figure 2. Combined CB/TAPS detector system.

3 Analysis

A Compton scattering event is identified in the analysis by selecting final states where a single neutral track and a single charged track are detected. Coincidence between the time of the neutral track and the time of a hit in the tagger allows for tagging the initial state photon. In a high flux environment, however, the tagger experiences a large amount of accidental coincidences, for which a sideband subtraction is necessary, as shown in Fig. 3. The charged track time is not utilized in the accidental subtraction due

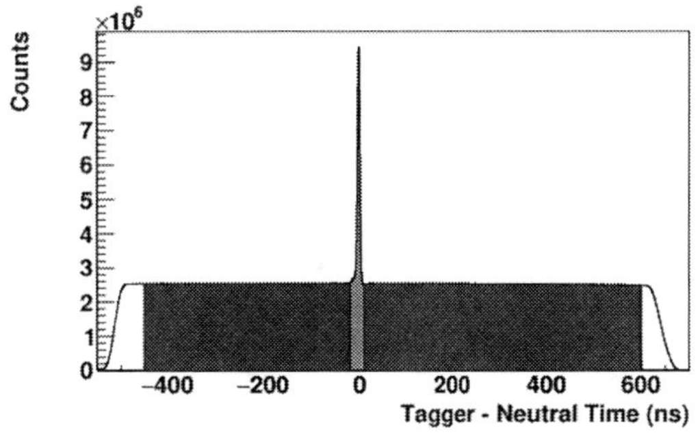

Figure 3. Timing difference between a neutral track and a hit in the Tagger. The sharp 'prompt' peak at zero sits on a large accidental background which is removed by integrating two sideband regions, shown in red, and subtracting it from the integration of the prompt region, shown in green, after scaling by the ratio of the widths of these regions.

to the Time of Flight (TOF) experienced by the heavier proton. While the distance from the target center to the center of a NaI crystal in the CB is only 45 cm, TAPS is three to four times that distance from the target center, depending on the inclusion of a Cherenkov detector. Protons detected in TAPS, a large fraction of the phase space in this analysis, therefore exhibit a notable TOF. As noted in Sec 1, the spin polarizabilities play larger roles at higher energies; unfortunately extraction of these parameters at higher energies is more problematic, notably as more reaction channels open up. A compromise of measuring just below the two-pion photoproduction threshold (308 MeV) was taken, and an additional cut is made on tagger channels within the region of interest.

Assuming an incoming photon with four-momentum $q_i = (E_i, \vec{p}_i)$ strikes an at rest proton $k_i = (m_p, 0, 0, 0)$ and results in a scattered photon $q_f = (E_f, \vec{p}_f)$, the 'missing' four-momentum is then $k_f = q_i + k_i - q_f$, and the missing mass for this reaction can then be calculated via

$$m_{miss} = \sqrt{(E_i + m_p - E_f)^2 - (\vec{p}_i - \vec{p}_f)^2}, \quad (8)$$

under which $m_{miss} = m_p$ for a Compton scattering event. While the detection of the proton is required, energy losses suffered by it before detection prevent a total missing energy calculation. These losses will be discussed in Sec. 4.4. Instead, the vector of the track is compared to the vector of k_f, and an opening angle cut on this greatly reduces background, notably from coherent and incoherent scattering on carbon, oxygen, and helium. Any remaining background from these reactions is then removed by subtracting out the separate data with the carbon target mentioned in Sec. 2.2.

Given that neutral pion (π^0 photoproduction has a cross section about 100 times larger than Compton scattering in the $\Delta(1232)$ region, an appreciable background signal is generated by cases where one of the decay photons ($\pi^0 \rightarrow \gamma\gamma$) goes undetected, and in some kinematic regions the reconstruction can survive the opening angle cut. This background can be studied either through a careful simulation, or directly through the data itself. Similar to the sideband analysis of the tagger accidentals, one can take neutral pion events, which are fully detected in the system, where one of the decay photons is close to the known holes in the detector phase space. By neglecting this photon and analyzing the other decay photon, and the recoil proton, as if they were a Compton scattering event, one can mimic the background in the missing mass distribution. This distribution, along with the various backgrounds, is shown in Fig. 4.

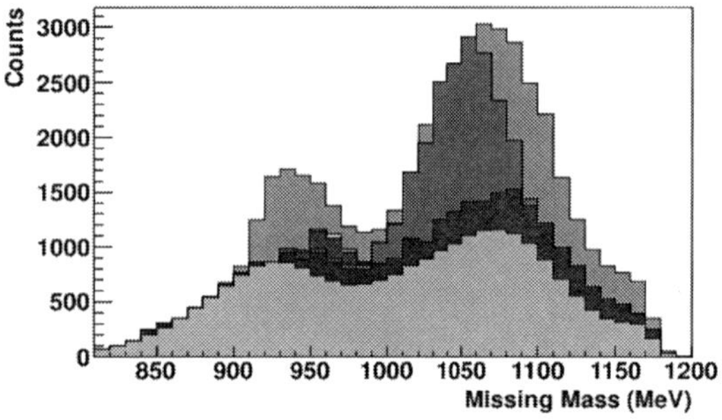

Figure 4. Missing mass distribution showing the tagger accidental (light blue), carbon (dark blue), and neutral pion (magenta, red, and yellow) background contributions, leaving the Compton signal in green.

Subtracting out the various background contributions results in the distribution shown in Fig. 5. Missing mass distributions like this one are then integrated up to conservative values near the proton mass, to ensure that any remaining neutral pion background is not included.

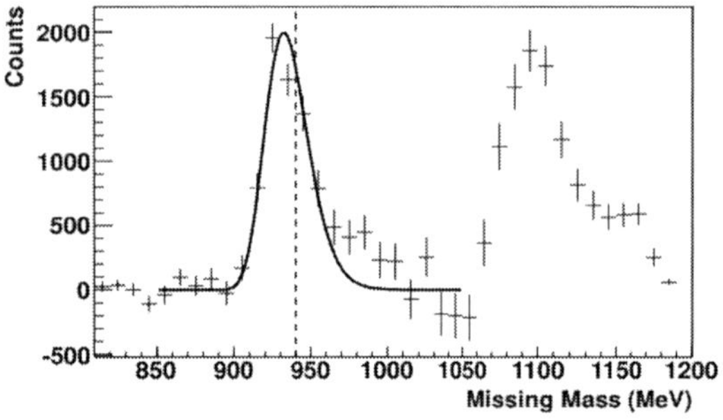

Figure 5. Missing mass distribution after background subtraction, including a fit of the simulated line-shape.

4 Results

The A2 collaboration has completed its initial program of Compton scattering experiments with the goal of extracting the spin polarizabilities. The results, in some cases preliminary, are given below for each of the sub-projects.

4.1 Σ_{2x}: Transverse Target, Circular Beam

Measuring Σ_{2x} required the use of the polarized target, with the polarization direction transverse to the incoming beam of circularly polarized photons. The asymmetry was then constructed either by flipping the helicity of the photon beam, or by flipping the polarization of the target. While it takes roughly a day to flip the direction of the target polarization, the beam helicity is flipped regularly at a rate of about 1 Hz. This eliminates potential systematic errors due to a change in detection efficiency between positive and negative target polarization data. However, to check such systematics, the target polarization was additionally flipped, albeit only once or twice during a beamtime. The missing mass distributions for each helicity-polarization combination were integrated up to the proton mass, and then applied into Eq. 7, to produce the points shown in Fig. 6. This reflects data in the energy range of $E_\gamma = 273 - 303$ MeV [23].

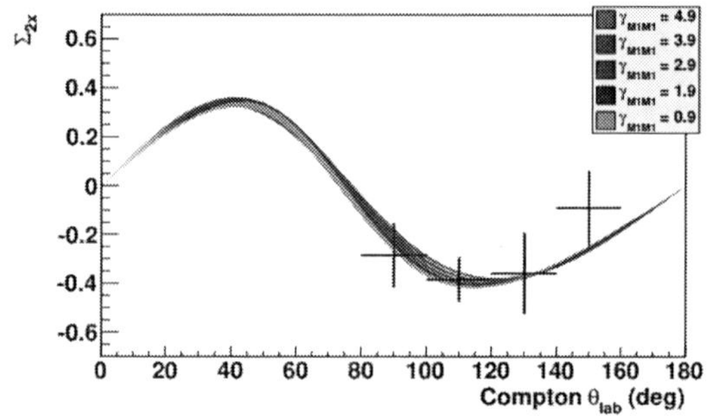

Figure 6. Σ_{2x} for $E_\gamma = 273 - 303$ MeV. The theoretical curves are from the HDPV dispersion code using different values for γ_{M1M1} as differentiated by color, while α_{E1} and β_{M1} are assigned the PDG2012 values, $\gamma_0 = -1.0$, $\gamma_\pi = 8.0$, and $\gamma_{E1E1} = -4.3$. The width of each band is generated by allowing α_{E1}, β_{M1}, γ_0, and γ_π to vary by their experimental errors [23].

The curves shown in this figure were generated with the use of a dispersion relation code, provided by B. Pasquini [12]. The 2012 PDG numbers [4] were used for the initial scalar polarizabilities values, and the HDPV numbers [10], as shown in Tab. 1, were used for the initial spin polarizability values. In Fig. 6, the initial value for γ_{M1M1} was then perturbed by ± 1 and ± 2, in the standard units, producing the different colored lines. The spread of each line reflects an additional step of allowing α_{E1}, β_{M1}, γ_0, and γ_π, to vary by their experimental errors, while keeping γ_{E1E1} fixed, and summing the resulting shifts in quadrature. What Fig. 6 then depicts is a rough idea of the sensitivity of this asymmetry to γ_{M1M1}, which is clearly small. Fig. 7 was constructed in a similar way except for varying γ_{E1E1} and keeping γ_{M1M1} fixed instead. Here one can see a much greater sensitivity to this particular spin polarizability. With just this one observable, one can extract values of $\gamma_{E1E1} = -4.6 \pm 1.6$ and $\gamma_{M1M1} = -7 \pm 11$ [23]. Obviously the value for γ_{M1M1} is poorly fit, as predicted by the sensitivity curves, and both values are constrained

DOI: 10.1051/epjconf/201714201021

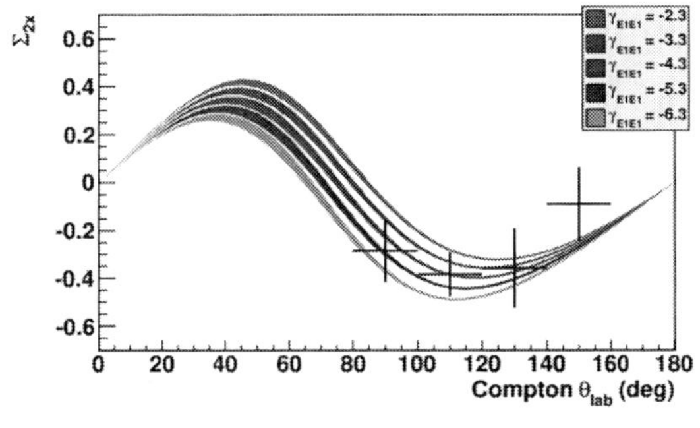

Figure 7. Σ_{2x} for $E_\gamma = 273 - 303$ MeV. Similar to Fig. 6, except now γ_{E1E1} is allowed to vary, while one fixes $\gamma_{M1M1} = 2.9$ [23].

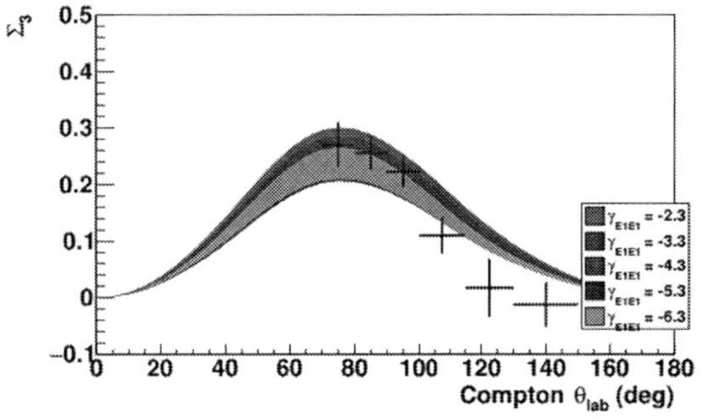

Figure 9. Preliminary Σ_3 for $E_\gamma = 286.9 - 307.1$ MeV. Similar to Fig. 7.

in the fit by the four other polarizabilities. For these reasons, more data on different observables was required.

4.2 Σ_3: Unpolarized Target, Linear Beam

While measuring Σ_3 only required the use of the simpler unpolarized LH_2 target, it then required the use of a diamond radiator to generate a linearly polarized photon beam. The plane of polarization is adjustable, and was flipped by 90° roughly each hour. Instead of constructing the asymmetry by the events with each beam helicity, it was then determined by the polarization plane. The asymmetry was generated as a function of ϕ in addition to E_γ and θ. The distribution over ϕ was fit with a $\cos(2\phi)$ function, with the amplitude then representing the magnitude of the asymmetry at that E_γ and θ. This was done for two different energy ranges, $E_\gamma = 267.0 - 287.2$ MeV and $E_\gamma = 286.9 - 307.1$ MeV, the latter of which is shown in Figs. 8 and 9, with similar sensitivity curves as in Sec. 4.1.

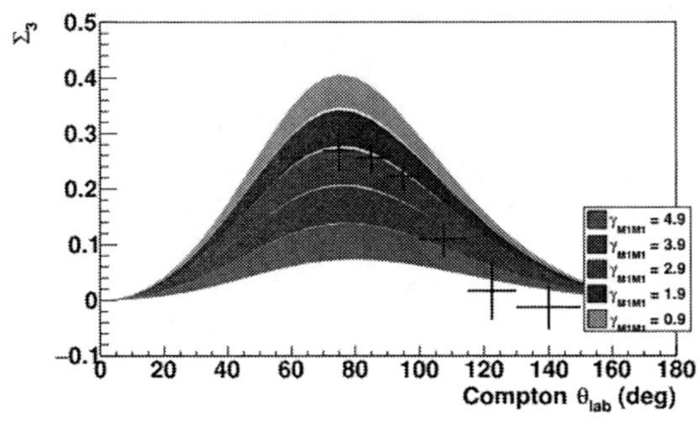

Figure 8. Preliminary Σ_3 for $E_\gamma = 286.9 - 307.1$ MeV. Similar to Fig. 6.

These results obtained at MAMI [24] complement those from a previous experiment at LEGS [25]. Each set of Σ_3 data can then be taken with the Σ_{2x} data, from which the polarizabilities can be fit while still utilizing the four constraints. The results of this inquiry are given in Tab. 2. The results from both MAMI and LEGS, for both energy ranges, are shown in Figs. 10 and 11. The theoretical

Table 2. Dispersion relation fitted to Σ_{2x} along with either Σ_3^{MAMI} [24] or Σ_3^{LEGS} [25]. Scalar polarizabilities in units of 10^{-4} fm^3, spin polarizabilities in units of 10^{-4} fm^4.

	Σ_{2x} and Σ_3^{LEGS}	Σ_{2x} and Σ_3^{MAMI}
$\bar{\gamma}_{E1E1}$	-3.5 ± 1.2	-5.0 ± 1.5
$\bar{\gamma}_{M1M1}$	3.16 ± 0.85	3.13 ± 0.88
$\bar{\gamma}_{E1M2}$	-0.7 ± 1.2	1.7 ± 1.7
$\bar{\gamma}_{M1E2}$	1.99 ± 0.29	1.26 ± 0.43
γ_0	-1.03 ± 0.18	-1.00 ± 0.18
γ_π	9.3 ± 1.6	7.8 ± 1.8
$\bar{\alpha} + \bar{\beta}$	14.0 ± 0.4	13.8 ± 0.4
$\bar{\alpha} - \bar{\beta}$	7.4 ± 0.9	6.6 ± 0.7
χ^2/dof	1.05	1.25

curves using the values given by the respective fits are also shown in Figs. 10 and 11. Additionally, the theoretical

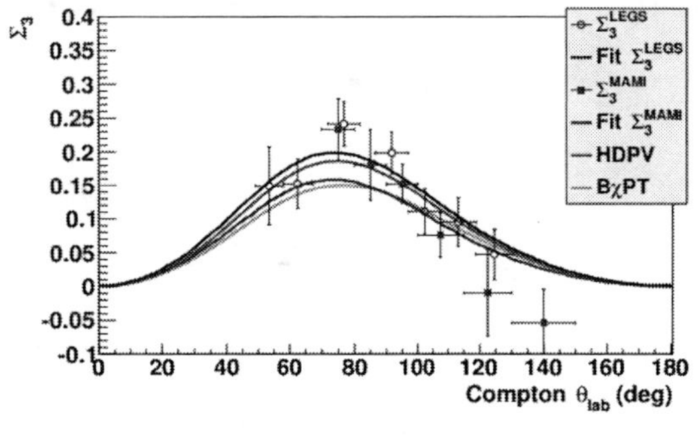

Figure 10. Preliminary Σ_3 for $E_\gamma = 267.0 - 287.2$ MeV. The open circle data are those taken at MAMI [24], while the closed square data are those from LEGS [25]. The black and blue curves represent fitting the Σ_{2x} data from MAMI along with the Σ_3 data from either MAMI or LEGS, respectively, as shown in Tab. 2. The red and green curves show the theoretical predictions from HDPV and BχPT, respectively.

curves from the HDPV and BχPT codes, using their values from Tab. 1, are plotted in those figures, depicting a better relationship between the HDPV theory and the fit-

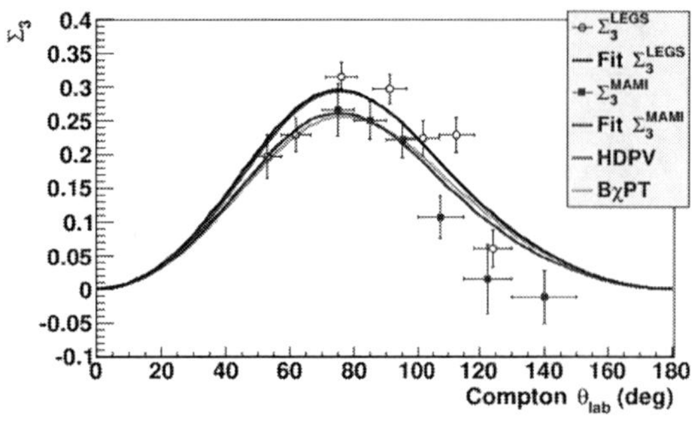

Figure 11. Preliminary Σ_3 for $E_\gamma = 286.9 - 307.1$ MeV. Similar to Fig. 10.

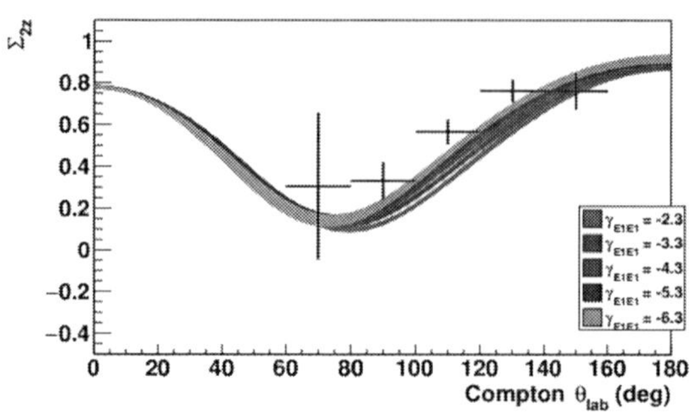

Figure 13. Preliminary Σ_{2z} for $E_\gamma = 273 - 303$ MeV. Similar to Fig. 7.

ted values from Σ_3^{LEGS} [25], and between the BχPT theory and the fitted values from Σ_3^{MAMI} [24].

4.3 Σ_{2z}: Longitudinal Target, Circular Beam

Completing the picture, and aiding the extraction of the spin polarizabilities without the use of the γ_0 and γ_π constraints, requires an additional lever-arm. This was accomplished in the form of Σ_{2z}, another polarized target asymmetry. In this case, the polarization of the target is set longitudinally with respect to the incoming beam of circularly polarized photons. These data are the most recent taken in this A2 Compton program, and as such are still in the process of being analyzed. Some preliminary asymmetry values are shown in Figs. 12 and 13, also for the energy range $E_\gamma = 273 - 303$ MeV. They only include about half of the data taken, were analyzed with only rough calibrations applied, and only use some average polarization values. As such, these will certainly change upon completion of the analyses of Paudyal and Rajabi. However, from the preliminary values one can at least observe both reasonable agreement with the curves as well as indications of reasonably high constraining power of these data.

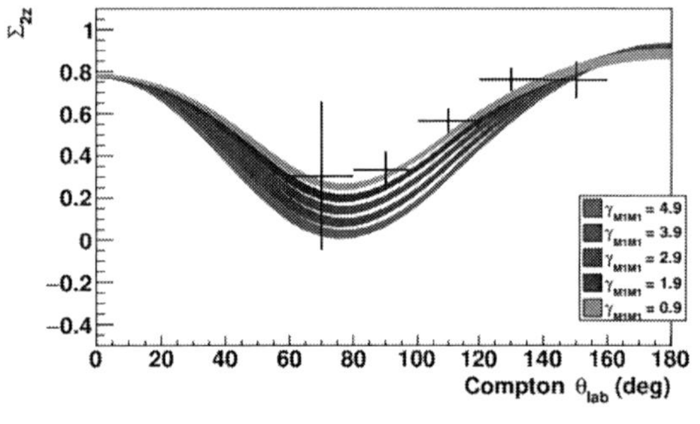

Figure 12. Preliminary Σ_{2z} for $E_\gamma = 273 - 303$ MeV. Similar to Fig. 6.

4.4 Active Target

As noted in Sec. 3, these analyses suffer in the detection of the recoil proton to the degree that its energy can not be absolutely determined. The energy and angular phase space is also limited by the energy loss of the proton, as lower energy recoils will simply not be detected. This can be verified by determining the proton efficiency from π^0 photoproduction, as shown in Fig. 14. This shows that the

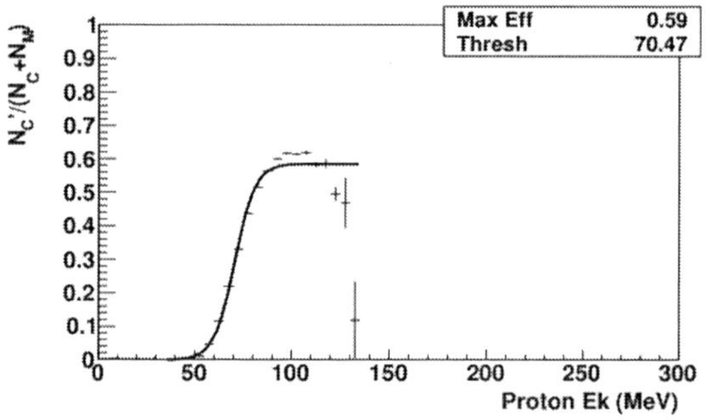

Figure 14. Proton efficiency $\epsilon = N_C'/(N_C + N_M)$ for the frozen spin target, where N_C' is the number of reconstructed $\gamma p \to \pi^0 p$ events that have a charged particle which satisfies an opening angle cut, N_C is the number of events that have a charged particle, and N_M is the number of events that missed the recoil particle.

threshold for detection in the FST, determined at the half-max of the distribution, is approximately 70 MeV, and by 50 MeV the efficiency is basically zero. The recoil proton energy for the Compton scattering phase space is shown in Fig. 15. With a threshold of 70 MeV, only the region in the upper right hand corner is accessible. For this reason it is difficult to go below 270 MeV or forward of 80°. The case is only slightly improved for the LH_2 target, where the threshold is about 50 MeV. To reach much lower energies, however, requires a different approach.

For this purpose an active polarized target has been developed, composed of discs of scintillating, yet polarizable, polystyrene as schematically shown in Fig. 16. When a charged particle passes through the material, it produces scintillation light. This light is then propagated along the plane of each disc and captured by the target nose, which then propagates it down the length of a light guide tube before reaching a ring of Silicon Photomultipliers (SiPMs). Since these SiPMs can detect single photons, this target

DOI: 10.1051/epjconf/201714201021

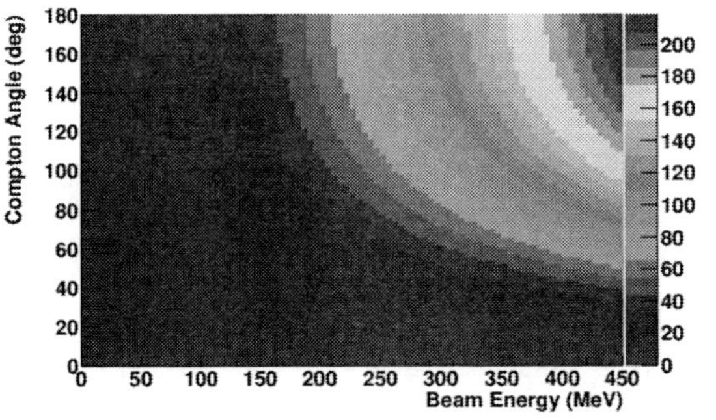

Figure 15. Compton kinematics, showing the recoil proton energy as color gradients as a function of the Compton scattering angle and the incoming beam energy.

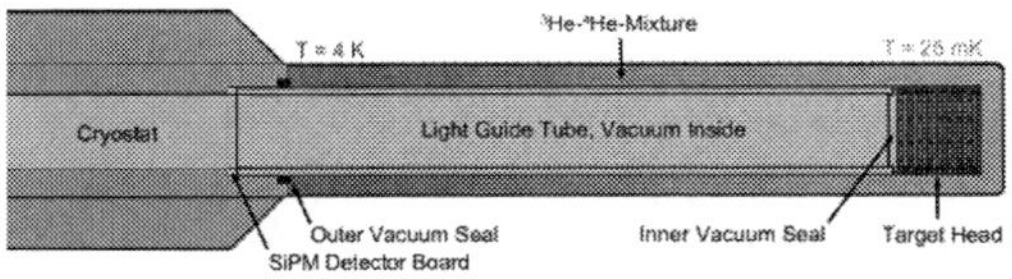

Figure 16. Schematic of the active target [26].

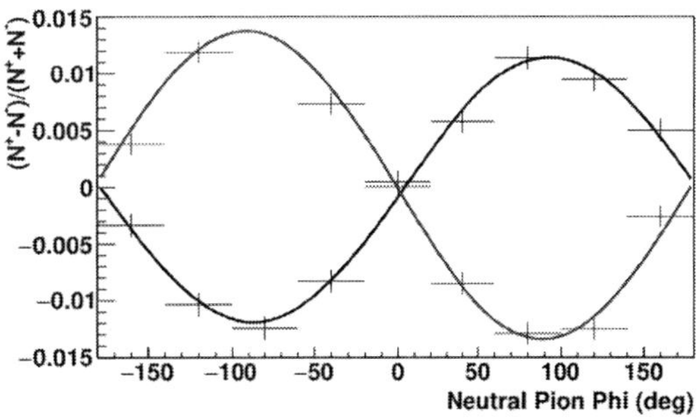

Figure 17. Preliminary π^0 target asymmetry from the active target, with the black points from the positively polarized target and the red points from the negatively polarized target, along with fits to the respective data.

can detect recoil protons immediately after the reaction, but before they have lost energy traversing the target material, cryostat, and various other sections before reaching the main detector system. Such a system will enable analyses at much lower energies, especially around the π^0 photoproduction threshold. One can already analyze the LH_2 target data in this region as, both below and slightly above threshold, other reactions can be kinematically separated from Compton scattering. With the polarized target however, as described in Sec. 3, the inclusion of non-hydrogen nuclei complicate things. With the detection ability of the active target, the dominant coherent background from these nuclei can be eliminated as the entire nucleus recoils, rather than individual protons. A first run with the active target, in a transversely polarized state, was performed in 2016, and the data is in the process of being analyzed. Two major achievements of this run are already evident however, the first being the detection of the scintillations within the target, and the second being the polarization of the target. This latter was clear from the NMR readings during polarization, but also confirmed in the analysis by looking at a simple π^0 target asymmetry, as shown in Fig. 17. The clear flip in the asymmetry between target polarization states proves the target was indeed polarized. The asymmetry is small since the analysis had not factored in the actual value of the target polarization, or removed contributions from the carbon which act to dilute the asymmetry.

4.5 Scalar Polarizabilities

In addition to studying the spin polarizabilities of the proton, the A2 collaboration undertook to improve the situation on the scalar polarizabilities. Initial studies, also of the Σ_3 observable with an LH_2 target, although at incident photon energies below the π^0 photoproduction threshold have been recently published [27]. As noted in Sec. 4.4, Compton scattering events from LH_2 can still be analyzed at these energies, despite the absence of recoil proton detection. These asymmetries are shown in Fig. 18, as presented in the paper [27].

With these data, fits were performed in the BχPT [2][29][14] and HBχPT [3] frameworks, which led to magnetic polarizability values of $\beta_{M1} = 2.8^{+2.3}_{-2.1} \times 10^{-4}\,\mathrm{fm}^3$ and $\beta_{M1} = 3.7^{+2.5}_{-2.3} \times 10^{-4}\,\mathrm{fm}^3$, respectively [27]. These data represent a proof-of-principle run taken in June 2013. A longer run, with improvements in both the tagging system and the linear beam polarization stability, is planned for the second half of 2017. This run will provide both a reduction in the asymmetry errors by a factor of about 3.5, and a set of cross section measurements, that together will enable a single extraction of α_{E1} and β_{M1} at level of the PDG errors.

5 Conclusions

The Compton scattering program at MAMI has included experiments on transversely and longitudinally polarized targets with a circularly polarized photon beam, as well as on an unpolarized liquid hydrogen target with a linearly polarized beam, with the intention of extracting the proton polarizabilities. Much of the data has already been acquired, and the analyses of these data are almost completed. From those analyses that are already completed, preliminary extractions of the scalar and spin polarizabilities have been performed, with the results presented here. The first run with a transversely polarized active target took place in 2016, and results are eagerly anticipated given the observed success of both the scintillation and polarization components of the target. A future run to extend the scalar polarizability study is in the preparation stage, and upon completion of these analyses, an unconstrained fit of the full data set will yield the first definitive determination of these important nucleon structure terms.

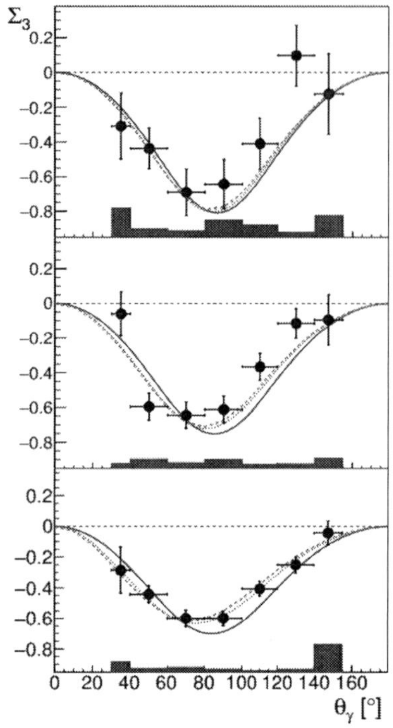

Figure 18. Beam asymmetry Σ_3 for three energy ranges (uppermost: 79 − 98 MeV, middle: 98 − 119 MeV, lowermost: 119 − 139 MeV). The errors represent statistical errors, the red bars indicate the systematic error. Green dashed curve: BχPT calculation [2], magenta dashed-dotted: DR calculation [28][11], blue dotted: HBχPT [3], all with $\alpha_{E1} = 10.65 \times 10^{-4}\,\mathrm{fm}^3$ and $\beta_{M1} = 3.15 \times 10^{-4}\,\mathrm{fm}^3$; brown solid: Born term (curves correspond to the central values of the shown energy bins) [27].

Acknowledgments

We wish to thank the accelerator group and operators of MAMI for their outstanding support; H.W. Grießhammer, J.A. McGovern, V. Pascalutsa, B. Pasquini, and D.R. Phillips for theory support; and the Messina conference organizers for the talk invitation.

References

[1] V. Olmos de Leon *et al.*, Eur. Phys. J. A **10**, 207 (2001).

[2] V. Lensky and V. Pascalutsa, Eur. Phys. J. C **65**, 195 (2010).

[3] J.A. McGovern, D.R. Phillips, H.W. Grießhammer, Eur. Phys. J. A **49**, 12 (2013).

[4] J. Beringer *et al.* (Particle Data Group), Phys. Rev. D **86**, 010001 (2012).

[5] C. Patrignani *et al.* (Particle Data Group), Chin. Phys. C **40**, 100001 (2016).

[6] J. Ahrens *et al.* (GDH/A2), Phys. Rev. Lett. **87**, 022003 (2001).

[7] H. Dutz, K. Helbing, J. Krimmer, T. Speckner, and G. Zeitler *et al.* (GDH), Phys. Rev. Lett. **91**, 192001 (2003).

[8] M. Camen *et al.* (A2), Phys. Rev. C **65**, 032202 (2002).

[9] S. Kondratyuk and O. Scholten, Phys. Rev. C **64**, 024005 (2001).

[10] B.R. Holstein, D. Drechsel, P. Pasquini, and M. Vanderhaeghen, Phys. Rev. C **61**, 034316 (2000).

[11] B. Pasquini, D. Drechsel, M. Vanderhaeghen, Phys. Rev. C **76**, 015203 (2007).

[12] D. Drechsel, B. Pasquini, and M. Vanderhaeghen, Phys. Rep. **378**, 99 (2003).

[13] A.M. Gasparyan, M.F.M. Lutz, and B. Pasquini, Nucl. Phys. A **866**, 79 (2011).

[14] V. Lensky and J.A. McGovern, Phys. Rev. C **89**, 032202 (2014).

[15] K.-H. Kaiser *et al.*, Nucl. Instrum. Methods Phys. Res., Sect. A **593**, 159 (2008).

[16] V. Tioukine, K. Aulenbacher, and E. Riehn, Rev. Sci. Instrum. **82**, 033303 (2011).

[17] J. McGeorge, J. Kellie, *et al.*, Eur. Phys. J. A **37**, 129 (2008).

[18] A. Thomas, Eur. Phys. J. Special Topics **198**, 171 (2011).

[19] D. G. Crabb and W. Meyer, Annu. Rev. Nucl. Part. Sci. **47**, 67 (1997).

[20] A. Starostin *et al.* (CB), Phys. Rev. C **64**, 055205 (2001).

[21] R. Novotny, IEEE Trans. Nucl. Sci. **38**, 379 (1991).

[22] C. M. Tarbert *et al.* (CB@MAMI/A2), Phys. Rev. Lett. **100**, 132301 (2008).

[23] P.P. Martel, R. Miskimen, *et al.* (A2), Phys. Rev. Lett. **114**, 112501 (2015).

[24] C. Collicott, Ph.D. thesis, Dalhousie University (2015).

[25] G. Blanpied *et al.* (LEGS), Phys. Rev. C **64**, 025203 (2001).

[26] M. Biroth, P. Achenbach, E. Downie, and A. Thomas (A2), PoS PSTP **2015**, 005 (2015).

[27] V. Sokhoyan, E.J. Downie, E. Mornacchi, J.A. McGovern, N. Krupina, *et al.* (A2), Eur. Phys. J. A **53**, 14 (2017).

[28] D. Drechsel, M. Gorchtein, B. Pasquini, M. Vanderhaeghen, Phys. Rev. C **61**, 015204 (1999).

[29] V. Lensky, J.A. McGovern, V. Pascalutsa, Eur. Phys. J. C **75**, 604 (2015).

Recent results on search for new physics at BaBar

Benjamin Oberhof[1,a]

[1] *Laboratori Nazionali di Frascati dell'INFN, Frascati, Italy*

Abstract. We present some recent measurements for the search of New Physics using 514 fb^{-1} of e^+e^- collisions collected with the BABAR detector at the PEP-II e^+e^- collider at SLAC. First we present a search for the decay $\Upsilon(1S) \to \gamma A^0$, $A^0 \to c\bar{c}$, where A^0 is a candidate for the CP-odd Higgs boson of the next-to-minimal supersymmetric standard model. No significant signal is observed and we set 90% confidence-level upper limits on $B(\Upsilon(1S) \to \gamma A^0) \times B(A^0 \to c\bar{c})$. We report the search for a light non-Standard Model gauge boson Z' coupling only to the second and third lepton families. Our results significantly improve current limits and further constrain the remaining region of the allowed parameter space. Finally, we present a search for a long-lived particle L that is produced in e^+e^- annihilations and decays into two oppositely charged tracks. We do not observe a significant signal and we and set 90% confidence level upper limits on the product of the L production cross section, branching fraction, and reconstruction efficiency as a function of the L mass. In addition, upper limits are provided on the branching fraction $B(B \to X_s L)$, where X_s is an hadronic system with strangeness -1.

1 Introduction

The Standard Model (SM) of particle physics has proven to be very successful in explaining the fundamental laws of nature, and most of the theoretical results of the SM agree with the experimental data. The SM, however, cannot be quantified as a complete theory of fundamental interactions as it shows theoretical drawbacks and it is unable to explain some experimental observations. The so called "hierarchy problem" requires the fine tuning of the Higgs-boson mass in order that the quadratical divergences occurring in loop corrections are systematically canceled and is deemed unnatural by many theorists. Some extensions of the SM account for this problems and naturally stabilize the hierarchy between the electroweak and the Plank scale [1]. Experimentally, astrophysical and cosmological observations indicate that a fraction of energy density in the universe is due to non-baryonic matter, usually called dark matter (DM). The nature of DM is currently unknown, and the SM does not have viable candidate for DM particles. Many models of physics beyond the SM predict the existence of a new gauge groups with gauge boson masses below 10 GeV which can typically interact with other SM elementary particles through so-called portals [2].

We herein present three different searches for new physics candidates using data recorded by the BABAR detector at the PEP-II asymmetric-energy e^+e^- storage rings operated at the SLAC National Accelerator Laboratory. The data sample consists of 424 fb^{-1} of e^+e^- collisions recorded at the center-of-mass (CM) energy of the $\Upsilon(4S)$ resonance, 27.8 fb^{-1} of data recorded at the $\Upsilon(3S)$ and 13.6 fb^{-1} of data collected at the $\Upsilon(2S)$ resonance. A detailed description of the BABAR detector is given elsewhere [3].

2 Search for a light Higgs in radiative decays of the $\Upsilon(1S)$ with a charm tag

An appealing extension of the Standard Model is represented by the next-to-minimal supersymmetric standard model (NMSSM) since it solves both the μ problem of the minimal supersymmetric standard model (MSSM) as well as the hierarchy problem of the SM [4, 5]. The NMSSM has an extended Higgs sector composed of two charged, three neutral CP-even, and two neutral CP-odd bosons. The mass of the lightest NMSSM Higgs boson A^0 could be low enough to be produced in the decay of the $\Upsilon(1S)$ [6]. The branching fractions of the A^0 to SM particles will depend on the A^0 mass and the NMSSM parameter $\tan\beta$ [7]. BABAR has already performed searches below the charm mass threshold searching for $A^0 \to \mu^+\mu^-$ [8, 9] and for $A^0 \to gg$ or $s\bar{s}$ [10] as well as above the charm mass threshold in $A^0 \to \tau^+\tau^-$ searches [11]. None of these searches have observed a significant signal and most of the NMSSM parameter space has been ruled out. Here we report a search for the decay $\Upsilon(1S) \to \gamma A^0, A^0 \to c\bar{c}$ with A^0 masses ranging from 4.00 to 9.25 GeV/c^2. For this search we use 13.6 fb^{-1} of data collected at the $\Upsilon(2S)$ resonance, corresponding to $(98.3 \pm 0.9) \times 10^6$ $\Upsilon(2S)$ mesons [12], which includes an estimated $(17.5 \pm 0.3) \times 10^6$ $\Upsilon(2S) \to \pi^+\pi^-\Upsilon(1S)$ decays [13]. The non-$\Upsilon(2S)$ backgrounds are studied using 1.4 fb^{-1} of "off-resonance" data collected 30 MeV below the $\Upsilon(2S)$ resonance. A Monte Carlo

[a] e-mail: benjamin.oberhof@lnf.infn.it

(MC) simulation is used to estimate the signal efficiency and optimize the search as well as for background studies. The signal decay chain $e^+e^- \to \Upsilon(2S) \to \pi^+\pi^-\Upsilon(1S)$, $\Upsilon(1S) \to \gamma A^0$, $A^0 \to c\bar{c}$ is simulated with EVTGEN [14], for A^0 masses between 4.0 and 9.0 GeV/c^2 in 0.5 GeV/c^2 steps and for A^0 masses of 9.2, 9.3, and 9.4 GeV/c^2. The A^0 decay width is assumed to be 1 MeV. The hadronization of the $c\bar{c}$ system is simulated using the JETSET [15] program. The detector response is simulated with GEANT4 [16]. $\Upsilon(1S)$ decays are tagged by the presence of a pion pair from $\Upsilon(2S) \to \pi^+\pi^-\Upsilon(1S)$. To tag the $A^0 \to c\bar{c}$ decay at least one charmed meson such as a D^0, a D^+, or a $D^*(2010)^+$ is required. In this analysis the A^0 is not fully reconstructed, instead the search for the A^0 is performed in the spectrum of the invariant mass of the system recoiling against the $\pi^+\pi^-\gamma$ system. Events are required to contain at least one photon candidate with $E > 30$ MeV. Each photon candidate is taken in turn to represent the signal candidate in the $\Upsilon(1S) \to \gamma A^0$ decays, allowing multiple candidates per event. D meson candidates are reconstructed in five channels: $D^0 \to K^-\pi^+$, $D^+ \to K^-\pi^+\pi^+$, $D^0 \to K^-\pi^+\pi^-\pi^+$, $D^0 \to K^0_S\pi^+\pi^-$, and $D^*(2010)^+ \to \pi^+D^0$ with $D^0 \to K^-\pi^+\pi^0$. Events are required to have at least one dipion candidate; m_R, the invariant mass of the system recoiling against the dipion in the $\Upsilon(2S) \to \pi^+\pi^-\Upsilon(1S)$, is calculated as

$$m_R^2 = M^2_{\Upsilon(2S)} + m^2_{\pi\pi} - 2M_{\Upsilon(2S)}E_{\pi\pi} \quad (1)$$

where $M_{\Upsilon(2S)}$ is the nominal $\Upsilon(2S)$ mass, $m_{\pi\pi}$ and $E_{\pi\pi}$ are the measured dipion mass and energy in the center-of-mass (CM) frame. Signal candidates must satisfy $9.45 < m_R < 9.47$ GeV/c^2. The mass of the A^0 candidate, m_X , is determined from the mass of the system recoiling against the dipion and photon through

$$m_X = (P_{e+e-} - P_{\pi+\pi-} - P_\gamma) \quad (2)$$

where P denotes four-momentum measured in the CM frame. Simulated background include $\Upsilon(2S)$ and $e^+e^- \to q\bar{q}$ events, where q is a u, d, s, or c quark. For m_X greater than 7.50 GeV/c^2, events with low-energy photons constitute an important background source, for this reason, the analysis is divided into a low mass region, for $4.00 < m_{A^0} < 8.00$ GeV/c^2 and a high mass region, for $7.50 < m_{A^0} < 9.25$ GeV/c^2. The overlap of the two regions is motivated by the need to have sufficient statistical precision for the background determination in each region. We train one Boosted Decision Tree (BDT) classifier [17] taking 24 different variables as input to separate background from signal candidates for each of the five D channels and for each mass region. To train the BDTs samples of simulated signal events, simulated generic $\Upsilon(2S)$ events, and the off-resonance data are used. For each channel in the two mass intervals, we require the BDT output to exceed a value determined maximizing the quantity $S/(1.5 + \sqrt{B})$ [18], where S is the expected number of signal events and B the expected number of background events, respectively. After selection, 9.8×10^3 and 7.4×10^6 candidates satisfy the selection criteria in the low- and high-mass regions, respectively. The corresponding distributions of m_X are shown in Fig. 1.

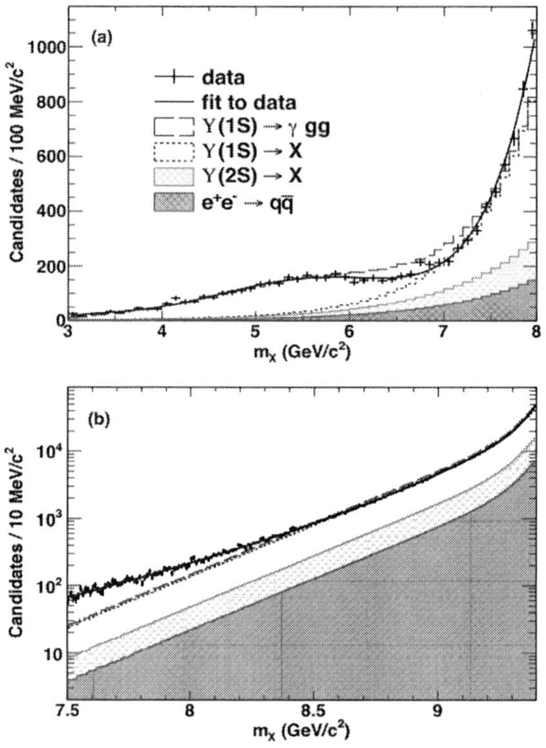

Figure 1. The m_X distribution of signal candidates in the low- (a) and high- (b) mass regions after applying all selection criteria. The points indicate the data. The solid curve shows the result of a fit to the data under a background- only hypothesis. The coloured histograms show the cumulative background contributions from $e^+e^- \to q\bar{q}$ (magenta dense-dot filled), $\Upsilon(2S) \to X$ (green sparse-dot filled), $\Upsilon(1S) \to X$ (blue dotted), and $\Upsilon(1S) \to \gamma gg$ (red dashed) events.

The backgrounds consist of $\Upsilon(1S) \to \gamma gg$, other $\Upsilon(1S) \to X$ decays, $\Upsilon(2S) \to X$ decays without a dipion transition and $e^+e^- \to q\bar{q}$ events. The corresponding contributions are 35%, 34%, 15% and 16% respectively in the low-mass region and 1%, 66%, 18%, and 15% respectively in the high-mass region. The contribution from $\Upsilon(1S) \to \gamma gg$ decays reaches a maximum near 5.5 GeV/c^2 and decreases above 7GeV/c^2. We search for the A^0 resonance as a peak in the m_X distribution; the scan is performed by means of extended maximum likelihood fits as a function of test-mass values, denoted m^0_A. For these fits the shape of the signal distribution are fixed while the parameters of the background PDF, the number of signal events N_{sig}, and the number of background events are determined in the fit. The signal PDF is modeled with a Crystal Ball function [19]. The background PDF is modeled with a second-order polynomial. The fits are performed in steps of 10 and 2 MeV/c^2 in the low- and high-mass regions, respectively. We use a local fitting range of $\pm 10\sigma_{CB}$ around test-mass values, where σ_{CB} is the width of the Gaussian component of the Crystal Ball function. The σ_{CB} parameter varies between 120 and 8 MeV/c^2 for values of m_A^0 between 4.00 and 9.25 GeV/c^2. In the high-mass region we exclude the interval $8.95 < m_A^0 < 9.10$ GeV/c^2 because of a large background from $\Upsilon(2S) \to \gamma\chi_{bJ}(1P), \chi_{bJ}(1P) \to \gamma\Upsilon(1S)$ decays. The fitting procedure is validated with background-only pseudo-experiments; a fifth-order polynomial PDF is used to fit the data and generate the pseudo-experiments in the low-mass region, while a sum of four exponential functions and six Crystal Ball functions is used for the high-mass region. The Crystal Ball functions describe the $\Upsilon(2S) \to \gamma\chi_{bJ}(1P)$ and $\chi_{bJ}(1P) \to \gamma\Upsilon(1S)$ transitions while the exponential terms describe the non-resonant background. The fitting procedure is found to require a correction to N_{sig} for values of m_{A^0} near 4.00 and 9.25 GeV/c^2. The corrections are determined from the average signal yield found in the fits to the background-only pseudo-experiments. The maximum correction reaches 50 candidates in the high-mass region; the uncertainty of the correction is taken to be half its value. The overall efficiencies range from 4.0% to 2.6% between 4.00 and 9.25 GeV/c^2. Potential bias introduced by the fitting procedure is evaluated using pseudo-experiments which use different values of $B(\Upsilon(1S) \to \gamma A^0) \times B(A^0 \to c\bar{c})$; depending on m_{A^0}, the extracted product branching fraction is found to be up to (4 ± 3)% higher than the value used to generate the events. The systematic uncertainties associated with the reconstruction efficiencies are dominated by the differences between data and simulation, including the BDT output modelling, $c\bar{c}$ hadronization, D-candidate mass resolution, dipion recoil mass and likelihood modeling, and photon reconstruction. Other systematic uncertainties are associated with the fit bias, the dipion branching fraction, the finite size of the simulated signal sample, and the $\Upsilon(2S)$ counting. The BDT output distributions in off-resonance data and $e^+e^- \to q\bar{q}$ simulation, are slightly shifted from one another; the associated systematic uncertainty is estimated by shifting the simulated distributions so that the mean values agree with the data, and then recalculating the efficiencies. The reconstruction efficiencies decrease by 7% and 2% in the low- and high-mass regions, respectively. Uncertainty associated with $c\bar{c}$ hadronization is evaluated by comparing D meson production in off-resonance data and $e^+e^- \to c\bar{c}$ simulation; the difference varies from 1% to 9% for the five D decay channels. We conservatively assign a global multiplicative uncertainty of 9% that includes effects due to the hadronization modeling, particle identification, tracking, π^0 reconstruction, and luminosity determination of the off-resonance data. The uncertainty due to the discrepancy between the reconstructed D mass resolution in data and simulation is estimated by gaussian smearing of the D mass input in simulation to match the data and measuring the difference in the reconstruction efficiency. Further corrections to account for data and simulation differences in reconstruction efficiencies are estimated with similar methods. Corrections are applied to account for the dipion recoil mass reconstruction, the dipion likelihood modeling, and the photon reconstruction [20].

The highest observed local significance in the low-mass region is 2.3σ, at 4.145 GeV/c^2, and 2.0σ at 8.411 GeV/c^2 in the high-mass region. Such fluctuations occur in 54% and 80% of pseudo-experiments, respectively, hence we conclude that our data are consistent with the background-only hypothesis [21]. Bayesian upper limits on the product branching fraction $B(\Upsilon(1S) \to \gamma A^0) \times B(A^0 \to c\bar{c})$ at 90% confidence level (C.L.) are determined assuming a uniform prior, with the constraint that the product branching fraction be greater than zero. The distribution of the likelihood function for N_{sig} is assumed to be gaussian with a width equal to the total uncertainty on N_{sig}. The upper limits obtained from the low-mass region are combined with those from the high-mass region to define a continuous spectrum. The results are shown in Fig. 2.

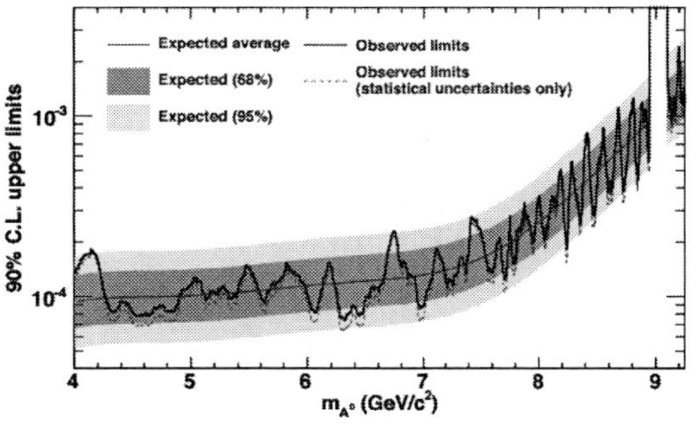

Figure 2. 90% CL upper limits for product branching fraction $\mathcal{B}(\Upsilon(1S) \to \gamma A^0) \times \mathcal{B}(A^0 \to c\bar{c})$ including all uncertainties (thick line) and statistical uncertainties only (thin dashed line). The inner and outer bands represent the 68% and 95% expected upper limits. The bands are calculated using all uncertainties. The solid line in the center of the inner band is the expected value of the upper limit.

3 Search for a muonic dark force

In the SM particles and interactions are insufficient to explain cosmological and astrophysical observations of dark matter. A possible scenario is represented by new hidden sectors that are only feebly coupled to the SM. In the simplest case of a hidden U(1) gauge interaction, such a sector contains it's own gauge bosons Z' and SM fields may directly couple to the Z', or alternatively the Z' boson may mix with the SM hypercharge boson [22]. In the latter case, the Z' couplings are proportional to the SM gauge couplings; however, due to large couplings to electrons and light-flavor quarks, such scenarios are strongly constrained by existing searches [23]. If SM fields are directly charged under the dark force instead, the Z' may interact preferentially with heavy-flavor leptons, greatly

DOI: 10.1051/epjconf/201714201022

reducing the sensitivity of current searches. Such interactions could account for the experimentally measured value of the muon anomalous magnetic dipole moment [24], as well as the discrepancy in the proton radius extracted from measurements of the Lamb shift in muonic hydrogen compared to observations in non-muonic atoms [25, 26]. In the following we report a search for dark bosons Z' with vector couplings only to the second and third generations of leptons [27] in $e^+e^- \to \mu^+\mu^- Z'$, $Z' \to \mu^+\mu^-$.

For this search we use the full $\Upsilon(4S)$ data sample plus about 28 fb^{-1} data at $\Upsilon(3S)$, 14 fb^{-1} data at $\Upsilon(2S)$, and 48 fb^{-1} off-resonance data. About 5% of the data set is used to validate and optimize the analysis method, the rest of the data was only examined after finish finalizing the analysis method. For the background study we use MC samples. Signal MC events are generated using MadGraph 5 [28] and hadronized in Pythia 6 [29] for Z' mass hypotheses from the dimuon mass threshold to 10.3GeV/c^2. Background samples include the direct processes of $e^+e^- \to \mu^+\mu^-\mu^+\mu^-$ generated with Diag36 [30], which includes the full set of the lowest order diagrams. The events of the process of $e^+e^- \to e^+e^-(\gamma)$ are generated using BHWIDE [31] while of $e^+e^- \to \mu^+\mu^-(\gamma)$ and $e^+e^- \to \tau^+\tau^-(\gamma)$ are generated using `KK2f` [32]. The off-resonance data samples, $e^+e^- \to q\bar{q}$ ($q = u, d, s, c$), are simulated using `JETSET`. The events processes of $e^+e^- \to \psi(2S)\gamma$ followed by $\psi(2S) \to \pi^+\pi^- J/\psi$ and $J/\psi \to \mu^+\mu^-$ were generated using **EVTGEN** with appropriate phase-space model. Finally the detector acceptance and reconstruction efficiency are determined using GEANT4.

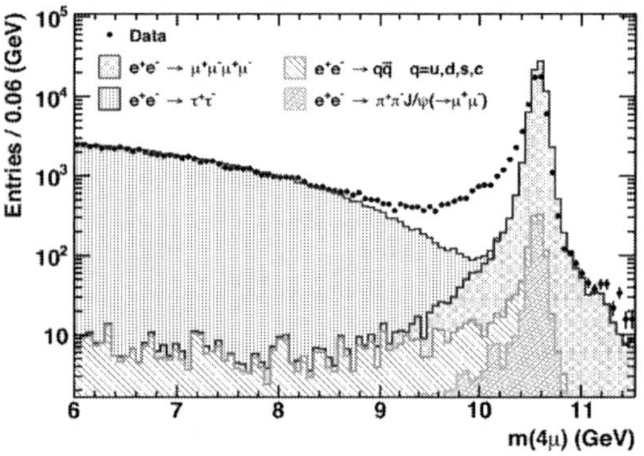

Figure 3. The distribution of the four-muon invariant mass, $m_{4\mu}$, for data taken at the $\Upsilon(4S)$ peak together with Monte Carlo predictions of various processes normalized to data luminosity. The $e^+e^- \to \mu^+\mu^-\mu^+\mu^-$ MC does not include ISR corrections.

We select events with exactly two pairs of oppositely charged tracks, consistent with the $e^+e^- \to \mu^+\mu^- Z'$, $Z' \to \mu^+\mu^-$ final state. The muons are identified by multivariate particle identification algorithms for each track. We require the sum of energies of the deposits in the electromagnetic calorimeter that are not associated to any track to be less than 200 MeV. We reject events coming from the $\Upsilon(3S)$ and $\Upsilon(2S)$, where $\Upsilon(2S, 3S) \to \pi^+\pi^-\Upsilon(1S)$, $\Upsilon(1S) \to \mu^+\mu^-$ decays if the dimuon combination lies within 100 MeV/c^2 of the $\Upsilon(1S)$ mass. The distribution of the four-muon invariant mass after all selections is shown in Fig. 3. The lower part of the four-muon invariant mass spectrum, $m_{4\mu} < 9$ GeV/c^2, is well reproduced by the Monte Carlo simulation while the MC simulation overestimates the full energy peak by about 30% and fails to reproduce the radiative tail. This, however, is expected because the Diag36 simulation does not simulate the initial state radiation (ISR). We select $e^+e^- \to \mu^+\mu^-\mu^+\mu^-$ events by requiring a four-muon invariant mass distribution within 500 MeV/c^2 of the nominal center-of-mass energy. We also require the tracks to originate from the interaction point to within its uncertainty and constraining the center-of-mass energy of the system to be within the beam energy spread. We do not attempt to select a single $Z' \to \mu^+\mu^-$ candidate per event, but instead consider all possible combinations. The most important contribution on the invariant mass peak besides the QED process $e^+e^- \to \mu^+\mu^-\mu^+\mu^-$ comes from $\Upsilon(2S) \to \pi^+\pi^- J/\psi$, $J/\psi \to \mu^+\mu^-$ as can be seen in Fig. 3. We extract the signal yield by a series of unbinned likelihood fits to the spectrum of the reduced dimuon mass m_R, defined as $m_R = \sqrt{m^2_{\mu^+\mu^-} - 4m^2_\mu}$, within the range of $0.212 < m_R < 10$ GeVc^2 and $0.212 < m_R < 9$ GeVc^2 for the $\Upsilon(4S)$ resonance data and $\Upsilon(2S)$ and $\Upsilon(3S)$ resonances data, respectively. We exclude a region of ± 30 MeVc^2 around the nominal known J/ψ mass. We probe a total of 2219 mass hypothesis. The signal efficiency at low masses is about 35% and it rises to about 50% around $m_R = 6 - 7$ GeV/c^2 to drop again at higher values of the reduced dimuon mass. The signal efficiencies include a correction factor of 0.82, which accounts for the impact of ISR not included in the simulation, as well as differences between data and simulation in trigger efficiency, charged particle identification, and track and photon reconstruction efficiencies. This correction factor is obtained from the ratio of the m_R distribution in simulated $e^+e^- \to \mu^+\mu^-\mu^+\mu^-$ events to the observed distribution in the mass region 1-9 GeV/c^2. We also assign a systematic uncertainty of 5% to cover the small variations between the uncertainties on the $e^+e^- \to \mu^+\mu^-\mu^+\mu^-$ and data taking period. We calculate the ISR contribution based on the quasi real electron approximation [38].

The cross section of $e^+e^- \to \mu^+\mu^- Z'$, $Z' \to \mu^+\mu^-$ is extracted as a function of Z' mass. The gray band indicates the excluded region. We find the largest local significance is 4.3σ around Z' mass of 0.82 GeV/c^2 that is corresponding to the global significance of 1.6σ and it is consistent with the null-hypothesis [33]. We also derive 90% confidence level (CL) Bayesian upper limit on the cross section of $e^+e^- \to \mu^+\mu^- Z'$, $Z' \to \mu^+\mu^-$ as shown in Fig. 4. We consider all uncertainties to be uncorrelated except for the uncertainties of the luminosity and efficiency. We finally extract the corresponding 90% CL on the coupling parameter g' by assuming the equal magnitude vector couplings muons, taus and the corresponding neutrinos together with the existing limits from Borexino and neutrino experiments as shown in Fig. 5. We set down to 7×10^{-4} near the dimuon threshold.

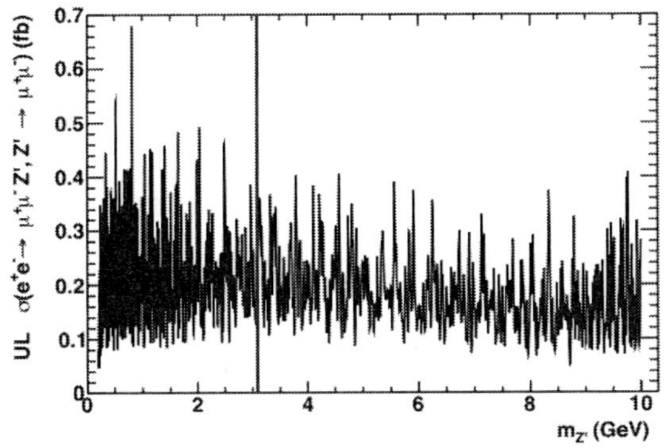

Figure 4. The 90% CL upper limits on the cross-section $\sigma(e^+e^- \rightarrow \mu^+\mu^- Z', Z' \rightarrow \mu^+\mu^-)$ as a function of the Z' mass. The dark gray band indicates the region excluded from the analysis.

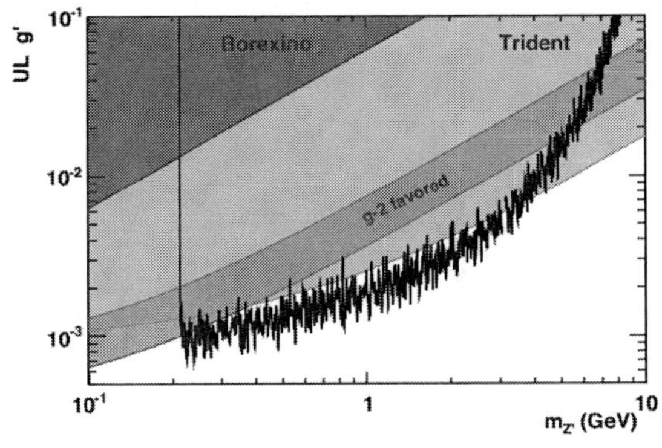

Figure 5. The 90% CL upper limits on the new gauge coupling g' as a function of the Z' mass, together with the constraints derived from the production of a $\mu^+\mu^-$ pairs in ν_μ scattering ("Trident" production) [34]. The region consistent with the discrepancy between the calculated and measured anomalous magnetic moment of the muon within 2σ is shaded in red.

4 Search for long lived particles

Anomalous astrophysical observations [35]-[37] have generated some interest in GeV-scale hidden-sector long-lived states [38]. *BABAR* offers an ideal laboratory to search for such a particle in the O(1GeV/c^2) mass range. We thus search for a neutral, long-lived particle L, which decays into one of the following final states $f = e^+e^-$, $\mu^+\mu^-$, $e^\pm\mu^\mp$, $\pi^+\pi^-$, K^+K^-, or $K^\pm\pi^\mp$. The signature for such a decay is a displaced vertex consistent with a two-body decay. The search is performed by fitting the L-candidate mass distribution. We provide "model-independent" limits on the product $\sigma(e^+e^- \rightarrow LX)(L \rightarrow f)\epsilon(f)$ of the inclusive production cross section $\sigma(e^+e^- \rightarrow LX)$, the branching fraction $B(L \rightarrow f)$, and efficiency $\epsilon(f)$ for each of the two-body final states f, where X is any set of particles. Detailed tables for efficiency as a function of L mass m, transverse momentum, and proper decay distance $c\tau$ are given in [39]. We also provide "model-dependent" limits on the branching fraction for the decay $B \rightarrow X_s L$, where X_s is an hadronic system with strangeness -1. To determine signal mass resolution and reconstruction efficiency we use MC simulations; signal events are produced with the EVTGEN generator, assuming the L spin to be zero, in two different ways: for the "model-independent" search the L is produced at 11 different masses, $m_{MC} = 0.5, 1, 2, 3, 4, 5, 6, 7, 8, 9$, and 9.5 GeV/$c^2$; for $m_{MC} \leq 4$ GeV/c^2, the L is created in the process $e^+e^- \rightarrow B\bar{B}$, with one B meson decaying to $L + N\pi$ ($N = 1$, 2, or 3) and the other B meson decaying generically. while at higher masses the production process is assumed to be $\Upsilon(4S) \rightarrow L + N\pi$. The production processes are chosen in order to populate properly the phase space and do not reflect any preferred production hypothesis; in both regions, the L is produced uniformly throughout the available phase space, with an average transverse decay distance of 20 cm. Efficiencies for other decay lengths are obtained by reweighting the events. For the "model-dependent" presentation we generate $B \rightarrow X_s L$ decays for the seven mass values $m_{MC} = 0.5$, 1, 2, 3, 3.5, 4, and 4.5 GeV/c^2. The composition of the X_s is taken to be 10% K, 25% $K^*(892)$, and 65% $K^*(1680)$, with the high-mass tail of the X_s spectrum suppressed by phase-space limitations. The other B meson in the event decays generically. In addition to the signal MC samples, we use the following background MC samples in order to optimize the event selection criteria and study the signal extraction method: $e^+e^- \rightarrow B\bar{B}$ (produced with EVTGEN), $\tau^+\tau^-$, $\mu^+\mu^-$ (KK2f), e^+e^- (BHWIDE), and $q\bar{q}$ events (JETSET), where q is a u, d, s, or c quark. The detector response is simulated with GEANT4.

Each track must satisfy $d_0/\sigma_{d_0} > 3$, where d_0 is the transverse distance of closest approach of the track to the e^+e^- interaction point (IP), and σ_{d_0} is its uncertainty. The two tracks have to originate from a common vertex, and the χ^2 value of the fit is required to be smaller than 10 for one degree of freedom. We require the distance r between the IP and the vertex in the transverse plane to be in the range $1 < r < 50$ *cm*, and the uncertainty on r is required to satisfy $\sigma_r < 0.2$ *cm*. We also require the angle α between $\vec{r}$ and the L-candidate reconstructed transverse momentum vector to be $\alpha < 0.01$ rad. The uncertainty σ_m on the measured L-candidate mass m must be less than 0.2 GeV/c^2. The L candidate is discarded if either of the tracks has associated hits in the detector between the IP and the vertex, or if the vertex is within the material of the beampipe wall, or the detector bulk material. Depending on the specific final state, the candidates must satisfy the following invariant-mass criteria: $m_{e^+e^-} > 0.44$ GeV/c^2, $m_{\mu^+\mu^-} < 0.37$ GeV/c^2 or $m_{\mu^+\mu^-} > 0.5$ GeV/c^2, $m_{e^\pm\mu^\mp} > 0.48$ GeV/c^2, $m_{\pi^+\pi^-} > 0.86$ GeV/c^2, $m_{K^+K^-} > 1.35$ GeV/c^2, and $m_{K^\pm\pi^\mp} > 1.05$ GeV/c^2. These criteria reject most background from $K^0_S \rightarrow \pi^+\pi^-$ and $\Lambda \rightarrow p\pi^-$ decays. In addition, except for the $\mu^+\mu^-$ mode, they exclude low-mass regions in which the mass distributions of background MC events are not smooth, and therefore incompatible with the background description method outlined below. We require at least one of the tracks of $L \to \mu^+\mu^-$ candidates with $m \geq 8$ GeV/c^2 to have an SVT hit. This rejects candidates that decay into $\mu^+\mu^-$ within the material of the final-focusing magnets, and thus have poor mass resolution. These selection criteria are found to yield near-optimal signal sensitivity given the broad range of m and r values of this search. For each decay mode, we determine the full efficiency ϵ for different values of m_{MC} , $c\tau$, and p_T, including the impact of detector acceptance, trigger, reconstruction, and selection criteria. The dominant source of background consists of hadronic events with high track multiplicity, where large-d_0 tracks originate mostly from K^0_S, Λ, $K^\pm$, and $\pi^\pm$ decays, as well as particle interactions with detector material. Random overlaps of such tracks comprise the majority of the background candidates.

We extract the signal yield for each final state as a function of L mass with unbinned extended maximum-likelihood fits of the m distribution. The procedure is based on the fact that signal MC events peak in m while the background distribution varies slowly. The fit probability density functions (PDFs) for signal and background are constructed separately for each mode and each data sample. We scan the data in search of an L signal, varying m_0 in steps of 2 MeV/c^2. At each scan point, we fit the data in the full mass range using the PDF $n_S P_S + n_B P_B$, where the signal and background yields n_S and n_B are determined in the fit. The statistical significance S =sign(n)2 log($\mathcal{L}_S / \mathcal{L}_B$), where $\mathcal{L}_S$ is the maximum likelihood for n_s signal events over the background yield, and $\mathcal{L}_B$ is the likelihood for $n_S = 0$, is calculated for each scan point. We find two scan points with a significance S greater than 3, both in the $\mu^+\mu^-$ mode in the $\Upsilon(4S)$+off-resonance sample. The highest significance is $S = 4.7$, with a signal yield of 13 events at the low-mass threshold of $m_0 = 0.212$ GeV/c^2. The second-highest significance of $S = 4.2$ occurs at $m_0 = 1.24$ GeV/c^2, corresponding to a signal yield of 10 events. To obtain the p-values for these significances, we perform a large number of pseudo experiments on $m_{\mu^+\mu^-}$ spectra generated according to the background PDF, obtained from the data. We find that the probability for $S \geq 4.7(4.2)$ anywhere in the $\mu^+\mu^-$ spectrum with $m_{\mu^+\mu^-} < 0.37$ GeV/c^2 ($m_{\mu^+\mu^-} > 0.5$ GeV/c^2) is 4×10^{-4} (8×10^{-3}). Further study provides strong indication for material interaction background in the 0.212 GeV/c^2 region. Specifically, most of the 34 $\mu^+\mu^-$ vertices with $m_{\mu+\mu-} < 0.215$ GeV/c^2 occur inside or at the edge of detector-material regions, including 10 of the vertices that also pass the e^+e^- selection criteria and 10 that pass the $\pi^+\pi^-$ criteria. Thus, the peak is consistent with misidentified photon conversions and hadronic interactions close to the mass threshold. We conclude that no significant signal is observed.

Systematic uncertainties on the signal yields are calculated for each scan fit separately. The dominant uncertainty is due to the background PDF; this uncertainty is a few signal events on average, and generally decreases with mass. The uncertainty due to the signal mass resolution is evaluated by comparing the mass pull distributions of K^0_S mesons in data and MC, whose widths differ by 5%. A conservative uncertainty of 2% on the signal reconstruction efficiencies is estimated from the K^0_S reconstruction efficiency in data and MC. Smaller uncertainties on the efficiency, of up to 1%, arise from particle identification, and signal MC statistics.

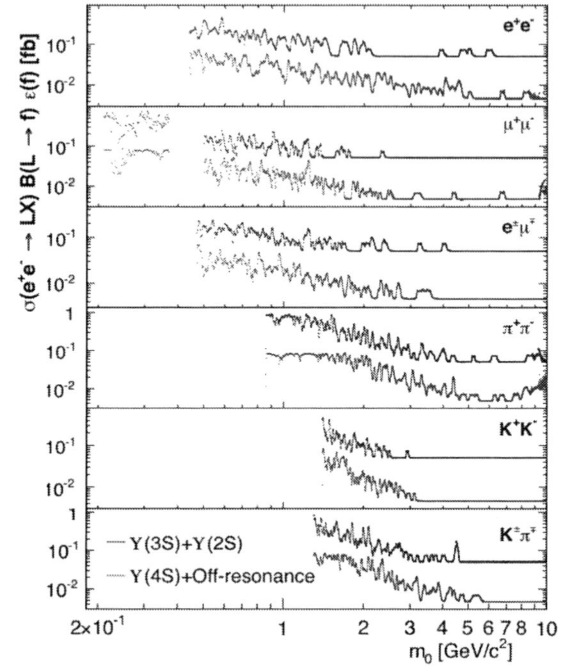

Figure 6. The 90% confidence level upper limits on $\sigma(e^+e^- \to LX)\mathcal{B}(L \to f)\epsilon(f)$ as a function of L mass for the $\Upsilon(4S)$ + off-resonance sample (red lower points) and for the $\Upsilon(3S) + \Upsilon(2S)$ sample (blue upper points). The limits include the systematic uncertainties.

The likelihood $\mathcal{L}_S$ which is a nearly normal function of the signal yield, is convolved with a Gaussian representing the systematic uncertainties in n_S, to obtain the modified likelihood function $\mathcal{L}'_S$. The 90% confidence level upper limit U_S on the signal yield is calculated from $\int_0^{U_S} \mathcal{L}'_S dn_S / \int_0^{\inf} \mathcal{L}'_S dn_S = 0.9$. Dividing U_S by the luminosity yields an upper limit on the product $\sigma(e^+e^- \to LX)B(L \to f)\epsilon(f)$. This limit is shown for each final state as a function of m_0 in Fig. 6 [39]. Determining the efficiency from the $B \to X_s L$ signal MC sample, we obtain upper limits on the product of branching fractions $B(B \to X_s L)B(L \to f)$ for each of the final states f. These limits are shown in Fig. 7.

5 Conclusion

In summary, we search for a resonance in radiative decays of the $\Upsilon(1S)$ with a charm tag. We do not observe a signif-

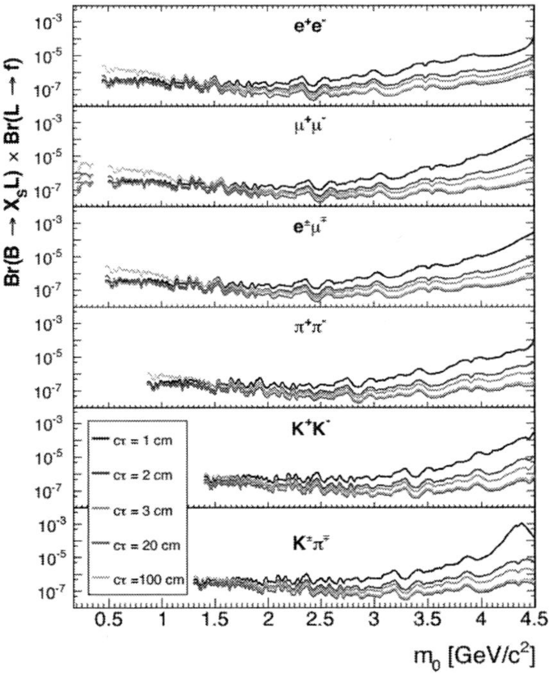

Figure 7. Implications of the results for Higgs-portal scenarios, showing the 90% confidence level upper limits on the product of branching fractions $\mathcal{B}(B \rightarrow XsL)B(L \rightarrow f)$ as a function of L mass for each final state f and for different values of $c\tau$. The limits include all systematic uncertainties.

icant signal and set upper limits on the product branching fraction $B(\Upsilon(1S) \rightarrow \gamma A^0) \times B(A^0 \rightarrow c\bar{c})$ ranging from 7.4×10^{-5} to 2.4×10^{-3} for A^0 masses from 4.00 to 9.25 GeV/c^2, excluding masses from 8.95 to 9.10 GeV/c^2 because of irreducible background from $\Upsilon(2S) \rightarrow \gamma\chi_{bJ}(1P)$, $\chi_{bJ}(1P) \rightarrow \gamma\Upsilon(1S)$ decays [21]. These results will further constrain the NMSSM parameter space.

We have performed the first direct measurement of Z' production from the decay of $e^+e^- \rightarrow \mu^+\mu^- Z', Z' \rightarrow \mu^+\mu^-$. No significant signal is observed for Z' masses in the range of 0.212 – 10 GeV/c^2. We set limits on the coupling parameters g' down to 7×10^{-4}. We set a strongest bounds for many parameter space below 3 GeV. We exclude most of the remaining parameter space preferred by the discrepancy between the calculated and measured anomalous magnetic moment of the muon above the dimuon threshold [33].

We have performed a search for long-lived particles L produced in e^+e^- collisions. No signal is observed, and upper limits on $\sigma(e + e^- \rightarrow LX)\mathcal{B}(L \rightarrow f)\epsilon(f)$ and on $\mathcal{B}(B \rightarrow XL)\mathcal{B}(L \rightarrow f)$ are set at 90% confidence level for six two-body final states f. We provide detailed efficiency tables to enable application of our results to any specific model [39].

References

[1] E. Witten, Phys. Lett. B **105**, 267 (1981).
[2] R. Essig *et al.*, arXiv:1311.0029 [hep-ph].
[3] B. Aubert *et al.* (BABAR Collaboration), Nucl. Instrum. Meth. A **479**, 1 (2002); B. Aubert *et al.* (BABAR Collaboration), Nucl. Instrum. Meth. A **729**, 615 (2013).
[4] R. Dermisek, J. F. Gunion, and B. McElrath, Phys. Rev. D **76**, 051105 (2007).
[5] M. Maniatis, Int. J. Mod. Phys. A **25**, 3505 (2010).
[6] F. Wilczek, Phys. Rev. Lett. **39**, 1304 (1977).
[7] R. Dermisek and J. F. Gunion, Phys. Rev. D **81**, 075003 (2010).
[8] B. Aubert *et al.* (BABAR Collaboration), Phys. Rev. Lett. **103**, 081803 (2009).
[9] J. P. Lees *et al.* (BABAR Collaboration), Phys. Rev. D **87**, 031102 (2013).
[10] J. P. Lees *et al.* (BABAR Collaboration), Phys. Rev. D **88**, 031701 (2013).
[11] J. P. Lees *et al.* (BABAR Collaboration), Phys. Rev. D **88**, 071102 (2013).
[12] J. P. Lees *et al.* (BABAR Collaboration), Nucl. Instr. Methods Phys. Res. Sect. **A** 726, 203 (2013).
[13] K. A. Olive *et al.* (Particle Data Group), Chin. Phys. C **38**, 090001 (2014).
[14] D. J. Lange, Nucl. Instr. Methods Phys. Res. Sect. A **462**, 152 (2001).
[15] T. Sjöstrand, Comp. Phys. Comm. **82**, 74 (1994).
[16] S. Agostinelli *et al.*, (GEANT4 Collaboration), Nucl. Instr. Methods Phys. Res. Sect. A **506**, 250 (2003).
[17] A. Höcker *et al.*, PoS ACAT, **040**, (2007), physics/0703039.
[18] G. Punzi, preprint physics/0308063 (2003).
[19] M. J. Oreglia, Ph.D. Thesis, report SLAC-R-**236** (1980), Appendix D.
[20] P. del Amo Sanchez *et al.* (BABAR Collaboration), Phys. Rev. Lett. **107**, 021804 (2011).
[21] J. P. Lees *et al.* (BABAR Collaboration) Phys. Rev. **D** 91, 071102 (2015).
[22] B. Holdom, Phys. Lett. B **166**, 196 (1986).
[23] A. Anastasi *et al.* [KLOE-2 Collaboration], Phys. Lett. B 757, 356 (2016).
[24] M. Pospelov, Phys. Rev. D **80**, 095002 (2009).
[25] V. Barger, C. W. Chiang, W. Y. Keung and D. Marfatia, Phys. Rev. Lett. **106**, 153001 (2011).
[26] D. Tucker-Smith and I. Yavin, Phys. Rev. D **83**, 101702 (2011).
[27] X. G. He, G. C. Joshi, H. Lew and R. R. Volkas, Phys. Rev. D **43**, 22 (1991).
[28] J. Alwall *et al.*, JHEP **1407**, 079 (2014).
[29] T. Sjostrand, S. Mrenna, and P. Z. Skands, JHEP **0605**, 026 (2006).
[30] F. A. Berends, P. H. Daverveld and R. Kleiss, Nucl. Phys. B **253**, 441 (1985).
[31] S. Jadach, W. Placzek, and B. F. L. Ward, Phys. Lett. B **390**, 298 (1997).

[32] S. Jadach, B. F. L. Ward, and Z. Was, Phys. Rev. D **63**, 113009 (2001).

[33] J. P. Lees *et al.* (*BABAR* Collaboration), Phys. Rev. D **94**, 011102 (2016).

[34] A. Kamada and H. B. Yu, Phys. Rev. D **92**, 113004, (2015).

[35] M. Aguilar *et al.* [AMS Collaboration], Phys. Rev. Lett. **110**, 141102 (2013).

[36] O. Adriani *et al.* [PAMELA Collaboration], Nature **458**, 607 (2009).

[37] M. Ackermann *et al.* [Fermi LAT Collaboration], Phys. Rev. Lett. **108**, 011103 (2012).

[38] B. Batell, D. McKeen and M. Pospelov, Phys. Rev. Lett. **107**, 011803 (2011).

[39] J. P. Lees *et al.* (*BABAR* Collaboration), Phys. Rev. Lett. **114**, 171801 (2015).

EPJ Web of Conferences **142**, 01023 (2017) DOI: 10.1051/epjconf/201714201023
Advances in Dark Matter and Particle Physics

Measuring gravitational effects on antimatter in space

Giovanni Maria Piacentino[1,2,3,a], Antonio Gioiosa[1,2], Anthony Palladino[4], and Graziano Venanzoni[5]

[1] *University of Molise*
[2] *INFN Lecce*
[3] *INAF Milano*
[4] *Boston University*
[5] *INFN Pisa*

Abstract. A direct measurement of the gravitational acceleration of antimatter has never been performed to date. Recently, such an experiment has been proposed, using antihydrogen with an atom interferometer and an antihydrogen confinament has been realized at CERN. In alternative we propose an experimental test of the gravitational interaction with antimatter by measuring the branching fraction of the CP violating decay of K_L in space. In fact, even if the theoretical Standard Model explains the CPV with the presence of pure phase in the KMC Kobaiashi-Maskava-Cabibbo matrix, ample room is left for contributions by other interactions and forces to generate CPV in the mixing of the neutral K and B mesons. Gravitation is a good candidate and we show that at the altitude of the International Space Station, gravitational effects may change the level of CP violation such that a 5 *sigma* discrimination may be obtained by collecting the K_L produced by the cosmic proton flux within a few years.

1 Introduction

Even if the Standard Model has proved capable of explaining all phenomena with which its predictions have been compared, many unanswered questions, or at least with an unsatisfactory answer, still remain in the official description of the Physical Universe.

First of all we know that in the observed Universe the ordinary matter prevails on antimatter even if both are always created together.

Cosmic Microwave Background is not anisotropic nor inhomogeneous enough to be compatible with the Big Bang model without the introduction of a still unknown interaction driving the inflation.

Furthermore, given the gravity we expect a negative acceleration of the expansion of the Universe. On the contrary it seems to accelerate. The gravitational field of Galaxies, clusters and even of the Solar system seems much stronger of the one due to the visible matter.

The suggested possible solutions are not modules of a general theory nor are capable of explaining the phenomena without an empirical tuning. For example the Sakarov suggested mechanism for matter/antimatter asymmetry, connected to CP violation (CPV) is contradicted by the fact that experimentally such phenomenon is far too weak to generate the present asymmetry. Models have been proposed to justify inflation by supersymmetric vacuum energy and SuperSymmetry Breaking but at present no evidence for supersymmetry has been found yet. Dark Energy has been introduced by hand in order to give a motivation to the accelerated expansion of the Universe. Finally Dark Matter has been introduced in order to give a motivation to the observed discrepancies between theory and measurements of the orbital speed of the stars of the external part of the Galaxies.

Too many different and independent motivations:

From the point of view of the elegance the situation is far from being satisfactory: There are as many hypothesis as problems this sounds extremely artificial especially because most of them are just put by hand into the theory.

Dark Matter and Dark Energy hypotheses, as an example, are similar in nature to that of Luminiferous aether for the transmission of the electromagnetic signals. All of them would not survive to the "Ockham's razor" criterion. This "lex parsimoniae" is due to the English Franciscan Scholar William of Ockahm (1287-1347), who inspired the character of William of Baskerville in the Umberto Eco's novel "The name of the Rose". It can be presented as follow:

"If there are several competing Hypotheses in order to explain a phenomenon or create a theory, The one that needs the fewest assumptions and parameters should be selected".

In this paper we show how all these problems can all be mastered with a single Hypothesis and will suggest an experiment capable to settle or at least put a limit on the applicability of the same Hypothesis. In order to understand the motivation of this Hypothesis, let's start from considering matter-antimatter symmetry:

[a]e-mail: nanni.piacentino@gmail.com, giovanni.piacentino@unimol.it, nannip@le.infn.it

At a very short distance, matter is always produced with the corresponding antimatter but only matter seems to dominate the landscape of the Universe, no stable antimatter seems to populate our Galaxy nor the Universe in general at large distance. Strong and Weak Interactions are limited in range. At the scale of 10^6 m even the Electromagnetic interaction is mostly screened and the only residual interaction is the gravity. At this scale no significant presence of antimatter can be find. Is there any connection between absence of antimatter and presence of the dominion of gravitation as interaction? Antimatter particles correspond to negative energy solution. Could this correspond to a negative gravitational mass and to a consequent gravitational repulsion between matter and antimatter? We simply suppose that gravitational interaction between matter and antimatter is repulsive and suggest that this could explain matter-antimatter asymmetry and also the nature of Dark Energy.

Gravitational repulsion between matter and antimatter could have powered the Inflaction because an equal mix of matter and antimatter would give rice to a net repulsive force. Dark Matter also could originate from quantum gravitational vacuum [1–15].

Most of people could be worried about the compatibility of this hypothesis with General Relativity (GR) but GR is not explicitly concerned about antimatter. Furthermore GR and Quantum Mechanics are incompatible so that we already need to revise our theoretical system. Many new ideas are suggested as MOND theories [12], possible new forces i.e. Gravivector and Graviscalers, Kerr-Newman Universe, where, as suggested by G. Chardin, repulsion could be connected with a negative proper time of antiparticles [1–4].

2 The repulsive Hypothesis of the interaction between matter and antimatter

We consider a simple interaction of the kind:

$$F = -G\frac{m_1 m_2}{r^2}\hat{r} \tag{1}$$

$$m_{matter} > 0\,;\; m_{antimatter} < 0$$

$F < 0$, attractive for masses of the same sign

$F > 0$, repulsive for masses of opposite sign

2.1 Gravitational dipoles

In this hypothesis the quantum vacuum around a large gravitational mass, populated as it is by virtual couples of particle antiparticle, can develop a gravitational polarization. If we consider that the distance between particle and antiparticle in a dipole should be of the order of the Compton wavelength, we obtain that the dipolar charge of each pair is independent of the mass of the particle:

$$\overrightarrow{P_g} = m\overrightarrow{d}$$

$$|\overrightarrow{d}| \approx \chi = \frac{\hbar}{mc} \tag{2}$$

$$|\overrightarrow{P}| \approx m\chi = \frac{\hbar}{c}$$

2.2 Energy of the dipoles

If the vacuum is permeated by an external gravitational field of intensity $\overrightarrow{g_0}$ the interaction energy is given by:

$$\eta = -\overrightarrow{p}\cdot\overrightarrow{g_0}$$

$$\eta = -\frac{\hbar}{c}g_0 \tag{3}$$

or in the case of a field of a spherical mass M_0

$$\eta = -\frac{\hbar}{c}\frac{GM_0^2}{r^2}$$

The polarization is the result of the competing action of the external Field and of the thermal agitation and as usual is described by a Langevin-like equation:

$$P = NP\left[coth\theta - \frac{1}{\theta}\right] \tag{4}$$

Where N is the number of dipoles/volume and

$$\theta = \frac{pg_0}{KT} = \frac{\hbar G M_0}{cr^2}\frac{1}{KT} \tag{5}$$

Of course in the Langevin equation the reference energy KT must be replaced by the one related to the phenomenon. That is:

$$KT \approx \frac{\hbar}{c}g_{dip} = \frac{\hbar}{c}\frac{G\rho_{dip}}{r} \tag{6}$$

where $g_{dip} = \dfrac{G\rho_{dip}}{r}$ and $\rho_{dip} = \dfrac{dM_{dip}}{dr} = constant$
So that the polarization can be written as:

$$\overrightarrow{P} = -N\frac{\hbar}{c}\left[\coth\left(\frac{M_b}{\rho_{dip}r}\right) - \frac{\rho_{dip}r}{M_b}\right]\hat{r} \tag{7}$$

where $\hat{r}$ is the radial versor.

In order to estimate N, we can consider that at low energy the dipoles can be considered a pure pion gas so that the density can be written:

$$N = \frac{1}{\lambda_\pi^3} = \left[\frac{m_\pi c}{2\pi\hbar}\right]^3 \tag{8}$$

So for the mass density due to polarization we have:

$$\rho_P = \nabla\cdot\overrightarrow{P} = \frac{1}{3}N\frac{\hbar}{c}\frac{M_B}{\rho_{dip}r^2} \tag{9}$$

So the mass due to polarization is:

$$M_p(r) = \frac{4\pi}{3}N\frac{\hbar}{c}\frac{M_B}{\rho_{dip}}(r - R_B) \tag{10}$$

where R_B is the radius of the bare mass M_B that produces the polarization, ρ_{dip} is the density of dipoles at the saturation. This is the same behavior of the Dark Matter around a

 DOI: 10.1051/epjconf/201714201023

large mass. As it is well known the accelerated expansion of the universe can be accomplished introducing a positive cosmological constant:

$$R_{\mu\nu} - \frac{1}{2}g_{\mu\nu}R = \frac{8\pi G}{c^4}T_{\mu\nu} + \Lambda g_{\mu\nu} \quad (11)$$

Now in our model energy and mass density of gravitational dipoles of the vacuum can accomplish for the positive cosmological constant:

$$\Lambda = \frac{8\pi G}{c^4}\rho_{Edipol} \quad (12)$$

3 Antimatter gravitational experiments

The problem of the gravitational interaction of antimatter is not new. Whitteborn and Fairbanks attempted to measure gravitational force on positrons. Los Alamos-led team proposed (1986) to measure gravitational force on antiprotons ($\overline{H}$) at the CERN Low Energy Antiproton Ring (LEAR) [16]. Both the Projects ended inconclusively but the problem is interesting and several efforts are in progress at CERN in order to measure the gravitational interaction of $\overline{H}$. The experiments AEgIS [17], ALPHA [18], ATRAP [19], ASACUSA, GBAR [20] are all in advanced state but the way to have a final answer is still a long one.

4 CPV and Gravity

We consider a different approach to the problem, looking at phenomena connected to the gravitational interaction of antimatter but, even if indirect, more sensitive. In the 1958, eight years before the discovery of CPV, Philip Morrison [21] published on the American Journal of Physics a paper showing that a strong difference in the gravitational interaction for matter and antimatter could generate a CPV in neutral Kaons system. At present even if the phenomenon is well described by the CKM formalism, space is left to other contribution to CPV that could contribute to the matter-antimatter asymmetry in the Universe as suggested by Sakharov.

4.1 The Sakharov conditions

In 1967, a Russian physicist Andrei Sakharov outlined that CPV could explain the matter antimatter asymmetry of the Universe. We would expect any reaction to produce antibaryons at the same rate as baryons For overall baryon production, antibaryon-producing reactions must be suppressed. This implies different treatment of matter and antimatter, which is exactly what the mechanism of CP violation describes.

Let us restrict to CPV in the $K_S - K_L$ system considering the possible contribution to CPV in this system of neutral mesons. The gravitational field is described by the acceleration g so the components of antimatter and matter of a meson are divided by a distance growing with the time that can be written as:

$$\Delta\zeta = gt^2 \quad (13)$$

The time useful for the phenomenon is a fraction $\Omega^{-1/2}$ of the mixing time $\Delta\tau$ where:

$$\Delta\tau = \frac{\pi\hbar}{\Delta mc^2} \approx 5.9 \times 10^{-10}s \approx 6\tau_s \quad (14)$$

Where τ_s is the life time of K_S
The dimension of a K meson is about 0.5 fm or:

$$\Delta L_k = \frac{\hbar}{m_k c} \quad (15)$$

The ratio $\frac{\Delta\zeta}{\Delta L_k}$ is the adimensional constant that characterizes the phenomenon. So we obtain the CPV parameter as:

$$\epsilon = \Omega\frac{g\frac{\pi^2\hbar^2}{\Delta m^2 c^4}}{\frac{\hbar}{m_k c}} = \Omega\frac{\pi^2\hbar^2 g m_k}{\Delta m^2 c^3} = \Omega \times 0.88 \times 10^{-3} \quad (16)$$

This means that gravity could be responsible for most of the CPV in the neutral K seen on the Earth.

4.2 CPV and Gravity

In 1992 Gabriel Chardin [2] showed that gravity on Earth has the right intensity to generate CPV in the mixing of the neutral K and B mesons. He also demonstrated that the phenomenon of antigravity for antimatter could be compatible with the General Relativity and that it could be the motivation of an instability of quantum vacuum in the presence of strong gravitational fields, mimic of the Hawking radiation.

In order to understand if gravitational contribution is present in the CPV of the neutral Kaons we propose an experiment in orbit. In fact, on a circular LEO at 500 Kilometers, gravity is about 10% less than on Earth. On a GEO orbit the intensity of the of the Earth's gravitational field is of the order of few percent. This is likely to cause a large fluctuation of any gravitational contribution to the CPV in the 2-state system of the neutral K mesons. The production of the kaons should not be a problem because, in orbit, a large flux of energetic protons is present and is only mildly modulated by the Earth's magnetic field.

On a square target of 70 cm of side, about $1.4x10^4$ protons per second will impact, the energy of the cosmic protons ranges from a few MeV to 200 GeV with the maximum flux around 1 GeV and several smaller local maxima at 5, 13, and 31 GeV. This spectrum can produce the neutral Kaons. The total number of K mesons decays over the predicted mission lifetime (> 2 years) will yield the required physical measurement.

5 A dedicated Satellite

Our payload is aimed at performing a particle physics experiment in orbit and is composed of the following parts:

Active Target: we simulated [22, 23] the production of the K_L ("long" decay neutral mesons) and K_S ("short")

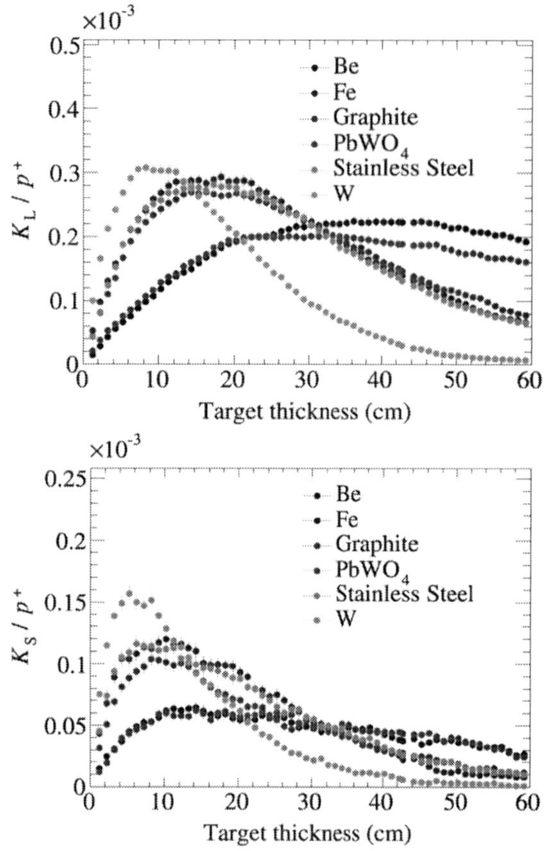

Figure 1. K_L production (Color on line). Number of K_L (top) and of K_S (bottom) which exit the downstream face of the target versus target thickness for several materials.

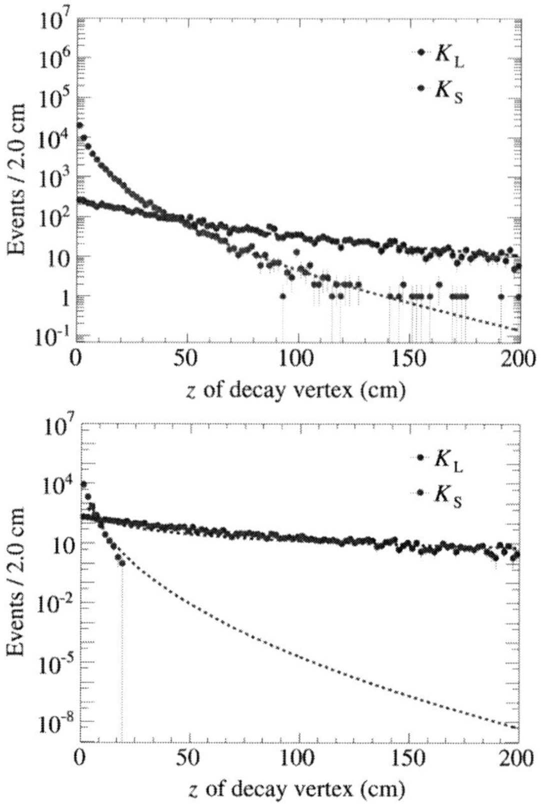

Figure 2. (Color on line) Axial position of the decay vertex (top) and the same with the additional constraint $pz < 0.5$ GeV/c (bottom).

mesons by the cosmic protons on an active target made of alternate layers of tungsten and scintillating fibers, read by photodiods, for a total thickness of few cm of W (see Figs. 1 and 2.

"Empty" structure: (necessary for the decay of the K_S) from our simulations we expect that 50 cm vacuum, in which the K_S may decay, are needed to separate the K_L from the K_S.

Spectrometer Magnet: will be made of neodymium/iron/boron permanent magnets, magnetized at 1.3 T. The spectrometer diameter will be 50 *cm*. The required 1.3 kGauss field needs a 25 mm × 500 mm cylindrical magnet.

Tracker: will consist of five silicon layers similar to those used in other satellites.

Calorimeter: the 4-radiation-length calorimeter uses a lead scintillating fibers array, and permits the absorption of about 90% of the electromagnetic shower.

The veto detector: consists of a layer of scintillator surrounding the entire detector.

6 Conclusions

We have proposed a possible test of the gravitational [24] behavior of antimatter by measuring the rate of the CP violating decay in space. We estimate that a 5σ measurement on a possible change in the CP violation parameter ε could be obtained within few years, depending on the detection efficiency, if one places a detector with thick tungsten target, a 1 m diameter by 1 m deep tracking region, a magnetic field for charged-particle identification, time-of-flight counters, and electromagnetic calorimeters for energy measurements, on a Leo or better on Geostationary orbit. Any difference between the amount of CP violation in orbit with respect to the level CP violation on the Earth's surface would be an indication of the nature of the gravitational interaction between matter and antimatter. A positive result may offer an explanation for the cosmic baryon asymmetry as well as a contribution to the observed effects thought to come from dark matter and dark energy.

ACKNOWLEDGMENTS

I WOULD LIKE TO THANK:
Gabriel Chardin CEA Saclay France

Erasmo Recami University of Milan Italy
Giovanni Fabrizio Bignami INAF
Thomas j. Phillips- Duke University NC USA
Anthony Palladino Boston University
Luca Latronico INAF Torino Italy
Patrizia Caraveo INAF Milano Italy
Graziano Venanzoni INFN LNF Italy
Dragan Slavkov Hajdukovic CERN
Dan Kaplan Illinois Institute of Technology

References

[1] G. Chardin, Nuclear Phys. A **558**, 477c (1993)

[2] G. Chardin, J.M. Rax, Phys. Lett. B **282**, 256-262 (1992)

[3] A. Benoit-Lévy, G. Chardin, Astron. Astrophys. **537** A78 (2012)

[4] A. Benoit-Lévy, G. Chardin, International Journal of Modern Physics: Conference Series. Antimatter and Gravity Conference (WAG 2013) **30**, 1460272 (2014)

[5] D.S. Hajdukovic, Astrophys. Space Sci. **334**, 215-218 (2011)

[6] D.S. Hajdukovic, Astrophys. Space Sci. **334**, 219-223 (2011)

[7] D.S. Hajdukovic, Space Sci. **337**, 9-14 (2012)

[8] D.S. Hajdukovic, Astrophys. Space Sci. **339**, 1-5 (2012)

[9] D.S. Hajdukovic, Phys. Dark Universe **3**, 34-40 (2014)

[10] D.S. Hajdukovic, Proceedings of the 3rd International Workshop on Antimatter and Gravity, WAG (2015) https://hal.archives-ouvertes.fr/hal-01254678v2

[11] M. Villata, Astrophys. Space Sci. **345**, 1-9 (2013)

[12] L. Blanchet, Classical Quantum Gravity **24**, 3529 (2007)

[13] L. Blanchet, A. Le Tiec, Phys. Rev. D **78**, Article 024031 (2008)

[14] L. Blanchet, A. Le Tiec, Phys. Rev. D **80**, 023524 (2009)

[15] L. Bernard, L. Blanchet, Phys. Rev. D **91**, Article 103536 (2015)

[16] CPLEAR Collaboration, Phys. Lett. B **452** 425-433 (1999)

[17] A. Kellerbauer (AEgIS Collaboration) et al., Nucl. Instrum. Methods Phys. Res. B **266** 351 (2008). http://dx.doi.org/10.1016/j.nimb.2007.12.010

[18] A.E. Charman (ALPHA Collaboration) et al., Nature Comm. **4**, 1785 (2013)

[19] G. Gabrielse (ATRAP Collaboration) et al., Phys. Rev. Lett. **108**, Article 113002 (2012)

[20] G. Chardin, P. Grandemange, D. Lunney, et al., Tech. Rep. CERN-SPSC-2011- 029 SPSC-P- 342

[21] P. Morrison, Journal of Physics **26**, 358 (1958).

[22] S. Agostinelli et al., Nucl. Instrum. Methods. Phys. Res. A **506** 250-303 (2003)

[23] Geant4 developments and applications, IEEE Trans. Nucl. Sci. **53**, (2006)

[24] G.M. Piacentino, A. Palladino, G. Venanzoni, Physics of the Dark Universe **13**, 162-165 (2016)

Recent result from CMD-3 detector

A. S. Popov[1,2,a] for collaboration CMD-3

[1] *Budker Institute of Nuclear Physics, SB RAS, Novosibirsk, 630090, Russia*
[2] *Novosibirsk State University, Novosibirsk, 630090, Russia*
[3] *Novosibirsk State Technical University, Novosibirsk, 630092, Russia*
[4] *University of Sydney, School of Physics, Falkiner High Energy Physics, NSW 2006, Sydney, Australia*
[5] *University of Tokyo, Department of Physics, 7-3-1 Hongo Bunkyo-ku Tokyo, 113-0033, Japan*

Abstract. Since December 2010, the CMD-3 detector has collected data at the electron-positron collider VEPP-2000. The sample of the accumulated data corresponds to about 60 pb^{-1} of integrated luminosity in the c.m. energy from 0.32 up to 2 GeV. Preliminary results of the analysis of various processes e^+e^- annihilation to hadrons are presented. It is shown the processes with multihadron events have several intermediate states which must be taken into account to correctly describe the angular and invariant mass distributions as well as cross section dependence versus energy.

1 Introduction

The electron-positron collider VEPP-2000 [1] operates at the Budker Institute of Nuclear Physics since December 2010. The collider is designed to provide luminosity up to $10^{32}\text{cm}^{-2}\text{s}^{-1}$ at the maximum center-of-mass energy $\sqrt{s}$ = 2 GeV. The current integrated luminosity accumulated by the CMD-3 is about 60 pb^{-1}. One of the main physical task is to measure the hadronic cross sections in the wide energy range, and searches for the new vector and scalar mesons and as well as for the exotic hadrons.

The precision data of the hadronic cross sections are required for various applications, in particular, to evaluate the anomalous magnetic moment (AMM) of muon, $a_\mu = (g-2)_\mu/2$. The VEPP-2000 energy range gives the major hadronic contribution to AMM ($a_\mu^{had} \sim 92\%$) both to the hadronic vacuum polarization itself and to its uncertainty [2].

The precision measurement of luminosity is a key ingredient of many experiments which study the hadronic cross sections at e^+e^- colliders. So far it is very important to have several QED processes such as $e^+e^- \to e^+e^-$, $\mu^+\mu^-$, $\gamma\gamma$ to have the cross check as it was done by CLEO [3].

The general purpose detector CMD-3 has been described in detail elsewhere [4]. Its tracking system consists of a cylindrical drift chamber (DC) and double-layer multiwire proportional Z-chamber, both also used for a trigger, and both inside a thin (0.2 X_0) superconducting solenoid with a field of 1.3 T. The liquid xenon (LXe) barrel calorimeter with 5.4 X_0 thickness has fine electrode structure, providing 1-2 mm spatial resolution, and shares the cryostat vacuum volume with the superconducting solenoid. The barrel CsI crystal calorimeter [5] with thickness of 8.1 X_0 is placed outside the LXe calorimeter, and the end-cap BGO calorimeter with a thickness of 13.4 X_0 is placed inside the solenoid. The luminosity is measured using events of Bhabha scattering and $\gamma\gamma$ at large angles with about 1% accuracy.

The beam energy has been monitored by measuring the current in the dipole magnets of the main ring, and for a few energy points by using the Back-Scattering-Laser-Light system [6]. Using measured average momentum from Bhabha events, and average momentum of proton-antiproton pairs from the $e^+e^- \to p\bar{p}$ process, we determine E_cm for each energy point with about 1 MeV accuracy.

To study a detector responce and to obtain a registration efficiency, we have developed a Monte Carlo (MC) simulation of our detector based on GEANT4 package, and all simulated events pass all our reconstruction and selection procedures. The MC simulation includes photon radiation by initial electron or positron, calculated according to Ref. [7].

2 Luminosity measurement

The energy range from 1 to 2 GeV was scanned upward and downward with a step of 50 MeV. At each energy point the integrated luminosity of about 500 nb^{-1} was accumulated. In the case of scanning downward, the energy points have been shifted by 25 MeV with respect to the previous case. The experiments were performed from January to June 2011 and, similarly, in 2012.

Two types the «CHARGED» and the «NEUTRAL» first level triggers were used while data have been taking.

[a] e-mail: al.s.popov@inp.nsk.su

DOI: 10.1051/epjconf/201714201024

A combinations of the signals from DC cells and ZC sectors, which roughly reproduce "track", start a special processor «TRACKFINDER» (TF). The combinations of signals from calorimeters with different energy thresholds actuated the «CLUSTERFINDER» (CF) processor. A positive decision of either processor allowed the recording of current event onto a hard disk with capacity about 2 TB. In the course of data accumulation, the mean trigger frequency ranged in $\sim 200 \div 400$ Hz. The sample of collinear Bhabha events e^+e^- were selected for luminosity determination.

The process $e^+e^- \to \gamma\gamma$ was also used, since it has essential advantages [8] with respect to the first one. It is free of radiation of the final state particles and its Coulomb interaction, the corresponding Feynman graphs do not contain photon propagators affected by the vacuum polarization effects. These advantages are the main motivation to exploit this process as an independent tool for luminosity.

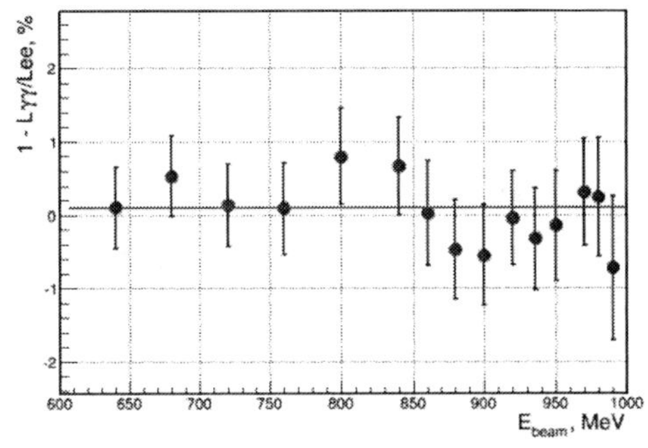

Figure 1. The ratio of the relative difference of the luminosities vs beam energy (scan 2011). Full circles - scan up, triangles - scan down. Line - fit: 0.2 ± 0.3%

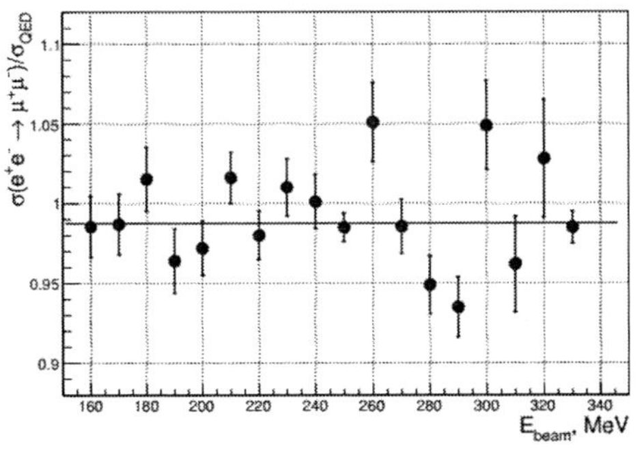

Figure 2. Result of the measurement of muon pair production in comparison with the QED prediction. Horizontal line - fit: 0.995 ± 0.005

The collected integrated luminosity above the ϕ mass is about 34.5 pb^{-1}, 8.3 and 8.4 pb^{-1} at the ω and ϕ resonances, respectively, and 9.4pb^{-1} from a scan below the ϕ. The peak luminosity $\sim 2 \cdot 10^{31}$ cm^{-2}s^{-1} is limited while by a positrons injection rate and it will gain by a factor of ten after upgrade of the injection facility.

The relative difference of the luminosities determined with two processes versus energy is presented in Fig. 1, where only statistical errors are shown (SCAN 2011). The horizontal line is a fit for this ratio and in average it is about 0.2±0.3%. The main sources which contribute to systematic error are: contribution due to the different angular resolutions for Bhabha events and $\gamma\gamma$ is estimated as ~0.8%; correction, which takes into account inclination of the beam axis with respect to detector is about ~0.4%; DC z-coordinate calibration contributes about 0.3%; radiative corrections are calculated according to [9] with the accuracy about 0.2%. Presently we estimate the current luminosity systematic accuracy as ~1% for energies higher than 1 GeV.

3 Processes with multihadrons in final states

One of the main goal of the CMD-3 experiment is to reduce a systematic uncertainty of the cross section of two pion production to 0.3-0.4%. The $\pi^+\pi^-$ events are separated either using the particles momentum or their energy deposition in EM-calorimeter. Two ways of event separation will provide cross-check and is expected allow to keep the systematic error under control. The first energy scan below 1 GeV was performed in 2013. The collected statistics is a few times higher than we had in the previous CMD-2 measurements and it is at the level of ISR statistics accumulated by the BaBar and KLOE. The process $e^+e^- \to \mu^+\mu^-$ is very importance since it provides overall systematic test of the event separation accuracy. Preliminary results for the cross section $\sigma(e^+e^- \to \mu^+\mu^-)$ are shown in Fig. 2 with respect to the QED prediction. Horizontal line is a fit for the double ratio $\sigma^{exp}_{\mu\mu}/\sigma^{QED}_{\mu\mu}/\sigma^{exp}_{ee}/\sigma^{QED}_{ee}$ which was found to be 0.995±0.005. At the moment this result demonstrates our potential power of the event separation procedure. Study of different systematic uncertaintices is ongoing.

3.1 Study of the processes $e^+e^- \to K_S K_L$ and $e^+e^- \to K^+K^-$.

It is known the CMD-2 and BaBar results for cross sections at the ϕ-peak region disagree at the level ~4% for charged channel, so a new measurement are required. The $e^+e^- \to K^0_S K^0_L$ and $e^+e^- \to K^+K^-$ cross sections were measured in the c.m. energy range 1.004-1.060 GeV at 25 energy points. The neutral mode detection is based on the search of two central tracks with common vertex in DC from the $K^0_S \to \pi^+\pi^-$ decay. The number of events is defined by the fit of two pions invariant mass distribution [10].

The registration of the charged mode is based on the search of two central collinear tracks of kaons with defined momentum in DC. Each track should has ionization losses significantly larger than mip due to relatively small velocity of kaons under study. After these requirements the level of remaining background is less than 0.5%. The detection efficiency of each kaon was determined with data and as well as with MC and deliver a deviation less 1.5%.

The obtained cross sections for the neutral (published) and charged mode (preliminary) are presented in Fig. 3 and Fig. 4, correspondingly. The measured cross section is approximated according to VDM model as a sum of ϕ, ω, ρ-like amplitudes and their excitations. The interference of non-resonant amplitudes with the amplitude of ϕ meson scales in ten times and shown too at the bottom of graphs. The neutral and charged channels were approximated simultaneously, as a result the following values of the ϕ meson parameters have been obtained: $m_\phi = 1019.464 \pm 0.060$ MeV/c^2, $\Gamma_\phi = 4.240 \pm 0.017$ MeV, $\frac{B_{\phi \to K^+K^-}}{B_{\phi \to K^0_S K^0_L}} = 1.573 \pm 0.06$. The obtained parameters have accuracy comparable or better than it was obtained in previous experiments.

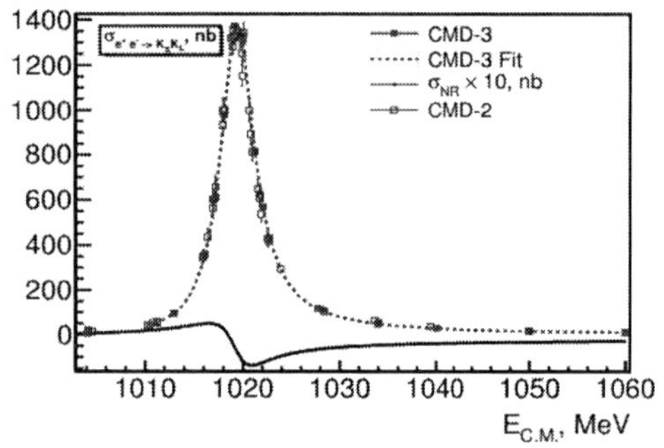

Figure 3. The cross section of the process $e^+e^- \to K_L K_S$ around ϕ-meson energy region. CMD-2, CMD-3 and BaBar data are presented.

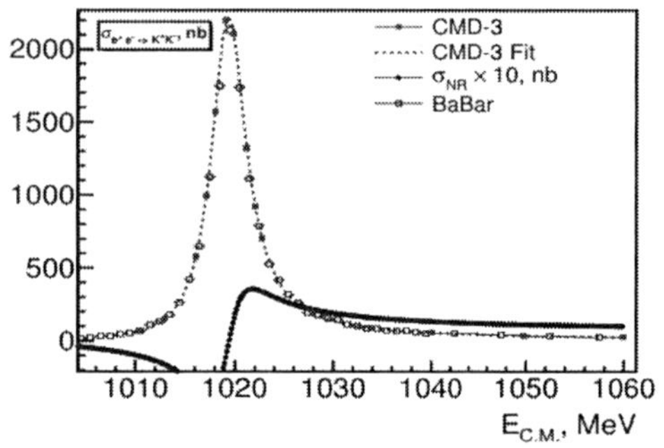

Figure 4. The cross section of the process $e^+e^- \to K^+K^-$ around ϕ-meson energy region. CMD-2, CMD-3 and BaBar data are presented.

3.2 Study of the process $e^+e^- \to K^+K^-\pi^+\pi^-$

The cross section measurement of the process $e^+e^- \to K^+K^-\pi^+\pi^-$ is based on the integrated luminosity of 22 pb^{-1} in the c.m. energy range from 1.5 to 2.0 GeV and early was measured by the BaBar via ISR. Nevertheless the direct measurements are very important, since some contributes to a_μ are based on isospin relations of various $K\bar{K} + n\pi$ final states. Any uncertainty of this approach will be crucial for the a_μ accuracy in the future calculations.

The signal events should have three or four tracks in the DC coming from interaction region and obey to the energy-momentum conservation. Two tracks corresponding to kaons should have specific ionization losses dE/dx in DC. Fig. 5 shows the difference between measured total

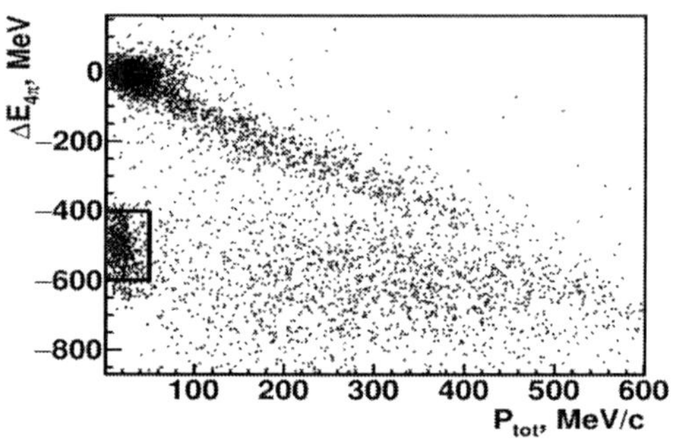

Figure 5. The difference between the total energy and c.m. energy (ΔE_4) versus the total momentum for the four-track events. The upper cluster of dots represents $\pi^+\pi^-\pi^+\pi^-$ while the lower one - $K^+K^-\pi^+\pi^-$ events.

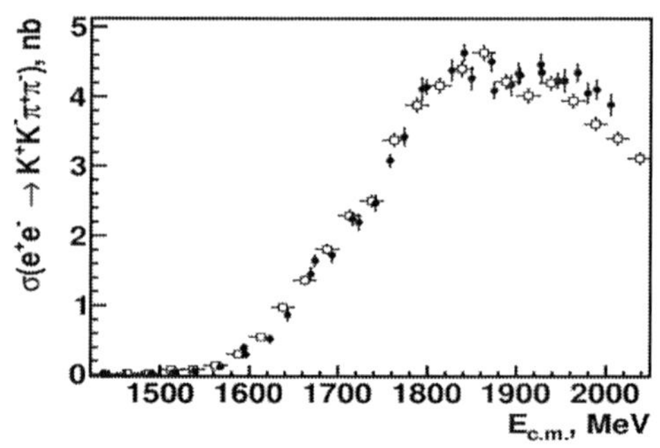

Figure 6. Dots - the $e^+e^- \to K^+K^-\pi^+\pi^-$ cross section measured with the CMD-3. The BaBar results are shown by open circles.

energy and c.m. energy $\Delta E_4 = E_{tot} - E_{c.m.}$ vs the total momentum for all events with four tracks. The signal events are located near in origin of coordinates. The cluster of events with a zero total momentum but shifted up along the vertical axis, corresponds to $\pi^+\pi^-\pi^+\pi^-$ events.

The similar procedure was used to select signal events with the three-track in DC. As a result, ~13300 four-track events and ~16000 three-track events were selected. To calculate a detection efficiency, the $K^+K^-\pi^+\pi^-$ events were simulated with a primary generator using the GEANT4 package and then reconstructed with the same software as experimental data.

The cross section as a function of energy shown in Fig. 6, and well agrees with the previous BaBar measurement [11] presented by open circles. Systematic error was studied in detail and currently is estimated as 6%

and mainly due to model dependence of the detection efficiency. More detail analyses can be found in publication [12].

3.3 Six pion production

Production of the six charged pions in e^+e^- annihilation was studied at DM2 [13] and BaBar [14]. The DM2 experiment observed a "dip" in the cross section of the process $e^+e^- \to 3(\pi^+\pi^-)$ near 1.9 GeV, confirmed later by the BaBar. The origin of the "dip" remains unclear, but the most popular explanation is based on a hypothesis of a presence of the under-threshold $(p\bar{p})$ resonance discussed in many theoretical papers [15].

The analysis is based on 22 pb^{-1} of the integrated luminosity collected in the c.m. energy range from 1.5 to 2 GeV. Candidates for the process under study are required to have five or six tracks in DC. For six- or five-track candidates the total energy and total momentum are calculated, assuming all tracks to be pions. To estimate the background MC simulation of the major processes $2(\pi^+\pi^-\pi^0)$ and $2(\pi^+\pi^-)\pi^0$ was performed and was found to be smaller than 1%. The analysis in detail for this channel can be found in [16].

To measure the cross section of the process $e^+e^- \to 2(\pi^+\pi^-\pi^0)$ the sample of events with the four charged and two neutral pions were selected. To select neutral pions the spectrum of invariant mass of all two photon combinations was studied inside energy gap from $60 < m_{\gamma\gamma} < 200$ MeV/c^2 and combination with the nearest to the pion mass is chosen. The number of events under study at each energy point was determined by fit of the distributions shown in Fig. 7 with a sum of three Gaussian functions for signal events and quadratic polynomial for background. The cross section calculates according to the number of determine events and takes into account RC and detection efficiency. The preliminary results for the cross section are presented in Fig. 8. The analysis of the data is in progress.

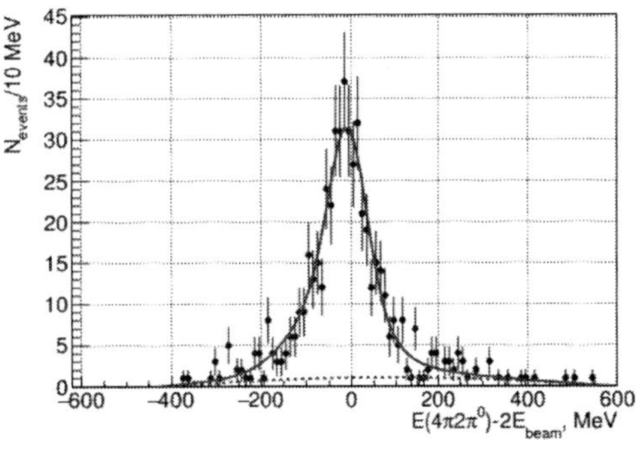

Figure 7. The number of events vs the total energy of the system $2(\pi^+\pi^-\pi^0)$. Signsl- fit with a sum of three Gaussian functions, background - fit with quadratic polynomial.

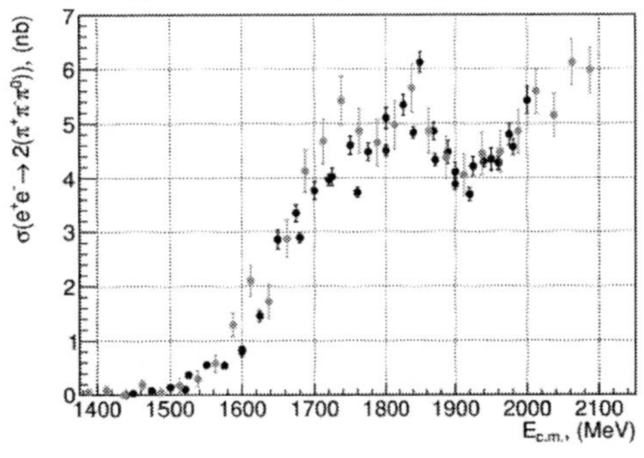

Figure 8. . Cross section of the process $e^+e^- \to 2(\pi^+\pi^-\pi^0)$ vs energy. Black points - CMD3, green - BaBar.

3.4 Cross sections measurement of the $e^+e^- \to \eta\pi^+\pi^-$ and $e^+e^- \to \omega\pi^+\pi^-$ processes

The process $e^+e^- \to \eta\pi^+\pi^-$ was studied in two decay modes of η: 2γ and $\pi^+\pi^-\pi^0$. The signal events should have two tracks and at least two photons in EM-calorimeter. The shape of two photons invariant mass distribution was fix from MC and used to determine the number of the $\eta\pi^+\pi^-$ events at each energy point. The result of such fit at 1500 MeV is shown in Fig. 9. The preliminary results for the Born cross section are presented in Fig. 10. The systematic uncertainty for this process is about 5.2% and mainly due to uncertainty of detection efficiency, which depends on angular distribution of the final particles, i.e. depends on the intermediate states (theoretical models) through this process goes.

These two processes were studied too when η and ω decay to the three pions: $\pi^+\pi^-\pi^0$. To determined the signal events the shape of the $\pi^+\pi^-\pi^0$ invariant mass distribution has been studied using Monte Carlo simulation and was used to count the number of the signal events. The preliminary results for the Born cross sections are consistent with other experiments. The current systematic uncertainty for these channels we estimated as 15% and mainly driven by the difference between angular distributions of charged particles in simulation and experiment. The analysis of systematic errors is ongoing.

3.5 Study of the processes $e^+e^- \to K^+K^-\eta$ and $e^+e^- \to K^+K^-\omega$

The analysis of the process $e^+e^- \to K^+K^-\eta$ was based on 19 pb^{-1} of an integrated luminosity collected by the CMD-3 detector in 2011–2012 in the $E_{c.m.}$ range from 1.59 to 2.01 GeV. On the base of these statistics we observed the contribution of $\phi(1020)\eta$ intermediate state only. Candidates for the events of the signal process were required to have two, three or four tracks in the DC, coming out of the beams intersection point. The kaon/pion separation was performed with the use of $f_{K/\pi}(p, dE/dx)$ functions [12], representing the probability density for charged kaon/pion with the momentum p to produce the energy losses dE/dx in the DC. We considered η-meson as a recoil particle, which allowed us to avoid the loss of statistics due to the selection of the specific η decay mode. But such an inclusive approach lead to the complication of the signal/background separation. Therefore

DOI: 10.1051/epjconf/201714201024

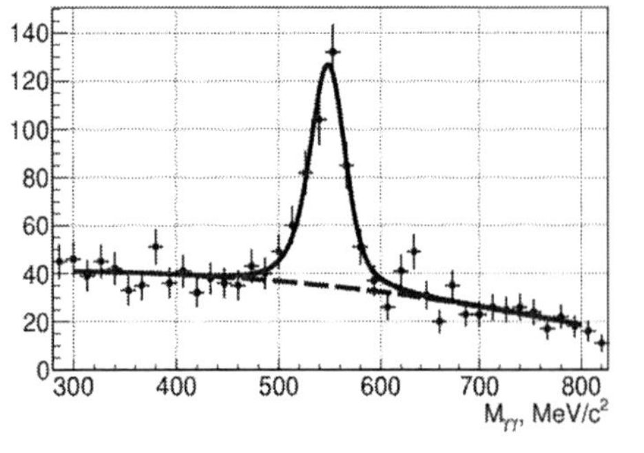

Figure 9. Two photons invariant mass at the energy of 1500 MeV.

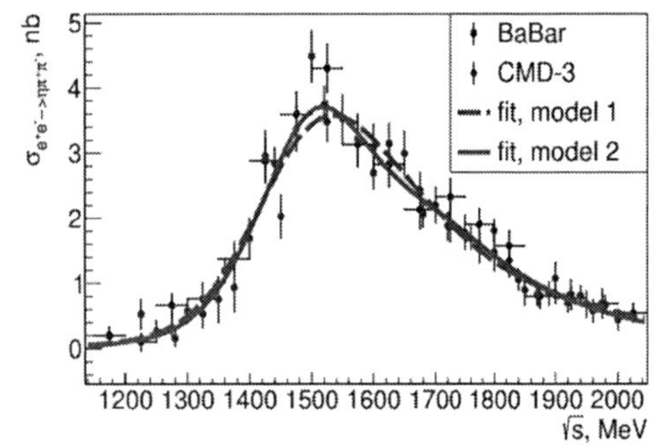

Figure 10. The $e^+e^- \to \eta\pi^+\pi^-$ Born cross section measured in the $\eta \to \gamma\gamma$ channel. The results are presented together with BaBar data.

the major background processes were studied and were found to be $e^+e^-\to K^+K^-\omega(782)$, $e^+e^-\to K^+K^-\pi^+\pi^-$, $e^+e^-\to\phi f_0(500)$, $K^{*\pm}(892)K^\mp\pi^0\to K^+K^-\pi^0\pi^0$, $e^+e^-\to 2\pi^+2\pi^-2\pi^0$. We perform the signal/background separation and 1296±43 signal events were extracted. The resulting $e^+e^-\to\phi(1020)\eta$ cross section is shown in the Fig. 11 along with the BaBar results. The overall systematic uncertainty of the cross section measurement was estimated to be 6%. Via the cross section approximation the $\phi(1680)$-meson parameters have been determined.

The analysis of the process $e^+e^-\to K^+K^-\omega(782)$ was based on 12 pb^{-1} of the integrated luminosity collected by the CMD-3 detector in 2011–2012 in the $E_{\rm c.m.}$ range from 1.8 to 2.01 GeV. Candidates for the events of the signal process were required to have three or four tracks in the DC, flying out of the area of the beams intersection. The kaon/pion separation was performed in the same way, as in the $e^+e^-\to K^+K^-\eta$ process analysis. We studied the process in the $\omega(782)\to\pi^+\pi^-\pi^0$ decay mode, considering π^0 as a recoil particle. The major background processes were found to be $e^+e^-\to K^+K^-\eta$ and $e^+e^-\to K^+K^-\pi^+\pi^-$, but their contribution was almost completely suppressed by the cuts on the invariant and missing masses of kaons and pions. After background suppression we performed a direct subtraction of the estimated number of background events and in total in the experiment we obtained 886±30

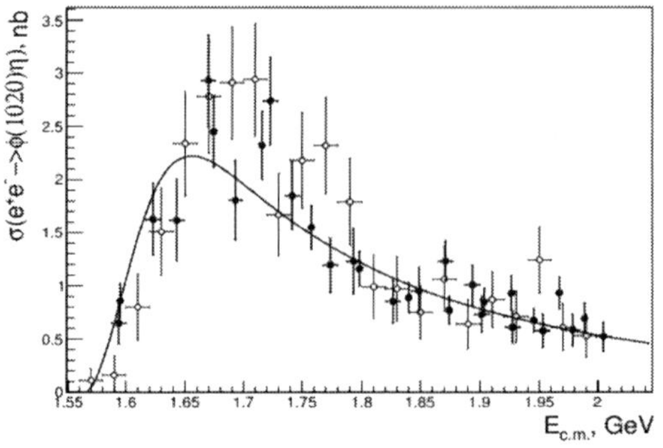

Figure 11. The BaBar (open circles) and CMD-3 (filled circles) results for the measurement of $e^+e^-\to\phi(1020)\eta$ cross section in the range from 1.59 to 2.01 GeV. The approximation of CMD-3 data together with the BaBar data in the range from 2.3 to 3.46 GeV is also shown.

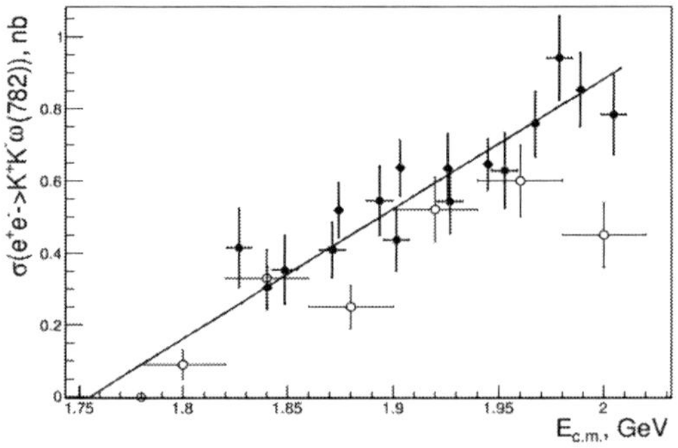

Figure 12. The BaBar (open circles) and CMD-3 (filled circles) results for the measurement of $e^+e^-\to K^+K^-\omega(782)$ cross section in the range from 1.8 to 2.01 GeV. The approximation of CMD-3 data is also shown.

signal events. The resulting $e^+e^-\to K^+K^-\omega(782)$ cross section is shown in the Fig. 12 along with the BaBar results. The overall systematic uncertainty of the cross section measurement was estimated to be 6%.

4 Summary and conclusion

CMD-3 detector will operate with a goal to get $\sim 1 fb^{-1}$ in 5-10 years and provides the new precise results on the hadron production. The current integrated luminosity was measured using two well known QED processes $e^+e^- \to e^+e^-$, $\gamma\gamma$ and systematic accuracy is estimated as 1%. Two type of the first level triggers «CHARGED» and «NEUTRAL» deliver the independent information that allowed to determine the detection efficiencies and to estimate their uncertainties. Data analysis is in progress, the already collected data sample provides the same or better statistical precision for the hadronic cross sections than in previous experiments were achieved.

Acknowledgements

The authors are grateful to A.I. Milstein for help with a theoretical interpretation and development of the models.

We thank the VEPP-2000 team for excellent machine operation.

This work is supported in part by the Russian Education and Science Ministry, by FEDERAL TARGET PROGRAM Scientific and scientific-pedagogical personnel of innovative Russia in 2009-2013, by agreement 14.B37.21.07777, by the Russian Fund for the Basic Research grants.

Part of this work related to the photon reconstruction algorithm in the electromagnetic calorimeter is supported by the Russian Science Foundation (project No. 14-50-00080).

RFBR 10-02-00695-a, RFBR 10-02-00253-a, RFBR 11-02-00328-a, RFBR 11-02-00112-a, RFBR 12-02-31501-mol-a, RFBR 12-02-31499-mol-a, RFBR 12-02-31498-mol-a, RFBR 12-02-01032-a, RFBR 13-02-00215-a.

References

[1] I. Koop *et al.*, Nucl.Phys.B, Proc.Suppl. **181**, 371 (2008)

[2] M. Davier *et al.*, Eur. Phys. J. C **31**, 503 (2003)
K. Hagiwara *et al.*, J. Phys. G **38**, 085003 (2011)

[3] G. Grawford *et al.*, Nucl. Instr. Meth. A **345**, 429 (1994)

[4] B. I. Khazin *et al.*, Nucl.Phys.B, Proc. Suppl. **181-182**, 376 (2008)

[5] V. M. Aulchenko *et al*, 2015 JINST 10 P10006 (2015)

[6] E. V. Abakumova *et al.*, Phys. Rev. Lett. **110**, 140402 (2013)

[7] E. A. Kuraev and V. S. Fadin, Yad. Fiz., vol.3 **41**, 733 (1984)

[8] S. I. Eidelman *et al.*, Eur. Phys. J. C **71**, 1597 (2011)

[9] A. B. Arbuzov *et al.* Eur. Phys. J. C **46**, 689 (2006)
S. Actis *et al.*, Eur. Phys. J. C **66**, 585 (2010)

[10] E. A. Kozyrev *et al.*, Phys. Lett. B **760**, 314 (2016)

[11] B. Aubert et al. (BaBar Collaboration), Phys. Rev. D **76**, 012008 (2007)
J. P . Lees *et al.* (BaBar Collaboration), Phys. Rev. D **86**, 012008 (2012)

[12] D. N. Shemyakin *et al.*, Phys.Lett. B **756**, 153-160 (2016)

[13] R. Baldini *et al.*, reported at the "Fenice" Workshop, Frascati (1988);
A. B. Clegg and A. Donnachie, Z. Phys. C **45**, 677 (1990);
M. R. Whalley, J. Phys. G **29**, A1 (2003)

[14] B. Aubert *et al.*, (BaBar Collaboration), Phys. Rev. D **73**, 052003 (2006)

[15] A. Sibirtsev and J. Haidenbauer, Phys. Rev. D **71**, 054010 (2005)
A. Antonelli et al. (FENICE Collaboration), Phys. Lett. B **365**, 427 (1989)

[16] R. R. Akhmetshin *et al.*, Phys.Lett. B **723**,82 (2013)

Status of the experiment for the search of a dark photon at VEPP–3

I. Rachek[1,a], D. Nikolenko[1], and B. Wojtsekhowski[2]

[1] *Budker Institute of Nuclear Physics, 630090 Novosibirsk, Russia*
[2] *Jefferson Laboratory, Newport News, VA, USA*

Abstract. The status of the experiment to search for a new gauge boson, A′, in e^+e^- annihilation by means of a positron beam incident on a gas hydrogen target internal to the bypass at the VEPP-3 storage ring is reported. The search method is based on a missing mass spectra in reaction $e^+e^- \to \gamma A'$. It allows observation of the A′ signal independently of its decay modes and life time. The projected result of this experiment corresponds to an upper limit on the square of coupling constant $\varepsilon^2 = (0.2 \div 1.0) \cdot 10^{-7}$ with a signal-to-noise ratio of two to one at A′ mass of 5-20 MeV.

1 Introduction

The search for an experimental signature of physics beyond the Standard Model is a major effort of modern particle physics. The existence of Dark Matter (DM) is supported by many astrophysical data. At first, weakly-interacting massive particles (WIMPs) were considered as main candidates for DM. However, up to now no evidence for WIMPs at LHC or in direct detection experiments has been reported. That has initiated a hunt in a wider area and sparked a renewed interest in searching for a new gauge boson, the so-called dark photon or A′-boson, which mediates interactions between DM particles, as well as between DM particles and Standard Model particles through the kinetic mixing mechanism [1]. In recent years, considerable efforts have been devoted to the re-analysis of existing data in attempts to observe a manifestation of a new boson. An extended review of the present status of such searches can be found in the recent report [2]. So far, no evidence of new particles has been found. In parallel, a series of new dedicated experiments has been proposed.

Here we present the status of the experiment to search for a dark photon, proposed to be conducted at the VEPP-3 electron/positron storage ring at the Budker Institute, Novosibirsk.

2 The concept

The concept of the method is partly described in [3]. A positron beam in a storage ring with an energy E_+ of a few hundred MeV and an internal hydrogen gas target make up an "e^+e^-–collider". In such a collider it is possible to search for the production of a light A′-boson with a mass of up to $m_{A'}[\text{MeV}] \sim \sqrt{E_+[\text{MeV}]}$. Unlike many other experiments with a fixed target, which are based on the detection of e^-e^+ or $\mu^+\mu^-$ pairs from A′-boson decay, in the described experiment no special assumptions about decay modes of the A′-boson are required.

Specifically for the VEPP-3 storage ring an experimental approach looks as follows:

- A positron beam in VEPP-3 repeatedly crosses an internal gas target. The target is a cooled windowless storage cell filled with hydrogen gas.
- VEPP-3 will work with the new injection complex, commissioned recently at the Budker Institute. The new injector is capable of providing 2×10^9 positrons per second. Besides linear accelerators, it contains a damping ring which can be used as a positron booster storage ring for VEPP-3. The switching between old and new injection facilities at VEPP-3 is ongoing.
- The beam energy will be equal to the injector energy, which is up to 500 MeV, so no energy ramping in VEPP-3 is envisaged. This ensures the highest luminosity.
- A six-bunch regime of VEPP-3 operation will be implemented. Bunches are equally spaced in the VEPP-3 orbit, and every 10 seconds the oldest bunch is replaced by a new one, which was accumulated in the injector dumping ring. Thus, a luminosity of up to $3 \times 10^{33}\text{cm}^{-2}\text{s}^{-1}$ can be achieved.
- A special magnet will be installed downstream from the target to turn away the positron beam and thus to allow a free flight for gamma-quanta from target to photon detector.
- A segmented EM-calorimeter will be used to detect photons.
- To control the systematics a run with the *electron* beam in VEPP-3 can be performed, thus making sure that a putative boson is seen only with a positron beam.

[a] e-mail: rachek@inp.nsk.su

DOI: 10.1051/epjconf/201714201025

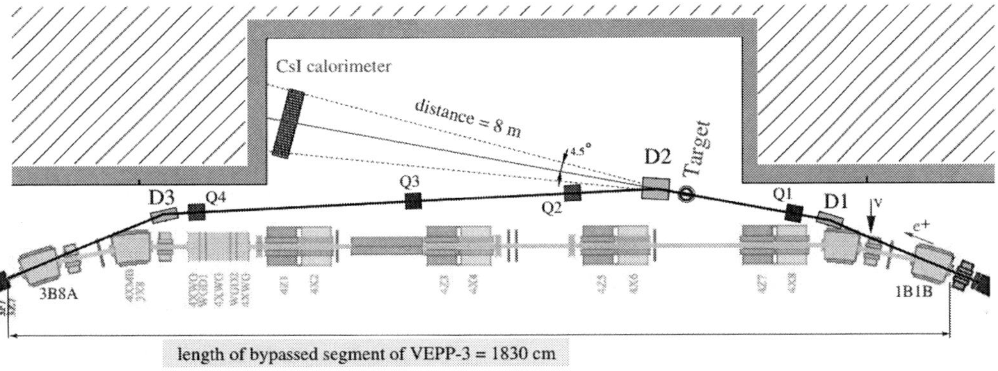

Figure 1. The layout of the proposed experiment with a bypass built along the straight section of the VEPP-3 ring. *D*1 – *D*3 - dipole magnets, *Q*1 – *Q*4 - quadrupole lenses.

3 The proposed experimental setup

VEPP-3 is a buster–ring, operating as an intermediate accelerator/storage ring of electrons and positrons for the VEPP-4 collider. There are time slots when VEPP-3 is available for other experimental programs, including the electro/photo-nuclear experiments, utilizing a super-thin internal target technique.

Usually the internal target is located in the first 12-meter-long straight section of the VEPP-3 ring. In the same straight section there are also two RF cavities, four quadrupole and one sextupole lens and elements of beam injection and extraction. The space available for the internal target equipment is 217 cm long. In our earlier proposal [4] we have described a possible configuration of the experiment based on a chicane magnet installed in this segment of VEPP-3. However, such a configuration has significant difficulties, both technical and organizational. The latter follows from the fact that when a chicane magnet is installed in the ring, VEPP-3 becomes completely inaccessible to other working regimes, including operation as a booster for VEPP-4, or as a synchrotron radiation source.

Therefore, for the proposed experiment it was decided to implement a special setup. The main element of this setup is a **bypass**, which has to be built along the second straight section of the VEPP-3 ring.

3.1 Bypass

The proposed configuration of the bypass is shown in Figure 1. It consists of *i)* a vacuum chamber with a total length of about 18 m; *ii)* vacuum pumps; *iii)* three dipole magnets with a combined bending angle of 45°; *iv)* a number of quadrupole lenses; *v)* elements of beam diagnostics; *vi)* an internal target section containing a thin-walled storage cell and set of powerful vacuum pumps.

The bypass will occupy the place which several decades ago was already used for another bypass, containing the Free Electron Laser. FEL has been dismounted, but some elements providing its operation at VEPP-3 (switching magnets, inlet/outlet vacuum channels) are still there. This should simplify significantly the commissioning of the new bypass.

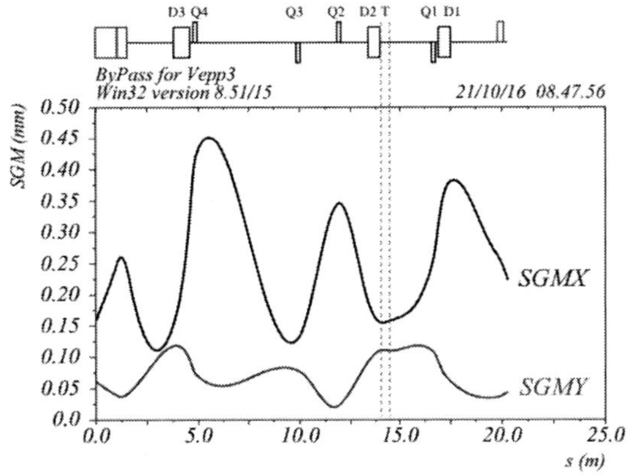

Figure 2. Calculated beam size (X and Y) along the bypass for the proposed electron optics. Vertical lines show the location of the internal target.

The R&D of the bypass electron optics has been recently completed, see Fig. 2, and detailed specifications of the magnetic elements of the bypass have been obtained.

3.2 Internal target

A thin-walled open-ended storage cell cooled to 25°K and filled with hydrogen gas will be used as an internal target. The magnetic structure of the bypass is designed in such a way as to provide small values of beta-functions in the location of the storage cell, Fig. 2, allowing the use of a small–opening cell. Together with cell cooling this permits us to obtain the required target thickness of about 10^{16} at/cm^2 with a smaller amount of hydrogen gas injected into the target. The gas leaks out of the cell ends into the ring vacuum chamber and must be pumped out promptly. A set of powerful turbomolecular and cryogenic pumps will be installed in the target chamber, as well as upstream and downstream from the target chamber.

DOI: 10.1051/epjconf/201714201025

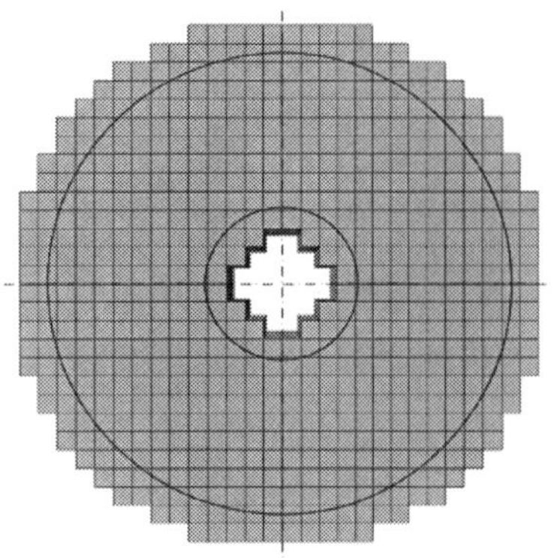

Figure 3. Calorimeter configuration: 624 CsI(Tl) crystals from the CLEO-II endcap calorimeter assembled in a ring shape.

3.3 Photon detector

The photon detector can be placed at a distance of up to 8 m from the target. The requirements for the detector are:

- Energy resolution on the level of $\sigma_E/E = 5\%$ for photons with energy $E_\gamma = 100 - 450$ MeV.
- Angular resolution on a level of 0.1°.
- Angular acceptance as defined by the requirement to detect both photons from two-photon annihilation:
  - in θ : symmetrical range in θ_γ^{CM} around 90°, e.g. $\theta_\gamma^{CM} = 60° - 120°$, which corresponds to $\theta_\gamma^{LAB} = 1.5° - 4.5°$.
  - in ϕ : total 2π coverage.
- The detector should be able to sustain a moderate total photon rate at a level of 1 MHz.

A calorimeter with sufficient energy resolution could be constructed using crystals from the discarded CLEO-II detector [5] from Cornell University. It is especially effective because, the Cornell group is considering a similar experiment with a 5 GeV positron beam [6]. The collaboration between the Budker Institute and Cornell University for development of such a detector is already organized [7].

The end-cap electromagnetic calorimeter of the CLEO-II detector consists of 1600 CsI(Tl) crystals of $5 \times 5 \times 30$ cm^3 size ($16.2X_0$). It was used to measure electron and photon energy in a wide range, and a direct measurement of its performance at a photon energy of interest for the proposed experiment, $E_\gamma = 180$ MeV, is available: energy resolution $\delta E/E = 3.8\%$; spacial resolution $\delta x = 12$ mm, which results in angular resolution of $\delta\theta = 1.5$ mrad for a distance of 8 m.

In this case it would take 624 crystals to cover the required angular range, see Fig. 3. A few notes on this detector option should be mentioned:

- The CLEO-II calorimeter assembly is clearly inappropriate for the proposed experiment, so a new mechanical support must be designed and constructed. This was recently completed.
- CsI(Tl) crystal has a long light emitting time. Therefore, its ability to work at a high background rate is limited. However, due to the high segmentation of the calorimeter, a long output pulse seems not to be a problem. Even for crystals covering the lowest polar angle, the expected rate of background photons is estimated to be at a level of a few tens kHz for the projected luminosity of 10^{33}cm^{-2}s^{-1}.

The experiment will require a careful account of the detector responses. The energy response will be calibrated by using $\gamma\gamma$ coincidence events produced with the hydrogen target. These data will also provide a detector line shape determination. The use of the electron beam instead of the positron beam provides the way to obtain the "white" photon spectra without the A′-boson signal and the two-photon line.

3.4 Positron veto counter

The main single–photon QED background comes from positron bremsstrahlung on hydrogen and from the radiative Bhabha process. Therefore, a rejection of such background events with an efficiency ϵ would result in an increase of the search sensitivity by a factor of $1/\sqrt{1-\epsilon}$. Since in these processes the positron loses energy and is swept out by the D2 dipole magnet, such background events can be vetoed by detecting the scattered positron. For this purpose, a compact sandwich will be installed downstream from the D2 magnet. The fraction of bremsstrahlung events detected by this counter depends on its geometry. The total rate of positrons emitting a photon with energy above a $E_\gamma = 50$ MeV threshold can be estimated to be 60 MHz for the luminosity of $L = 10^{33}$ cm^{-2}s^{-1}. Hence, the veto counter would be useless at this luminosity if it detects all positrons including those which emit photons outside the photon detector acceptance. Using the MC simulation we have chosen a configuration with a positron rate of $f_{e^+} \sim 2.5$ MHz, which gives only 10% event loss due to accidental coincidences – see Fig. 4. The simulation has shown that the veto efficiency is a function of missing mass range and it is generally higher for the bremsstrahlung from a hydrogen target process than for the radiative Bhabha one, with overall mean value of about 70%.

However, in the case of a visible decay of A′ to an electron/positron pair the use of such a veto counter results in a loss of signal events because positrons from the A' decay often hit the veto counter. We have considered a simple extension – a layer of scintillation strips placed in front of the veto sandwich, Fig. 5. This allows us to reconstruct a positron energy and, together with photon energy, reconstructed in the calorimeter, allows us to distinguish between the A′decay process and the bremsstrahlung on the proton. However, the veto efficiency for radiative Bhabha events is rather low in this case. Nevertheless, such a relatively inexpensive configuration of the veto counter does improve the search sensitivity and we are going to implement it.

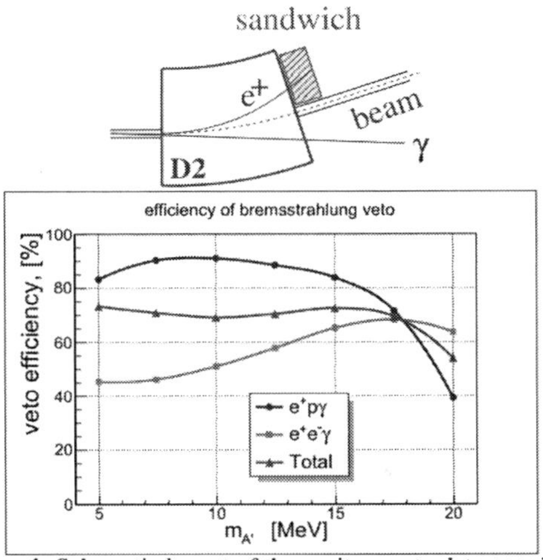

Figure 4. Schematic layout of the positron veto detector placed behind the D2 dipole magnet, and its efficiency as a function of missing mass.

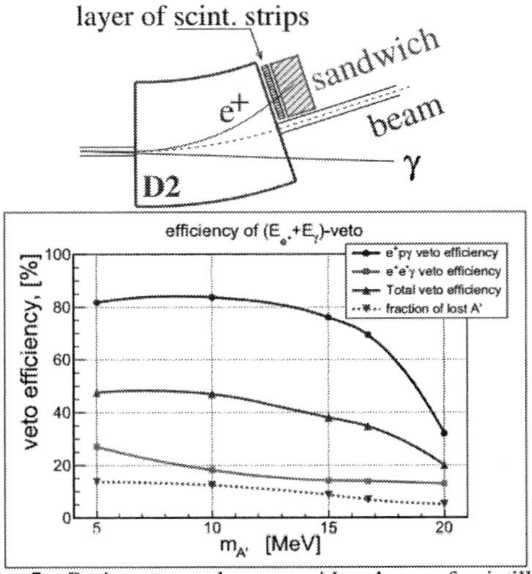

Figure 5. Positron veto detector with a layer of scintillation strips, capable to reconstruct positron energy, and its efficiency as a function of missing mass.

3.5 Background from QED annihilation process

Conventional QED e^+e^- annihilation processes have two or more gamma quanta in the final state, while in the signal process there is strictly one photon. This can be used to reject such background.

For this purpose the acceptance of the photon calorimeter is chosen to be symmetrical with respect to $\theta_\gamma^{cm} = 90°$ in the center of mass frame of an electron/positron pair. This means that for the 2-photon annihilation, the calorimeter will always detect either both photons or none. Therefore, the 2-photon annihilation events will be rejected very effectively. Three-photons events will also be largely rejected.

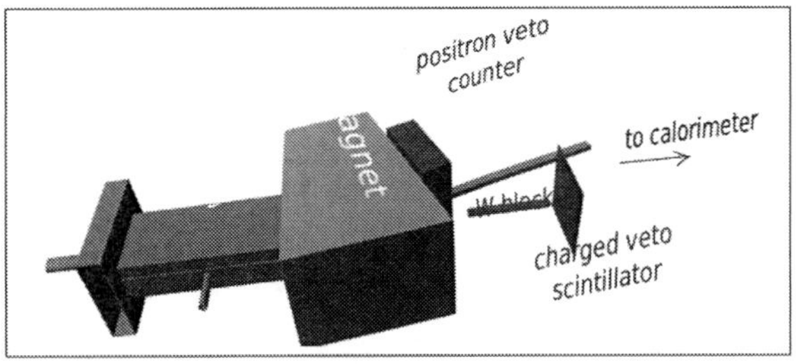

Figure 6. Charged veto scintillator behind the tungsten "blocker", covering the calorimeter acceptance.

3.6 Secondary background suppression

Secondary sources of background events should also be considered. Such events are produced when secondary particles (photons, positrons, electrons) hit the materials inside and outside the target area. Special efforts should be devoted to install shielding wherever it is applicable. Also, since most secondary showers, born outside the target area, contain electrons and positrons, an additional suppression of such background can be achieved by installation of charge veto counters.

In the proposed configuration we investigate an installation of a tungsten rod (a "photon blocker") to dump a high flux of photons, emitted from the target at small angles outside the acceptance of the photon detector, and a thin scintillator, covering the colorimeter acceptance, to veto charged particles. Both elements are installed close to the exit of the dipole magnet – see Fig. 6.

3.7 Data Acquisition

The expected rate of events with a threshold of minimum energy deposition in the calorimeter $E_{\rm cal} > 25$ MeV is about 450 kHz for the projected luminosity of the experiment. Assuming a conservative value for the on-line suppression factor of 3 for combined veto-channels and for events with two or more photons in the calorimeter, one obtains a 150 kHz final trigger rate.

It is clear that a conventional readout (based, e.g. on FASTBUS ADCs) does not meet the specifications. The front-end and digitizing electronics should be based on Flash ADC and FPGA logic to provide an on-line timing, cluster-finding and zero-suppression. Similar or substantially faster systems are now widely used or being designed for Data Acquisition of various experiments. However, at the moment the DAQ electronics for the described experiment is at an early stage of development.

4 Monte Carlo

In the proposed experiment we will search for a bump in a missing mass distribution on top of a QED-background. Some part of the background events can be identified and rejected; the remaining ones define the search sensitivity. For the study of the detector response to the background processes and for realistic

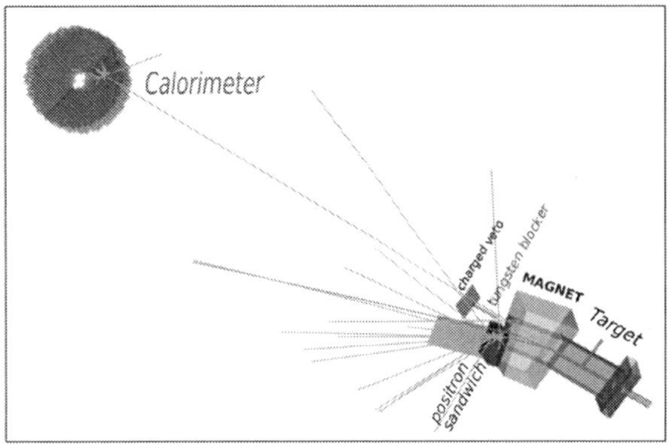

Figure 7. The final configuration accepted for the Monte Carlo simulation of the experiment.

estimation of the search sensitivity of the experiment, a detailed Monte Carlo simulation was performed using the GEANT4 toolkit. A set of dedicated event generators have been developed for the following processes:

1) positron bremsstrahlung $e^+ + H \to e^+ + X + \gamma$
2) radiative Bhabha scattering $e^+ + e^- \to e^+ + e^- + \gamma$
3) two-photon annihilation $e^+ + e^- \to \gamma + \gamma$
4) three-photon annihilation $e^+ + e^- \to \gamma + \gamma + \gamma$
5) signal process $e^+ + e^- \to \gamma + A'$

The final configuration accepted for Monte Carlo is shown in Fig. 7.

To implement a bump search procedure one needs to know the experimental resolution for a reconstructed missing mass. This was determined from Monte Carlo. The missing mass resolution depends on the energy and angular resolution of the photon detector. Measured resolutions of the CLEO calorimeter, see ref. [5], were reproduced in our Monte Carlo by adding an equivalent of "electronics" noise to simulated energy depositions in the crystals. Noise parameters were taken from [5]. The simulation results for 4 values of A′-boson masses: 5, 10, 15 and 20 MeV, when purely A′ production events were generated, are shown in Fig. 8. One can see that the missing mass resolution rapidly improves with increasing the masses of A′. The dependence can be fitted with an exponential function:

$$\sigma_{miss}\,[\text{MeV}] = 3.6 \cdot \exp\left(-0.13 \cdot M_A\,[\text{MeV}]\right), \quad (1)$$

which was used in the simulation of the search procedure. Note that such behavior means that a big tail from the two-photon annihilation ($M_A = 0$) is expected in a missing mass spectrum, and this indeed is observed in Monte Carlo. Therefore, for an efficient search at low M_A, a strong suppression of two-photon events is important.

Figure 9 shows the results of a full simulation for the same 4 possible masses of A′-boson. The kinetic-mixing parameter was set to $\epsilon = 0.1$. The shown distributions are obtained for invisible decay of A' boson after all cuts are applied. For a designed experimental luminosity of 10^{33}, each such spectrum can be collected in about 3 seconds.

5 Projected search sensitivity

The simulation data, presented in Fig. 9, allow us to carry out a realistic estimation of the search sensitivity. The search conditions are:

- Positron beam energy is 500 MeV.
- Luminosity is $10^{33}\text{cm}^{-2}\text{s}^{-1}$, which corresponds, for example, to a 30 mA beam current and a target thickness of 5×10^{15} atoms/cm^2.
- Run time is taken to be 10^7 seconds, which means a half-year run with 65% time utilization.
- The search is performed using a sliding missing mass window with a width of $\pm 1.5\sigma_{miss}$, where σ_{miss} is evaluated using Eq.(1).

As expected, the search sensitivity is a function of missing mass range, and it exhibits a moderate dependence on the decay mode of the A′-boson and on the regime of operation of the veto counter. The best search sensitivity is demonstrated for invisible decay of A′ and veto-ON regime; the worst sensitivity is for a veto-OFF regime, but the difference is only a factor ~2.

Graphical representations of the projected search sensitivity are shown in Figures 10, 11.

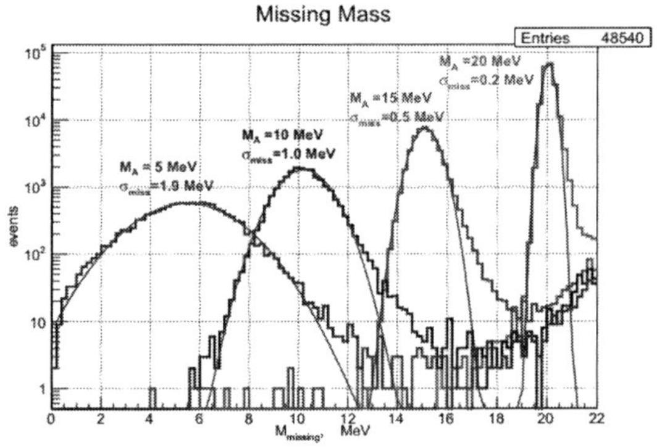

Figure 8. Missing mass reconstruction for 4 possible masses of A′-boson.

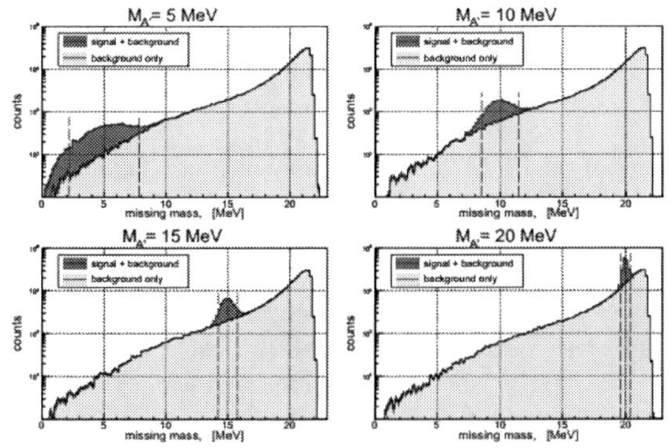

Figure 9. Full simulation of the experiment for 4 possible masses of A′-boson. Here the kinetic-mixing parameter ϵ was set to 0.1.

DOI: 10.1051/epjconf/201714201025

Table 1. Simulated search sensitivity.

$M_{A'}$	M_{mis} window	$\varepsilon^2 \times \mathbf{10^8}$ (95% CL)			
MeV	MeV	veto **ON** invisible	veto **ON** visible	$(E_{e^+} + E_\gamma)$-cut visible	veto **OFF** any
5	±2.8	**9.7**	**15**	**13**	**17**
10	±1.5	**8.7**	**15**	**12**	**15**
15	±0.8	**5.2**	**11**	**8.2**	**9.1**
20	±0.3	**2.4**	**5.9**	**3.3**	**3.5**

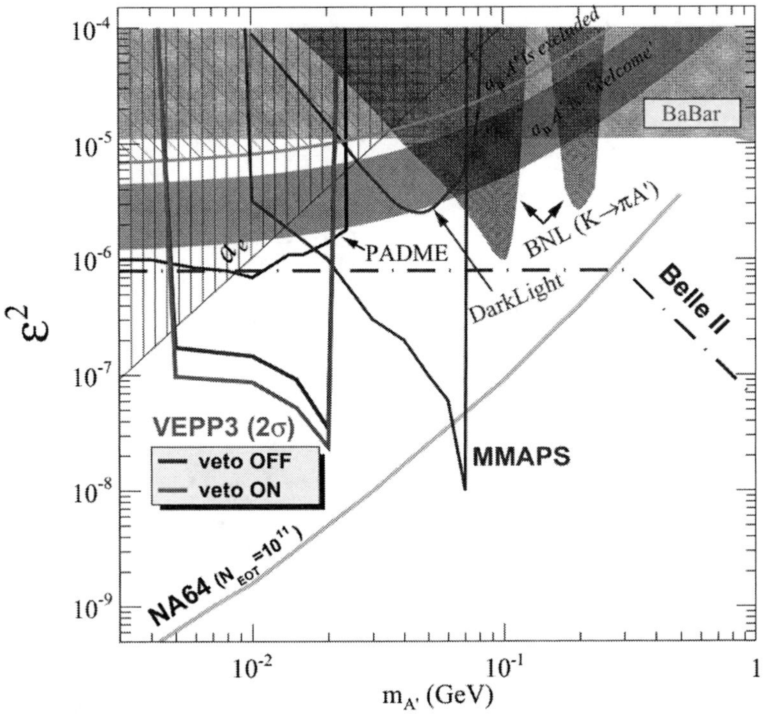

Figure 10. Projected parameter space and search sensitivity of the VEPP-3 experiment in the case of invisible decay of A′. Shaded areas show existing bounds [8–10]. Curves demonstrate the search regions of the proposed experiments [6, 11–14].

6 Conclusions

- A decay mode independent search for a dark photon is effective in a setup with an intense circulating positron beam and an internal hydrogen gas target.
- At VEPP-3 such an experiment could be performed at the ByPass, which is now under development.
- Crystals from the CLEO endcap EM-calorimeter would be a good choice for the photon detector, but for optimal performance they should be equipped with fast photo-sensors and a sophisticated DAQ .
- Monte Carlo simulation shows that with such a setup in a half-year run, one can conduct a sensitive, decay mode-independent search in an interesting unexplored area of parameter space, where the "Atomki boson" could show up.

References

[1] B. Holdom Phys. Lett. B **178**, 65 (1986).
[2] J. Alexander, *et al.*, arXiv:1608.08632 (2016).
[3] B. Wojtsekhowski, arXiv:0906.5265 (2009).
[4] B. Wojtsekhowski, D. Nikolenko and I. Rachek, arXiv:1207.5089 (2012).
[5] Y. Kubota, *et al.*, Nucl. Inst. and Meth. A **320**, 66 (1992).
[6] J. Alexander (2016), URL https://indico.cern.ch/event/507783/contributions/2150181/attachments/1266367/1874844/SLAC-MMAPS-alexander.pdf
[7] J. Alexander, private communication (2016).
[8] B. Aubert, *et al.*, (BaBar Collaboration), arXiv:0808.0017 (2008).
[9] S. Adler, *et al.*, (E787 Collaboration), Phys. Rev. D **70**, 037102 (2004).

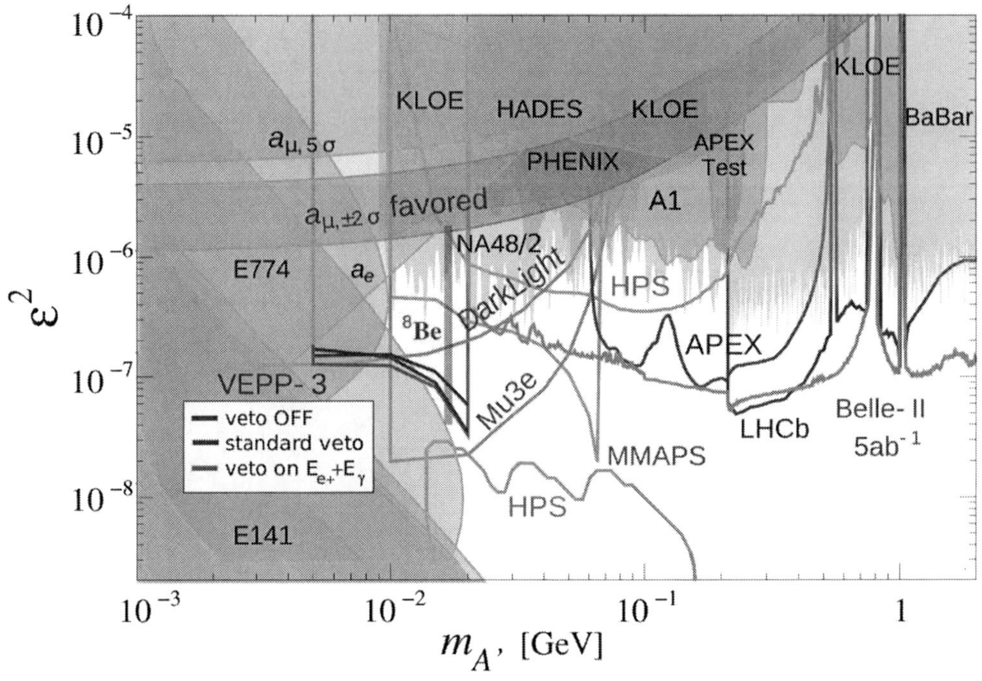

Figure 11. Projected parameter space and search sensitivity of the VEPP-3 experiment in the case of visible decay of A′ to (e^+e^-)-pair. Shaded regions show existing bounds; curves indicate the search areas of proposed experiments, see the review [2] and references therein. Note that the VEPP-3 search area covers almost the whole parameter space allowed for the putative "Atomki boson" [15], which is proposed to explain an anomaly in ^{8}Be* decay [16].

[10] A. V. Artamonov, *et al.*, (E949 Collaboration), Phys. Rev. D **79**, 092004 (2009).

[11] M. Raggi and V. Kozhuharov, Adv. High Energy Phys. **2014**, 959802 (2014).

[12] J. Balewski, *et al.*, arXiv:1412.4717 (2014).

[13] R. Essig, J. Mardon, M. Papucci, T. Volansky, and Y.-M. Zhong, JHEP **11**, 167 (2013).

[14] S. Andreas, *et al.*, arXiv:1312.3309 (2013).

[15] A. Krasznahorkay, *et al.*, Phys. Rev. Lett. **116**, 042501 (2016).

[16] J. L. Feng, *et al.*, Phys. Rev. Lett. **117**, 071803 (2016),

EPJ Web of Conferences **142**, 01026 (2017) DOI: 10.1051/epjconf/201714201026
Advances in Dark Matter and Particle Physics

The PADME experiment at LNF

M. Raggi[1], on behalf of the PADME Collaboration[a]

[1] *Istituto Nazionale di Fisica Nucleare, Sezione di Roma*
P.le A. Moro 2, I-00185 Roma (Italy).
E-mail: e-mail: mauro.raggi@roma1.infn.it

Abstract. The PADME experiment, approved by INFN at the end of 2015, aims to search for missing mass signals in the annihilation of positrons on a thin fixed target produced by invisible decays of the dark photon. The detector construction will be completed by the end of 2017 to be ready to run in spring of 2018. The collaboration aims at collecting about 10^{13} positron on target by the end of 2018 to reach a sensitivity down to 1×10^{-3} on the coupling of A' up to 23.7 MeV mass.

1 Introduction

The existence, the nature, and the interactions on the dark matter with Standard Model (SM) particles are among the most important open questions in modern particle physics. Despite an intense experimental activity, no direct evidence of dark matter interaction with ordinary matter through the known SM forces, except the gravity, has been reported. This experimental evidence suggests that a new force might be the solution to the dark matter puzzle. The idea of a new force and a new "dark sector" of particles, originally proposed in the '80[1], has been recently revitalised by the possibility of explaining the long standing discrepancy between theory predictions and experimental measurements of muon anomalous magnetic moment $(g-2)_\mu$[2]. The simplest "dark sector" model introduces a single new force with a gauge structure similar to the SM electromagnetic interaction, $U(1)_D$, and the corresponding mediator the so called "dark photon" (A'). The dark photon (DP) has the quantum numbers of both the SM electromagnetic interaction and the dark sector force. For this reason the DP can then mix with the ordinary photon and be produced in collisions of electrons, positrons, and protons on a target or in meson decays[4]. In this case the charges of SM fermions with respect to the $U(1)_D$ interaction will be proportional to the electric charge and the associated mixing term in the QED Lagrangian will be:

$$\mathcal{L}_{mix} = -\frac{\epsilon}{2} F^{QED}_{\mu\nu} F^{\mu\nu}_{dark}. \qquad (1)$$

[a]G. Chiodini (INFN Lecce), A. P. Caricato, M. De Feudis, M. Martino, G. Maruccio, A. Monteduro, S. Spagnolo (INFN Lecce and Dip. di Matematica e Fisica, Università del Salento); R. Bedogni, B. Buonomo, F. Bossi, R. De Sangro, G. Finocchiaro, L. G. Foggetta, A. Ghigo, M. Palutan, G. Piperno, B. Sciascia, T. Spadaro (INFN Laboratori Nazionali di Frascati); G. V. Georgiev, V. Kozhuharov (University of Sofia and INFN Laboratori Nazionali di Frascati); U. Dosselli (INFN Padova); F. Ferrarotto, E. Leonardi, P. Valente (INFN Roma); S. Fiore (INFN Roma and ENEA); G. C. Organtini, M. Raggi (INFN Roma and Dip. di Fisica, Sapienza Università di Roma).

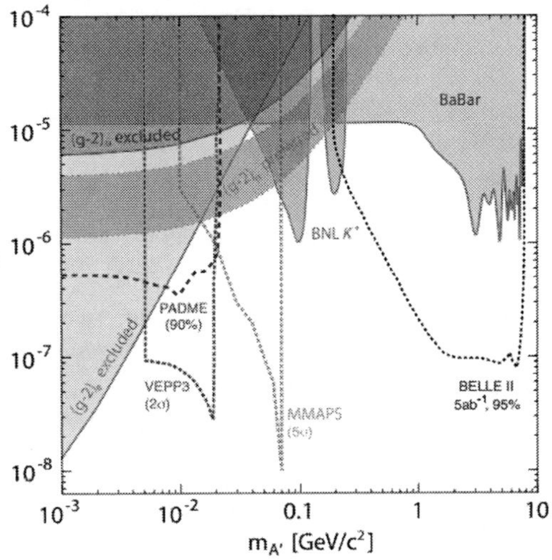

Figure 1. Status of the dark photon invisible decays searches.

The associated mixing coupling constant, ϵ, can be so small ($< 10^{-3}$) to hide the dark sector particle to experimental searches so far. An intense experimental activity in the last 5 years allowed to exclude a large fraction of the parameter space, including the $(g-2)_\mu$ favorite region, in the hypothesis that the dark photon decays to SM final states[3]. An exciting possibility is that dark matter states χ with mass smaller than $M_{A'}/2$ exist, thus allowing dark photon's invisible decays, $A' \rightarrow \chi\chi$. In this scenario almost all of the current constraints do not hold. Current limits on invisible dark photon's decays are summarised in fig.1 together with the expected reach of future missing mass proposals. The green band represents the region of the parameter space preferred by the $(g-2)_\mu$ explanation.

DOI: 10.1051/epjconf/201714201026

2 The PADME technique

The main goal of the PADME experiment is to search for dark photons produced by the annihilation of a positron beam with the electrons at rest in a thin active diamond target[5]. The resulting final state of the process $e^+e^- \rightarrow A'\gamma$ with $A' \rightarrow \chi\chi$, will be a single photon and nothing else. Measuring the incoming positron (P_{beam}) and recoil photon (P_γ) 4-momenta, the mass of the missing state (A') can be measured as:

$$M^2_{miss} = (P_{e^-} + P_{beam} - P_\gamma)^2. \quad (2)$$

being the electron in the target at rest. Once the number of A' candidates in each of the missing mass regions has been measured a natural choice for the A' production rate normalisation is the $e^+e^- \rightarrow \gamma\gamma$ process. The A' coupling constant ϵ can be determined using the formula

$$\frac{\sigma(e^+e^- \rightarrow A'\gamma)}{\sigma(e^+e^- \rightarrow \gamma\gamma)} = \frac{N(A'\gamma)}{N(\gamma\gamma)} * \frac{Acc(\gamma\gamma)}{Acc(A'\gamma)} = \epsilon^2 * \delta, \quad (3)$$

where $N(A'\gamma) = N(A'\gamma)_{obs} - N(A'\gamma)_{bkg}$ is the number of the signal candidates after the background subtraction, $N(\gamma\gamma)$ is the number of observed annihilation events, δ is the $e^+e^- \rightarrow A'\gamma$ cross section enhancement factor. The measurement of the ratio reduces the theoretical uncertainties just to the determination of the enhancement factor δ[4].

The PADME experiment setup uses the 550 MeV/c positron beam of the DAΦNE Beam Test Facility[7] impinging on a thin diamond target and is composed of the following parts:

- **Thin (100μm) active diamond target**, to measure the average position of the beam during a single BTF spill
- **Dipole magnet**, to deflect the primary positron beam out of the calorimeter acceptance and to convey any charged particle from interactions in the target toward the charged particle veto detectors
- **Charged particles veto**, to reject bremsstrahlung background.
- **Vacuum chamber**, to minimise the unwanted interactions of primary and secondary particles.
- **BGO Electromagnetic calorimeter**, to measure and/or veto final state photons.
- **Small angle fast photon veto**, to veto photons emitted at small angle with respect to the primary positron beam.

A schematic view of the experimental apparatus is shown in Figure 2.

A beam of 5000 positrons with very precise energy and angular distribution will cross the active diamond target 50 times per second. The target will measure the beam position and total charge providing information on the interaction point and luminosity of the experiment. In case a dark photon event occurs the 4-momentum of the recoil photon produced in the $e^+e^- \rightarrow A'\gamma$ annihilation will be measured by an electromagnetic calorimeter, placed at a distance of 3 m from the production target. The PADME calorimeter will be realised as a 600 mm diameter cylinder made up of 616 BGO crystals 21×21×230 mm^3, readout by 19mm diameter photo-multiplier tubes. The small size of the crystals, which are reshaped blocks of the endcap electromagnetic calorimeter of the L3 experiment, will allow the PADME calorimeter to reach ~ 5 mm position resolution for electromagnetic clusters allowing a precise measurement of the direction of the impinging photon. A 100×100 mm^2 central hole will allow the most part of Bremsstrahlung photons to reach a smaller and faster Cherenkov based veto detector, featuring a much better time resolution and smaller dead time. Thanks to CERN (TE-MSC-MNC) a dipole magnet from the SPS transfer line, modified to provide 0.55 T over a gap of 23 cm will be used to sweep away the non-interacting positrons.

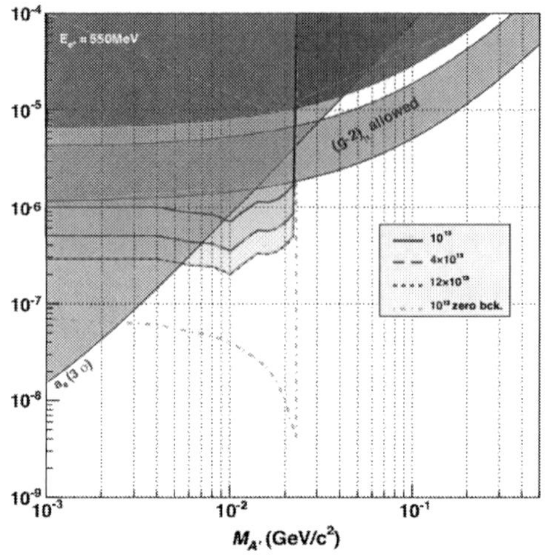

Figure 3. The PADME experiment projected sensitivity.

The rejection of events with a radiating positrons will mainly rely on charged particle veto detectors made of scintillating bars read out by silicon photomultipliers, placed inside the magnetic field of the dipole: a set of scintillators will be placed along each of the deflection sides vetoing lower momentum positrons/electrons, while another set will be placed outside the magnet, close to the nominal beam trajectory to detect higher momentum positrons. The entire volume from the target to the veto detectors and to the front face of the BGO calorimeter will be in vacuum, in order to minimize interactions of photons and charged particles, so that a relatively large and asymmetrically shaped vacuum vessel is realised. The PADME detector is under construction at the Laboratori Nazionali di Frascati of INFN and is expected to be operational by the end of 2017.

3 PADME sensitivity study

The PADME physics potential is extended beyond dark photon searches: presently a sensitivity estimate based on Monte Carlo simulations is available only for the invisible dark photon decays produced by annihilation process.

 DOI: 10.1051/epjconf/201714201026

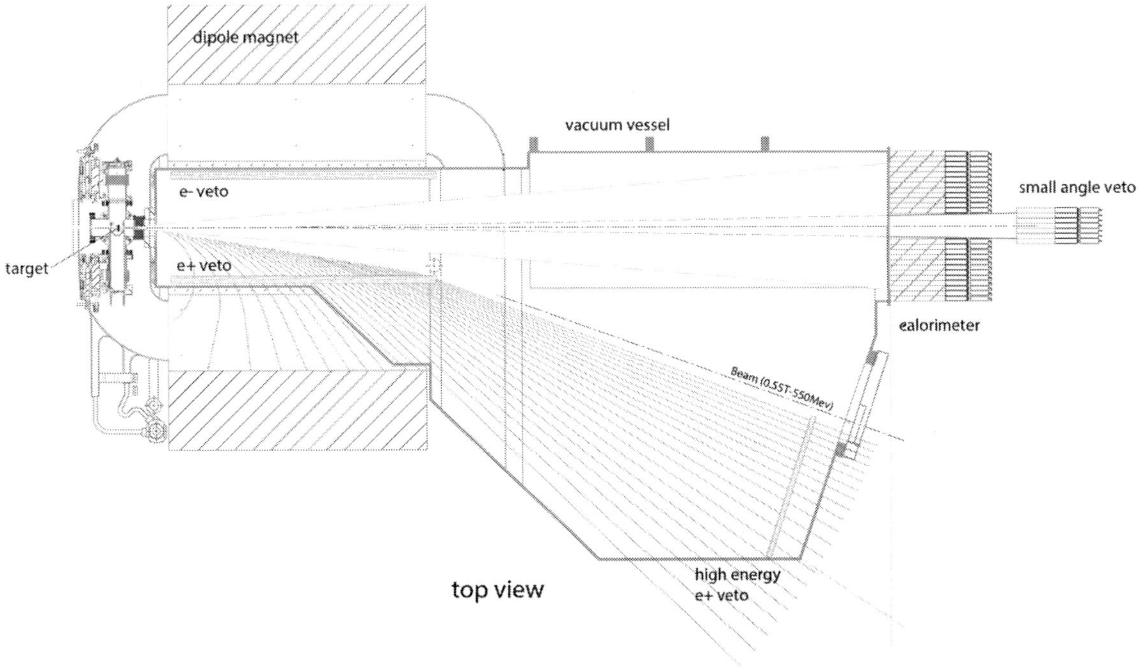

Figure 2. Top view of the PADME experiment.

The sensitivity of the PADME experiment has been estimated using a detailed Monte Carlo simulation, based on GEANT4, assuming 40 ns long beam bunches, each containing 5,000 positrons of 550 MeV/c momentum, with a momentum spread of 1%, an angular divergence of ~ 1 mrad, and a Gaussian beam spot with $\sigma_{x,y}$ 1 mm. This beam configurations allows to integrate a total luminosity of 1×10^{13} positrons in 3.15×10^{7} s corresponding to two years data taking at 50% efficiency. Recent development in the BTF extraction line allowed to reach bunch lengths up to 200 ns. Profiting of this possibility the PADME experiment running time can be significantly reduced.

To emulate the bunch structure, a simultaneous multi positron gun has been implemented, taking into account beam spot size, beam divergence, and energy spread in each single burst. The simulation includes backgrounds simulated by GEANT4 low-energy electromagnetic libraries, two photon annihilation, ionization processes, Bhabha and Moller scattering, and the production of δ-rays. A custom generator for the three photon annihilation background and to simulate the A' production and its decays into e^+e^- or invisible have been developed. Realistic performance of the detectors have been simulated: in particular for the reconstruction of the calorimeter clusters, the energy resolution obtained with the calorimeter prototype during test beams (2% / $\sqrt{E(GeV)}$ [8]), a minimum detectable energy of 1 MeV, and 1 ns timing resolution have been used.

The sensitivity of the PADME experiment to the invisible decays of the dark photon is shown in fig. 3 for various scenarios of recorded luminosity. The effect of the residual background is evident by comparing the sensitivity curve for 10^{13} positrons on target and the corresponding single event sensitivity (orange dashed curve). A margin of improvement of one order of magnitude in the full mass range still exist.

4 Conclusions

Searches for dark photons are well motivated within several new physics scenarios. The PADME experiment will be able to search for invisible decays of the dark photon using the present Beam Test Facility of the DAΦNE LINAC in just one-two years of data taking with a sensitivity on the A' mass up to 23.7 MeV/c^2 in a complete model independent way. The experiment's construction will be completed by the end of 2017 and the first beam for physics is expected in spring of 2018. An upgrade of the BTF positron beam energy will also allow to extend the sensitivity of PADME to higher mass values. Additional physics goals including APLs, are under investigation.

Acknowledgments

This work is supported by the project PGR-226 of the Italian Ministry of Foreign Affairs and International Cooperation (MAECI), CUP I86D16000060005.

References

[1] B. Holdom, Phys. Lett. B **166**, 196 (1986)
[2] M. Pospelov, Phys. Rev. D **80**, 095002 (2009).
[3] J. Alexander *et al.*, arXiv:1608.08632 [hep-ph].
[4] M. Raggi and V. Kozhuharov, Riv. Nuovo Cim. **38**, no. 10, 449 (2015). doi:10.1393/ncr/i2015-10117-9
[5] M. Raggi and V. Kozhuharov, Adv. High Energy Phys. **2014**, 959802 (2014)
[6] M. Raggi, V. Kozhuharov and P. Valente, arXiv:1501.01867 [hep-ex].
[7] P. Valente *et al.*, arXiv:1603.05651 [physics.acc-ph].
[8] M. Raggi *et al.*, arXiv:1611.05649 [physics.ins-det].

EPJ Web of Conferences **142**, 01027 (2017)
Advances in Dark Matter and Particle Physics
DOI: 10.1051/epjconf/201714201027

The End Point Tagger physics program at A2@MAMI

Oliver Steffen[1,a], on behalf of the A2-Collaboration

[1] *Institut für Kernphysik, Johannes Gutenberg-Universität Mainz, Johann-Joachim-Becher-Weg 45, 55099 Mainz*

Abstract. The A2-Collaboration uses a beam of real photons from the tagged photon facility at the electron accelerator MAMI in Mainz, Germany, to study photo-produced mesons. A new tagging device allows access to the higher photon beam energy range of 1.4 to 1.6 GeV. A large dataset containing more than 6 million η' and roughly 29 million ω decays has been obtained. Analyses are ongoing, including a study of the cusp effect and Dalitz plot in $\eta' \to \eta\pi^0\pi^0$, giving insight to the $\pi\pi$ scattering length and the structure of the $\eta\pi\pi$ system, as well as the measurement of the electromagnetic transition form factor in $\eta' \to e^+e^-\gamma$, a cross section measurement of $\gamma p \to 3\pi^0$, and branching ratio analyses of $\eta' \to \omega\gamma$ and $\omega \to \eta\gamma$.

1 Introduction

The A2-Collaboration uses the tagged photon facility at the Mainz Microtron MAMI [1] to study baryon resonances, meson photoproduction and meson decays.

After upgrading the existing experiment setup, a large dataset containing more than 6 million η' decays has been obtained during 10 weeks of beam time in 2014. It serves as a rich base for many physics topics to be studied. Analyses are ongoing, including the measurement of the electromagnetic transition form factor in $\eta' \to e^+e^-\gamma$ is relevant for the understanding of η/η' mixing and theory of the anomalous magnetic moment of the muon (see section 3.1). A study of the cusp effect and Dalitz plot in $\eta' \to \eta\pi^0\pi^0$, giving insight to the $\pi\pi$ scattering length and the structure of the $\eta\pi\pi$ system (see section 3.2). Branching ratio analyses of $\eta' \to \omega\gamma$ and $\omega \to \eta\gamma$ will provide input for chiral effective theories (see section 3.3). The cross section measurement of $\gamma p \to 3\pi^0$ will extend existing data from previous measurements (see section 3.4).

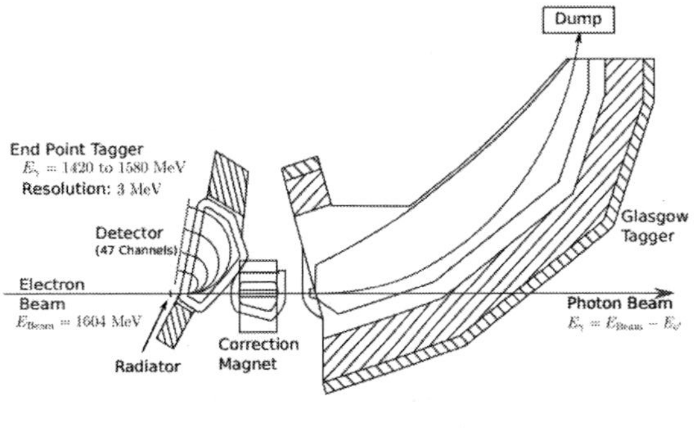

Figure 1. The photon tagging setup: Bremsstrahlung is created by shooting electrons on a thin metal foil. The energy of the scattered electrons is measured by a magnetic spectrometer. The specialized End Point Tagger is mounted in front of the main tagger, followed by a correction magnet.

2 Experiment

The electron beam produced by the MAMI multi-stage racetack microtron has an energy E_{beam} of up to 1.6 GeV [1]. It is converted into bremstrahlung photons by a thin copper foil. The electron, together with the unused portion of the beam, is bent away by the tagging spectrometer magnet (see Fig. 1). The energy of the scattered electron $E_{\text{scattered}}$ is measured by a hodoscope, and the photon energy can be calculated: $E_\gamma = E_{\text{beam}} - E_{\text{scattered}}$. Forming a time coincidence with the detector system of the experiment makes it possible to energy tag the photons.

A new tagging device, the End Point Tagger, was constructed in 2014. It focuses on the highest photon energies of the bremsstrahlung spectrum from 1.4 GeV to 1.6 GeV with a resolution of 3 MeV. It is an ideal tool to study rare and comparatively heavy mesons like the η' and the ω, produced in photo production reactions off the proton (see Fig. 2). Together with upgrades and fine-tuning of the readout system the existing experimental setup could be turned into an η' factory.

The photon beam impinges on a 10 cm long cell filled with liquid hydrogen, acting as a high density proton target. The target cell is contained in a cryostat and cooled down to temperatures below 20 K.

[a] e-mail: steffeo@uni-mainz.de

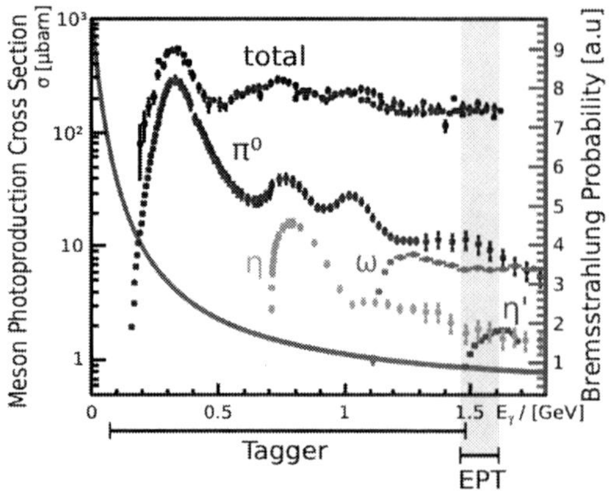

Figure 2. Sketch of photoproduction cross sections for prominent mesons (points) in the energy range covered by the Crystal Ball/TAPS experiment, together with the energy distribution of the bremsstrahlung spectrum (line). The End Point Tagger focuses on a narrow energy region around the η' threshold, while the main tagger covers a wide range of the lower energies. This allows rejection of the much more probable low energetic background events.

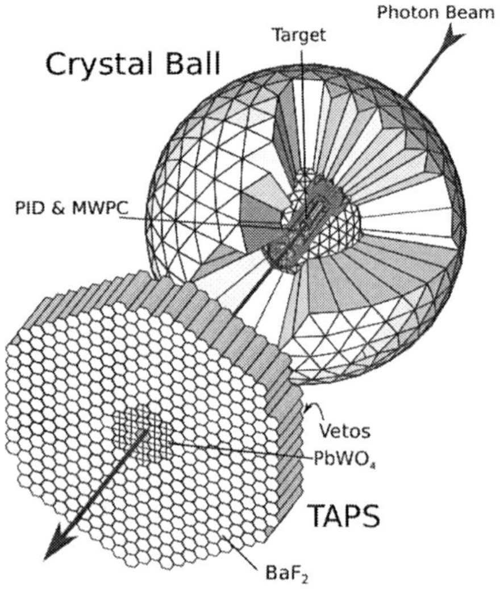

Figure 3. The detector system: The calorimeters Crystal Ball and TAPS cover over 96% of the solid angle. Additional particle identification detectors are located around the target and in front of TAPS.

The target is surrounded by the main calorimeter, the Crystal Ball [2]. It consists of 672 thallium doped sodium iodine crystals (NaI(Tl)), forming a sphere. The forward opening is covered by the TAPS detector [3], forming a wall of 366 barium fluoride (BaF_2) and additional 72 lead tungstate ($PbWO_4$) crystals. These detectors together cover over 96 % of the full solid angle (see Fig. 3) and are ideal for the measurement of neutral final states. Both calorimeters are equipped with additional detectors made out of plastic scintillators, allowing particle identification by measuring the energy loss of traversing charged particles. Additionally, a pair of cylindrical multi-wire proportional chambers (MWPC) is located inside the Crystal Ball. Due to the extremely high rates they could not be used in the beam times mentioned above. Table 1 summarizes the properties of the detector system.

Table 1. Properties of the Crystal Ball/TAPS detector system.

	CB	TAPS
Energy Resolution	$\frac{2\,\%}{\sqrt[4]{E_\gamma/\text{GeV}}}$	$\frac{0.8\,\%}{\sqrt{E_\gamma/\text{GeV}}} + 1.8\,\%$
Crystals	672	438
Material	NaI(Tl)	BaF_2, $PbWO_4$

3 Analyses

3.1 Electromagnetic Transition Form Factors

The electromagnetic transition form factors (TFF) of pseudo-scalar mesons are important input for solving the mystery of the anomalous magnetic moment of the muon [4] and better understanding the η/η' mixing [5].

Currently a 3.6 σ discrepancy exists between then experimental observation [6] and theoretical predictions of $a_\mu = (g-2)/2$. The theory is limited by QCD contributions [7, 8], consisting of hadronic vacuum polarizations and hadronic light-by-light scattering [9], where the transition form factors enter in the description primarily through the light-by-light scattering terms. The studies of meson decays allow the measurement of the q^2 dependent TFF in the time-like region ($q^2 > 0$). They appear as a modification of the QED prediction of the differential cross section for the point-like meson:

$$\frac{d\sigma}{d\Omega} = \left(\frac{d\sigma}{d\Omega}\right)_{\text{QED}} \cdot \left|F(q^2)\right|^2 \qquad (1)$$

Electromagnetic transition form factors of the π^0 [10] and η [11] meson have already been studied in previous experiments of the A2-Collaboration (see Fig. 4). This experiment gives the opportunity of extending the existing data set to higher energies and also study the η' TFF in the $\eta' \to e^+e^-\gamma$ decay, to complete the set of TFF of the light neutral mesons. The experiment covers a wide q^2 range in the time like region up to $q^2 \approx 0.7\,\text{GeV}^2$. This allows a fit containing the ρ peak region which is not covered by competing experiments [12]. A Monte Carlo simulation of the expected q^2 distribution is shown in Fig. 4. The analysis is still ongoing.

DOI: 10.1051/epjconf/201714201027

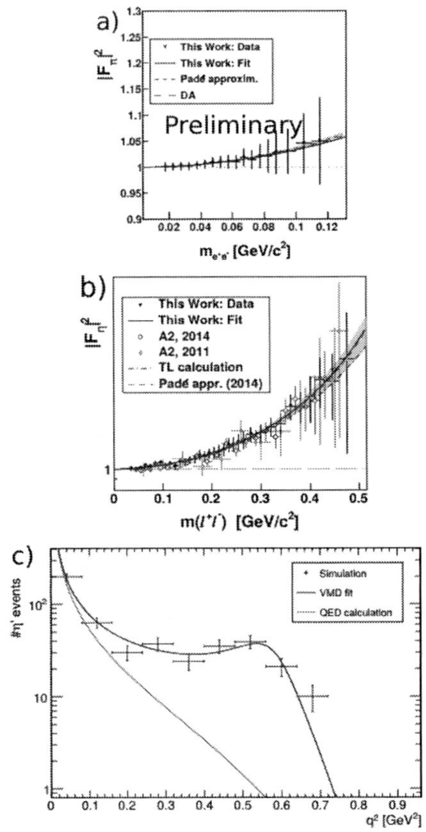

Figure 4. Electromagnetic transition form factors previously studied at A2: a) Preliminary π^0 TFF from $\pi^0 \to e^+e^-\gamma$ [10], b) η TFF from $\eta \to e^+e^-\gamma$ [11], c) Monte Carlo simulation of the q^2 distribution of $\eta' \to e^+e^-\gamma$, together with a VMD fit and the QED calculation.

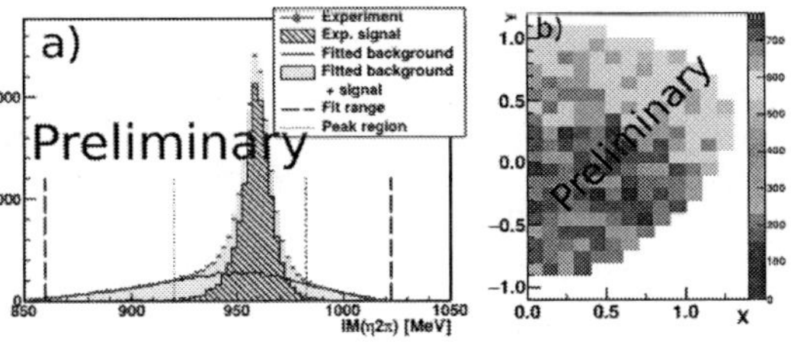

Figure 5. a) Invariant mass spectrum corresponding to one bin in the Dalitz plot. The peak from $\eta' \to \eta\pi\pi$ is clearly visible above the background. b) One half of the symmetric Dalitz plot, containing 1.3×10^5 events.

3.2 Cusp Effect and Dalitz Plot

The large amount of η' allows detailed studies of the $\eta' \to \eta\pi^0\pi^0$ decay. Two analyses of particular interest are the Dalitz plot analysis amd the search for the cusp effect.

A Dalitz plot analysis reveals the full information about the dynamics of a decay. The Dalitz plot variables X and Y for $\eta' \to \eta\pi^0\pi^0$ can be calculated from the kinetic energies T and masses m of the decay products as follows:

$$X = \frac{\sqrt{3}(T_{\pi_1} - T_{\pi_2})}{Q} \quad (2)$$

$$Y = \frac{(m_\eta + 2m_\pi)}{m_\pi}\frac{T_\eta}{Q} - 1 \quad (3)$$

$$Q = T_{\pi_1} + T_{\pi_2} + T_\eta = m_{\eta'} - m_\eta - 2m_\pi\,. \quad (4)$$

Since the two pions, π_1 and π_2, are indistinguishable both combinations have to be considered, making the plot symmetric in X. This Dalitz plot is well suited for testing extensions of chiral perturbation theory, such as large N_c and resonance models [13, 14, 15]. The plot obtained during this experiment contains $1.3 \cdot 10^5$ events (see Fig. 5). This is a significant increase in statistics compared to previous experiments of the GAMS-4π ($1.5 \cdot 10^4$ $\eta' \to \eta\pi^0\pi^0$ events) [16] and BES-III ($4.3 \cdot 10^4$ $\eta' \to \eta\pi^+\pi^-$ events) [17] collaborations. The Dalitz plot parameters a, b, and d can be extracted by fitting

$$|A|^2 = |N|^2\left[1 + aY + bY^2 + cX + dX^2\right] \quad (5)$$

to the plot. The parameter c, describing the slope in X, cannot be extracted since the plot in this case is symmetric in X. The preliminary values from this analysis are compared to previous measurements and theory calculations in table 2.

A cusp effect, due to $\pi^+\pi^- \to \pi^0\pi^0$ re-scattering, has been predicted for $\eta' \to \eta\pi^0\pi^0$ [18] and is also being studied with the data set. This kind of effect was first observed in $K^+ \to \pi^0\pi^0\pi^+$ by NA48/2 [19], and has also been predicted for $\eta/K_L^0 \to 3\pi^0$ as a few percent effect. The analysis of the cusp would allow the determination of the S-wave $\pi^0\pi^0$ scattering length [20, 21].

3.3 Branching Ratio Measurements

Reactions between pseudo-scalar and vector mesons with a photon at the same vertex contain valuable input for implementing vector mesons in chiral effective field theories. The goal is a consistent picture of pseudo-scalar and vector mesons within the theory and finding a way to set up a power counting for unstable particles which to fulfill additional constraints due to spin 1 nature [15, 22].

The decay $\eta' \to \omega\gamma$ has a branching ratio of $2.75 \pm 0.23\,\%$ [23], and was measured by the BESIII collaboration as $(2.55 \pm 0.03_{\rm stat} \pm 0.16_{\rm syst})\,\%$ [24]. A

Table 2. Preliminary results for the Dalitz plot parameters in the decay $\eta' \to \eta\pi^0\pi^0$ ($\chi^2/\text{dof} = 233.6/195 = 1.198$) with statistical errors, compared to the GAMS-4π measurement and predictions from large N_c and resonance extensions of chiral perturbation theory.

Result	a	b	d
A2 preliminary	-0.073 ± 0.078	-0.072 ± 0.013	-0.054 ± 0.009
GAMS-4π	$-0.067 \pm 0.016 \pm 0.004$	$-0.064 \pm 0.029 \pm 0.005$	$-0.067 \pm 0.020 \pm 0.003$
Large N_c extensions	-0.098 ± 0.048	-0.050 ± 0.001	-0.092 ± 0.008
Resonance ChPT	-0.098 ± 0.048	-0.033 ± 0.001	-0.072 ± 0.001

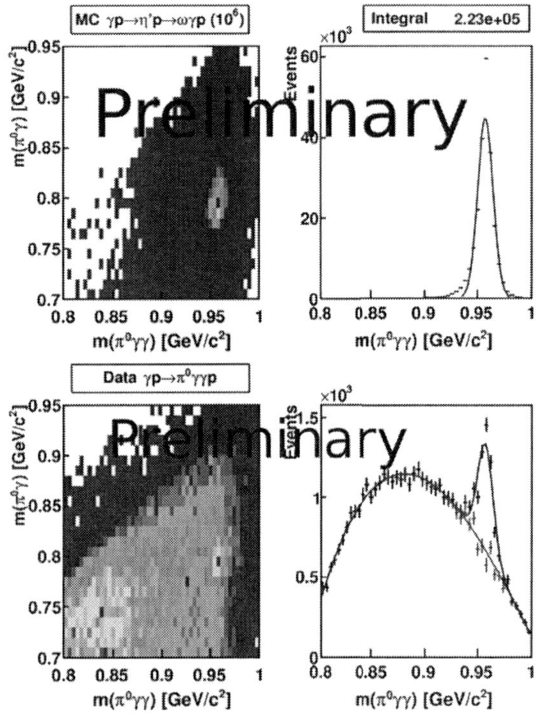

Figure 6. Branching ratio measurement of $\eta' \to \omega\gamma$: Preliminary invariant mass spectra from Monte Carlo (pure signal sample) and Data (including background).

branching ratio measurement of this decay relative to $\eta' \to \gamma\gamma$ is being performed on our data set with about 10^4 estimated signal events. Preliminary invariant mass spectra from Monte Carlo and experiment data are shown in Fig. 6.

Additionally about 29 million ω mesons were produced during this experiment, which allows the measurement of the $\omega \to \eta\gamma$ decay ($\Gamma_i/\Gamma = (4.6 \pm 0.4) \cdot 10^{-4}$ [23]). We are performing a relative measurement to $\omega \to \pi^0\gamma$, with an expected number of about 1500 signal events.

3.4 Photoproduction cross section of $3\pi^0$

The photo production process of three neutral pions off the proton, $\gamma p \to p3\pi^0$, is a major background for the other ongoing analyses. This reaction is also

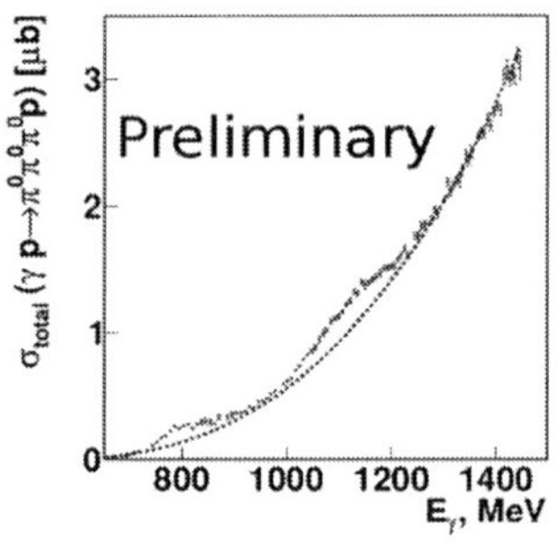

Figure 7. Preliminary result of a previous experiment measuring the three pion photoproduction cross section: $\gamma p \to p3\pi^0$ [25].

interesting to study as a production process. It can be described by the isobar model and involves a series of intermediate resonances. The cross section has been analyzed in previous experiments [25] (see fig 7). With the obtained data set we can extend and improve the previous results. Since the analysis is in an early state no preliminary results can be shown.

4 Summary

With the specialized tagging spectrometer and the upgraded detector readout of the Crystal Ball/TAPS experiment a large data set of η' and other mesons was recorded in 2014. Analyses are now ongoing within the A2-Collaboration: A study of the $3\pi^0$ photoproduction cross section, a Dalitz plot analysis of $\eta' \to \eta\pi\pi$, including the search for a cusp effect, as well as the measurement of the electromagnetic transition form factor in $\eta' \to e^+e^-\gamma$, and branching fraction measurements of $\eta' \to \omega\gamma$ and $\omega \to \eta\gamma$.

References

[1] M. Dehn *et al.*, EPJ Special Topics **198**, 19 (2011).

[2] M. J. Oreglia, PhD Thesis, Department of Physics of Stanford University (1980).

[3] R. Novotny, IEEE Trans. Nucl. Sci. **38**, 379 (1991).

[4] E. Czerwinski *et al.*, arXiv:1207.6556 [hep-ph], (2012).

[5] L. Ametller, Phys. Scripta **T99**. 45 (2002).

[6] G. W. Bennett *et al.*, Phys. Rev. D**73**, 072003 (2006).

[7] M. Davier *et al.*, Eur. Phys. J. C **71**, 1515 (2011). [Erratum: Eur. Phys. J.C **72**, 1874 (2012)].

[8] K. Hagiwara *et al.* J. Phys. G **38**, 085003 (2011).

[9] P. Masjuan, Nucl. Part. Phys. Proc. **260**, 111 (2015).

[10] P. Adlarson *et al.*, Phys. Rev. C **95**, 025202 (2017).

[11] P. Adlarson, *et al.*, Phys. Rev. C **95**, 035208 (2017).

[12] M. Ablikim *et al.*, Phys. Rev. D **92**, 012001 (2015).

[13] R. Kaiser and H. Leutwyler, EPJ C **17**, 623 (2000).

[14] R. Escribano *et al.*, JHEP **2011**, 94 (2011).

[15] G. Ecker *et al.*, Phys. Let. B **223**, 425 (1989)

[16] A. M. Blik *et al.*, PAN **72**, 231 (2009).

[17] M. Ablikim *et al.*, Phys. Rev. D **83**, 012003 (2011).

[18] Kubis, B., Schneider, S., EPJ C **62**, 511 (2009).

[19] J.R. Batley *et al.*, Phys. Let. B **633**, 173 (2006).

[20] N. Cabibbo and G. Isidori, , JEHP **03**, 21 (2005).

[21] C.-O. Gullström, A. Kupść and A. Rusetsky, , Phys. Rev. C **79**, 028201 (2009).

[22] C. Terschlüsen, S. Leupold, and M. F. M. Lutz, Eur. Phys. J. A **48**, 190 (2012).

[23] K. A. Olive *et al.*, Chin. Phys. **C38**, 1 (2014).

[24] M. Ablikim *et al.*, Phys. Rev. D **92**, 051101 (2015).

[25] A. Starostin *et al.*, arXiv:1101.3744 [nucl-ex], (2011)

EPJ Web of Conferences **142**, 01028 (2017) DOI: 10.1051/epjconf/201714201028
Advances in Dark Matter and Particle Physics

Status of positron beams for dark photons experiments

Paolo Valente[1,⋆]

[1] *INFN Sezione di Roma, P.le Aldo Moro, 2, I-00185 Rome, Italy*

Abstract. High energy positron beams are an important tool for fixed-target experiments searching for new particles produced in the annihilation on atomic electrons of a target. The status of existing or planned infrastructures is reviewed.

1 Introduction

The idea that the new dark particles are not directly connected with the Standard Model (SM) gauge fields, but only through mediators, linking a new, "secluded" sector to the ordinary one, has triggered a number of experiments, mainly devoted to search for a vector boson (or dark photon, A'), kinetically mixed with the ordinary photon (with coupling α' related to the electromagnetic one through the parameter $\varepsilon^2 = \alpha'/\alpha$).

Practically all the experimental results are based on the search for the decay into a SM fermion anti-fermion pair of the dark photon, being irradiated in the interaction of an electron beam with the nuclei of a target (in fixed-target experiments), or coming from the decay of a meson (in colliders).

In the so-called "visible" decays of the dark photon, detecting the decay of the A'-boson to a SM fermion pair (e^+e^- if $m_{A'} < 2m_\mu$) requires a significant branching of A' decay to the lepton pair. This is indeed the only available channel if the dark photon is the lightest particle in the new sector. A complication of this method is the high level of electromagnetic background in the e^+e^- mass spectrum. In addition, a large number or results have been obtained both in fixed-target experiments, using A'-strahlung production from high-intensity/high-energy electron beams, and in collider experiments, using meson decays to $\gamma A'$.

In the hypothesis that $m_{A'} < 2m_\chi$, being χ the lightest fermion in the dark sector, the most interesting range of the parameters space is already excluded [1]; in particular the mass range between 1 and 100 MeV/c^2 and the coupling down to $\varepsilon^2 \sim 10^{-6}$ that could explain the discrepancy between theory and observation for the muon anomalous magnetic moment [2]. However, removing any model-dependent hyphotesis on the A' decay modes and mass, the parameter space is much less constrained.

Following the proposal by B. Wojtsekhowski [3], an alternative approach is to look for the "invisible" decays of the A', i.e. without requiring SM fermions in the final state, and exploiting the kinematics of the annihilation of a positron beam of well-known momentum onto the electrons of a target. Measuring the momentum of the ordinary photon in events $e^+e^- \to \gamma A'$ would indeed allow the observation of a peak in the missing mass spectrum for $m^2_{miss} = m^2_{A'} \neq 0$, independently from the decay mode and life-time of the dark photon.

⋆e-mail: paolo.valente@roma1.infn.it

In this approach, a high energy positron beam, with well defined quadri-momentum impinges on a thin (i.e. not fully absorbing the incoming beam), low-Z target, in a well defined position and with the best possible reproducibility. Of course the sensitivity scales with the statistics, i.e. with the number of interactions of the incoming positrons with the target electrons.

The intensity of the beam is ultimately dictated by the pile-up probability in the downstream detector detecting the ordinary photon in the final state. Among the various background sources, mainly Bremsstrahlung on the target nuclei, $e^+(N) \to e^+(N)\gamma$, and two (three) photons process, $e^+e^- \to \gamma\gamma(\gamma)$ are the dominant ones.

The original idea proposed in Ref. [3] is to use the positron beam circulating in a collider ring (like the Novosibirsk VEPP-3 machine), striking it onto an internal gaseous Hidrogen target. This requires to have a internal target facility and the possibility of clearing the acceptance of the emitted ordinary photon in order to place a calorimeter at a suitable distance for measuring the angle and the energy with the required resolution.

A more flexible approach would be to use an extracted positron beam, in order to optimize both the experimental setup for the missing mass spectrum measurement, and the target, as proposed by the PADME [4, 5] and MMAPS [7] collaborations. It is important to note that the ideal time structure of the beam for the proposed annihilation experiments is to have a continuous positron beam, or at least a long beam pulse, in order to dilute the highest possible number of positron interactions, while keeping under control the probability of pile-up in the photon detector, taking into account both the $A'\gamma$ signal and the other photon sources (Bremsstrahlung and $2/3\gamma$ backgrounds).

The availability of a high-energy extracted beam would of course allow performing any other kind of positron fixed-target experiment, like searches of "visible" dark photon decays from A'-strahlung, dump experiments, or other light dark matter searches, opening also the possibility (with respect to the circulating beam) of further manipulation (collimation, focussing, etc.). In the following some of the available or planned accelerator facilities with the possibility of having an extracted, high-energy (>few tens of MeV) positron beam are briefly described.

2 Laboratori Nazionali di Frascati positron beam

2.1 Beam-Test Facility (BTF)

The Beam-Test Facility (BTF) is part of the LNF accelerator complex in operation since 2004 [8]. It makes use of the LINAC, used as injector of the Frascati e^+e^- high-luminosity collider DAΦNE, running at the Φ meson resonance (center of mass energy of 1020 MeV).

The DAΦNE linear accelerator [9] is a constant gradient, travelling wave, $2/3\pi$, S-band (2856 MHz) LINAC, composed by 15 SLAC-type, 3 m long, accelerating structures, fed by four radio-frequency power stations, each with a modulator supplying a 45 MWp klystron (Thales TH-2128C) at the maximum repetition rate of 50 Hz.

The electron source consists of a gridded triode gun with replaceable cathode. The LINAC has been designed on the model of the two-mile linear accelerator in SLAC. In order to get higher accelerating fields, each klystron is followed by a SLED (SLAC Energy Doubling) system using two coupled cavities, in order to compress the 4.5 μs long RF pulse.

The time duration of each LINAC pulse is dictated by the characteristics of the pulsing system, providing a rectangular waveform to the gun for the extraction of electrons emitted by the filament, kept at a fixed bias voltage. In the standard configuration [10], it is capable of delivering pulses of adjustable total width in the range 1.5 to 40 ns. After acceleration, of course the beam will have a micro-bunched structure at the RF frequency of 2856 MHz. For the injection into the DAΦNE rings, the optimal pulse duration is 10 ns, while the typical beam charge is of 0.5/1 nC for electron/positron pulses. Un upgraded pulsing system has been recently commissioned, allowing to deliver pulses of increasing length, up to 5 μs, so that – after having optimized the RF power and phases and the magnetic focussing in the LINAC – electron or positron beams could get accelerated close to the maximum energy, with a width of several hundreds of ns.

After the gun, electrons are bunched and brought to $v = c$ by pre-buncher and buncher cavities, followed by the five 3 m long, standard SLAC sections, accelerating them up to ~ 220 MeV. Positrons can then be produced by a two radiation lengths Tungsten-Rhenium converter, followed by solenoids for focussing and a magnetic chicane for electron separation and a high-field accelerating section (capture section); together with the remaining nine standard sections, positrons can be accelerated up to ~ 550 MeV.

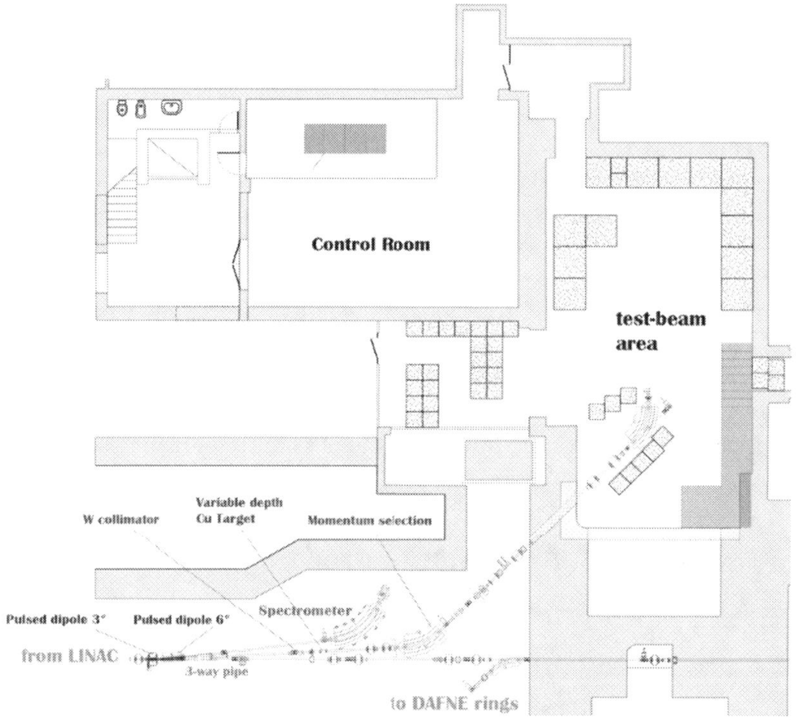

Figure 1. General layout of the Beam-Test Facility (BTF) of the Frascati DAΦNE accelerator complex.

As shown in the scheme of Fig. 1, at the end of the LINAC the electron/positron beam can be diverted in three different lines using two pulsed magnets:

- At 0°, to the transfer-line to the damping ring (and from there, back to the pulsed magnet driving it to the transfer-line towards the DAΦNE main rings);
- at 3°, to the BTF beam-line;
- at 6°, to the spectrometer line: a 60° dipole plus a segmented (metallic strip) detector allows measuring the beam energy with a < 0.5% accuracy.

The DAΦNE timing system can realize different sequences in which a portion of the 50 pulses available in one second are injected to the damping ring, while the remaining ones are deflected on the BTF line. At least one pulse (two during the extraction from the damping ring and injection to the main rings, which is performed at 2 Hz) is sent to the spectrometer line, in order to have a 1 Hz measurement of the LINAC beam energy. A dedicated sequence for the standalone operation of the BTF is available, sending all available 49 pulses to the test-beam, and is used when not injecting for the collider operation. The resulting time structure for the positron (or electron) beam extracted to the BTF line is shown in Fig. 2, yielding a duty-factor of 40 ns/20 ms = $2 \cdot 10^{-6}$.

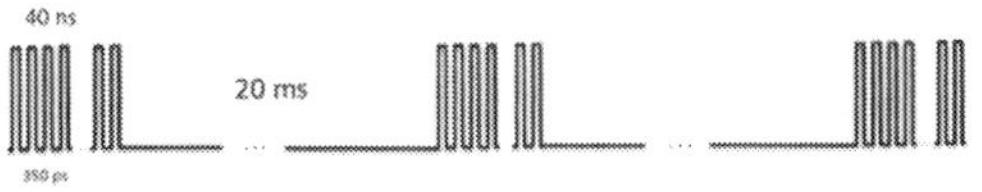

Figure 2. Time structure of the BTF extracted beam.

Since there is only one transfer line from the LINAC to the damping ring (also used for the reverse path from the damping ring to the main rings), it is necessary to reverse the polarity of all magnetic elements when switching from the injection of electrons to positrons, and viceversa. In this switching time the LINAC cannot deliver particles. This is a not negligible limitation of the duty-cycle of the BTF beam-line, when operating it together (in the so-called "parasitic mode") with the e^+e^- collider.

It is worth noticing that it is possible either to deliver to the BTF line the full LINAC beam, i.e. at high intensity and at the fixed energy of the LINAC (510 MeV during collider operation), or producing secondary particles intercepting the LINAC beam with a Copper target, and then selecting the charge and momentum of secondaries with a 43° dipole magnet plus collimators.

Together the focussing system of the BTF line, this also results in a strong and tunable attenuation of the beam, so that electrons or positrons can be delivered in the full energy range down to the "single particle" regime, i.e. of a Poisson distribution with adjustable average number of particles. As shown in Fig. 1, two FODO quadrupole doublets and a final 45° dipole, allow driving and focussing the beam in the BTF test-area, shielded by moveable concrete blocks. Taking into account the all different operation modes, the resulting sets of parameters for the BTF beam are shown in Tab. 1.

Parameter	**Values**			
Maximum flux	$3.125 \cdot 10^{10}\ s^{-1}$			
Spot size (hor.)	0.7–55 mm			
Spot size (vert.)	0.7–25 mm			
Divergence	1–2 mrad			
Parameter	**Parasitic mode**		**Dedicated mode**	
Pulse duration	10 ns		1.5–40 ns	
Repetition rate	Variable (depending on DAΦNE injection		$1–49\ s^{-1}$ (selectable)	
	With target	**Without target**	**With target**	**Withtaout target**
Particle type (e^+ or e^-)	selectable by users	alternating according to DAΦNE status	selectable by users	selectable by users
e^- Energy (MeV)	25–500	510	25–700	250–730
e^+ Energy (MeV)	25–500	510	25–500	250–530
Energy spread	1% at 500 MeV	0.5%	0.5%	0.5%
Intensity	$1–10^5$	$10^7–1.5 \cdot 10^{10}$	$1–10^5$	$10^3–3 \cdot 10^{10}$

Table 1. Beam parameters achievable in all possible operation modes at the end of the BTF line.

The present BTF target is shaped so that three different thickness values can be selected, corresponding to 1.7, 2 or 2.3 radiation lenghts. Positrons produced by high-energy primary electrons striking such a target exhibit a typical energy-angle anti-correlation: higher energy positrons will be produced mostly in the forward direction, while lower energy ones will be spread over a much wider angle distribution. Considering the effect of the collimation and the magnetic focussing, it is possible to optimize the yield of high energy positrons (close to E_0, the incoming primary electron energy) as a function of the thickness of the target, as shown in Fig. 3. This suggests that a thinner Copper target (of the order of one radiation length) with respect to the present one, can improve the number of produced positrons starting from primary electrons accelerated – exploiting the full length of the LINAC – to the maximum energy of about 750 MeV. Recently, an upgrade of the LINAC energy up to 1 GeV has been also proposed [11].

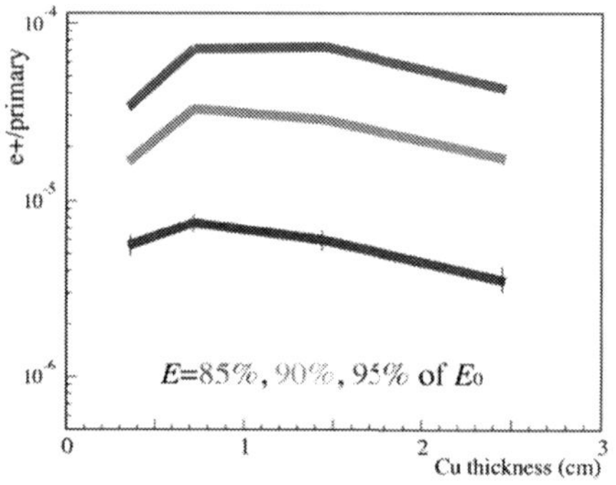

Figure 3. Positron yield as a function of the BTF target depth (Copper).

Studies are currently under way for extending the accelerated positron beam duration above 40 ns (up to 200 ns, resulting in a duty-factor of $\approx 10^{-5}$.), in order to increase the luminosity of fixed-target annihilation experiments while keeping the pile-up probability constant. The length of electron (and positron) pulses from the LINAC is limited by the peculiar, peaked shape of the accelerating voltage due to the compression of the RF power by the SLED system, so that a dedicated development for tuning the RF power, phase and delays in all the sections of the LINAC has to be performed, on the model of what achieved in Ref. [12].

2.2 Possible extracted beam from DAΦNE

An alternative to the extraction of the LINAC beam in the BTF line would be to perform a slow extraction from the DAΦNE collider ring, of 510 MeV positrons. The e^+e^- collider is composed by two separated rings (as schematically shown in Fig. 4), one for the electron and one for the positron beam, intersecting in one interaction region (IP1) and vertically separated on the opposite side of the machine. The RF frequency for the 97.7 m long rings is of 368.26 MHz, corresponding to up to 120 bunches, spaced by 2.7 ns. The positron stored current can exceed 1 A, with $\beta_x \simeq 1.6$ m, $\beta_y \simeq 17$ mm, and emittance $0.34 \cdot 10^{-6}$.

The slow extraction could be realized by using the third order resonance, excited by means of a sextupole and tuned by quadrupoles, together with RF knockout and septum magnet. Detailed studies are needed in order to design the extraction layout and thus estimate the parameters of the extracted beam; however good emittance and beam energy spread are expected, while the extraction efficiency and spill length has to be carefully calculated.

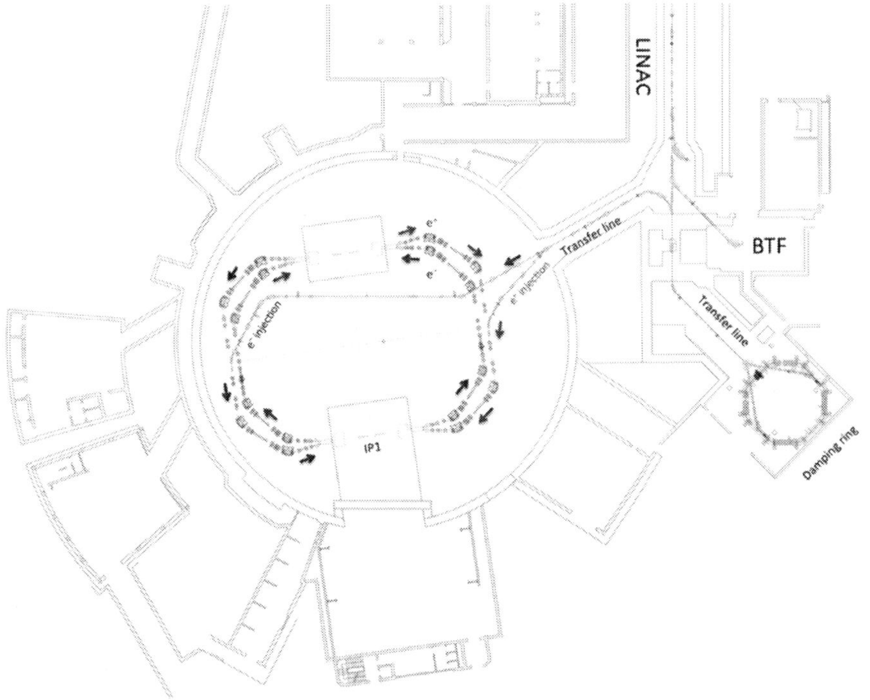

Figure 4. DAΦNE accelerator complex in Frascati, showing the two positron and electron intersecting rings.

3 Other facilities

At SLAC, FACET-II is a new test facility to develop advanced acceleration and coherent radiation techniques with high-energy electron and positron beams, representing a major upgrade over current FACET capabilities. The FACET-II project has two stages: first a new photoinjector and two bunch compressors in the LINAC will restore operation with electrons; then a new positron damping ring with injection and extraction lines will restore operation with positrons. Positrons are accelerated to 10 GeV, compressed and focused to a transverse size of <20 μm at the interaction point and dumped. Repetition rate of the LINAC will be 30 Hz, and the planned charge is 1 nC for positrons produced by 4 nC of electrons, using the existing target and capture section [6].

3.1 Proposed extracted beam at Cornell

The Cornell Electron Storage Ring (CESR) has been operational as e^+e^- collider until the late 2000's, and is now used as a synchrotron radiation source (CHESS). Electrons produced by a thermo-ionic gun are accelerated and then driven onto a Tungsten target for producing positrons, then accelerated by the LINAC to 150 MeV. Up to 16 positron bunches from the LINAC are loaded into the Cornell Synchrotron and accelerated up to 5.3 GeV. From there positrons are then injected into the external storage ring, which includes the magnets for the production of synchrotron radiation. The average circulating current is $\approx$ 10 nA.

By using the slow extraction at 60 Hz, quasi-continuous spills of $\approx$ 2 ms length, $\approx$ 2.5 mA current, of 5.3 GeV positrons (6 GeV after the planned upgrade of the complex, after the removal of the straight section for high-energy experiments like CLEO and CLEO-c) can be driven to a beam-line

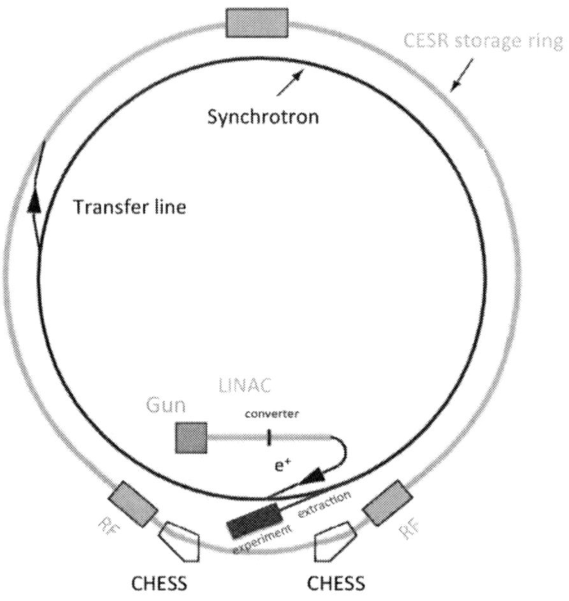

Figure 5. Proposed Cornell extracted beam for dark photon experiments.

for fixed-target experiments. Pulsed quadrupoles will shift the tune towards a third integer resonance, then pulsed sextupole will reduce the stable phase space pushing positrons away from the stable orbit (spiralizing outwards), where a septum can kick them out. The layout of the slowly extracted positron line from the synchrotron and the possible location of fixed-target experiments in between the two rings, is shown in Fig. 5, together with the production and acceleration chain (LINAC followed by synchrotron).

The beam time structure is made up by $\approx$ 12,000 bunches, spaced by 168 ns, repeated every 16.67 ms (60 Hz repetition rate), resulting in a duty-factor of $\approx 10^{-1}$.

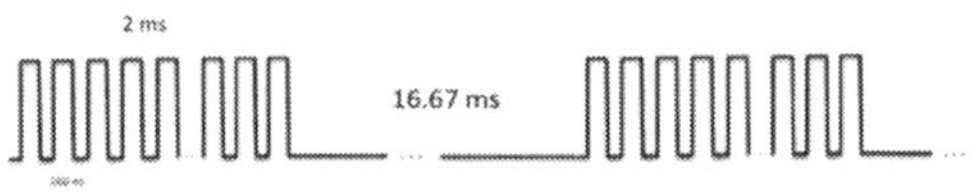

Figure 6. Time structure of the proposed Cornell extracted beam.

3.2 VEPP-3

Positrons are produced and accelerated up to $\approx$ 500 MeV in the VEPP-3 ring, part of the Budker Institute for Nuclear Physics (BINP) in Novosibirsk. In particular, 6 bunches, 1 ns long and spaced by 41.5 ns, of $\approx 2 \cdot 10^{10}$ positrons are injected and accumulated into the VEPP-3 storage ring. Positron annihilations can be realized by installing an internal gas target. The project is to use one of the straight sections of the machine for building a dedicated by-pass line (using three dipoles and six quadrupoles), as shown in the schematic layout of Fig. 7 and further discussed in Ref. [14].

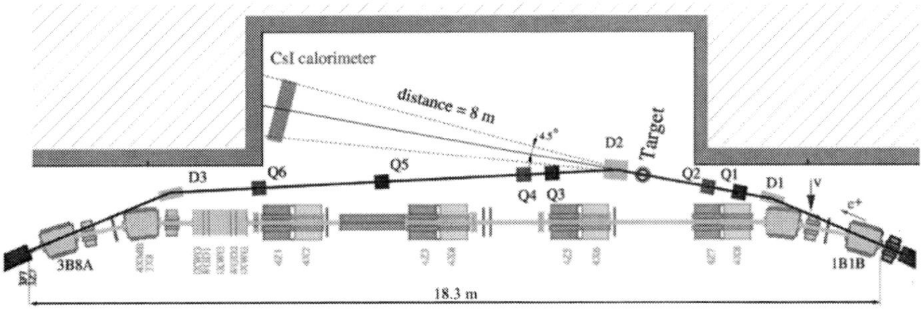

Figure 7. Proposed by-pass line at VEPP-3 in Novosibirsk.

Acknowledgments

This work is supported by the project PGR-226 of the Italian Ministry of Foreign Affairs and International Cooperation (MAECI), CUP I86D16000060005.

References

[1] J. Alexander *et al.*, "Dark Sectors 2016 Workshop: Community Report," arXiv:1608.08632 [hep-ph].
[2] M. Pospelov, Phys. Rev. D **80** (2009) 095002.
[3] B. Wojtsekhowski, D. Nikolenko and I. Rachek, "Searching for a new force at VEPP-3," arXiv:1207.5089 [hep-ex].
[4] M. Raggi and V. Kozhuharov, Adv. High Energy Phys. **2014** (2014) 959802.
[5] M. Raggi, V. Kozhuharov and P. Valente, EPJ Web Conf. **96** (2015) 01025.
[6] V. Yakimenko *et al.*, doi:10.18429/JACoW-IPAC2016-TUOBB02
[7] J. Alexander *et al.*, "Cornell Dark Photon Search", APS April Meeting (2015), 2015APS..APRU15006A.
[8] A. Ghigo, G. Mazzitelli, F. Sannibale, P. Valente and G. Vignola, Nucl. Instrum. Meth. A **515** (2003) 524.
[9] B. Buomomo and L. Foggetta, IPAC 2015, Richmond (USA), TUPWA057 (2015).
[10] B. Buomomo, L. Foggetta and G. Piermarini, IPAC 2015, Richmond (USA), TUPWA056 (2015).
[11] P. Valente *et al.*, "Linear Accelerator Test Facility at LNF: Conceptual Design Report," arXiv:1603.05651 [physics.acc-ph].
[12] F. J. Decker, Z. D. Farkas and M. G. Minty, SLAC-PUB-8111.
[13] S. Guiducci, DAΦNE Technical Note G-73 (2017).
[14] I. Rachek, D. Nikokenko and B. Wojtsekhowski, these Proceedings.

Study of the one-way speed of light anisotropy with particle beams

Bogdan Wojtsekhowski[1],*

[1] *Thomas Jefferson National Accelerator Facility, Newport News, Virginia 23606 USA*

Abstract. Concepts of high precision studies of the one-way speed of light anisotropy are discussed. The high energy particle beam allows measurement of a one-way speed of light anisotropy (SOLA) via analysis of the beam momentum variation with sidereal phase without the use of synchronized clocks. High precision beam position monitors could provide accurate monitoring of the beam orbit and determination of the particle beam momentum with relative accuracy on the level of 10^{-10}, which corresponds to a limit on SOLA of 10^{-18} with existing storage rings. A few additional versions of the experiment are also presented.

1 Introduction

A constant and isotropic speed of light is a key postulate of the special relativity theory, SR, as formulated by A. Einstein in 1905 [1]. In a search for a light related medium, ether, A. Michelson [2] developed an extremely sensitive method of measurement of the anisotropy of the two-way speed of light, c_2, which is an average of the speeds in two opposite directions. A modern experiment of this type [3] reached an extremely high precision, 10^{13} times higher than was obtained in the Michelson-Morley experiment [4], and found that c_2 is isotropic at least to the level of 10^{-18}. The one-way speed of light anisotropy tells us about the directional variation of a difference in the speed of light in two opposite directions. Current studies of SOLA are motivated by predictions in string and quantum gravity theories [5], as well as by general interest in physics beyond the Standard Model. The one-way speed of light, c_1, could be measured by using two synchronized clocks. However, there is a very significant difficulty in realizing clock synchronization at the required precision. If observed, variation of the speed of light could be an indication of a quantum gravity effect in the laboratory.

Several tests have been proposed for the study of c_1 which do not require clocks because they address the difference between the speed of light at different polarizations or the difference between the speed of light and the speed of electrons in the same direction. The latter could be explored using threshold reactions such as vacuum Cherenkov radiation and photon decay to an electron-positron pair [6]. The maximum attainable speed of electrons could also have directional anisotropy. A Compton process was used recently in an experiment [7], which provided a constraint on a sidereal time variation of a combination of the maximum attainable speed of electrons and the one-way speed of light. Many (but not all) the SOLA experiments were analyzed in terms of mSME developed by A. Kostelecky and collaborators [8].

*e-mail: bogdanw@jlab.org

2 Concept of the method

The trajectory of the high energy beam in a transverse magnetic field provides an exciting opportunity to investigate SOLA [9]. Indeed, according to SR, the momentum of the particle is defined by the difference between its speed and a maximum attainable speed for the particle (assumed from now on to be equal to the speed of light). The relativistic expression for the particle momentum, $p = mv/\sqrt{(c-v)\cdot(c+v)}$, shows that when v approaches c, the p value is very sensitive to the variation of $(c-v)$. For a charged particle moving in a transverse magnetic field, according to the conventional Lorentz force, we should have a constant absolute value of the velocity vector $\vec{v}$. If the speed of light, c_1, varies, the experimental observable, a particle momentum, will also change with sidereal time and its relative variation would be enhanced by a factor of γ^2.

Many modern accelerators are equipped with beam position monitors which have an accuracy of a few microns or better in a single measurement. Using a typical magnetic system with the dispersion function of a few meters, it should be possible to measure beam momentum to a few 10^{-6} precision many times per second. There are several accelerators with a beam gamma factor of 10^4, which provides a boost factor of 10^8 in sensitivity for the speed of light variation, which could as a result reach the quantum gravity domain whose possible onset is 10^{-17} [5].

In addition to a sidereal time variation, very short-term changes are also well motivated physics, e.g. transient-in-time effects [10], and they could be detected via particle momentum changes in a high energy accelerator. Correlated measurements at several facilities could provide a realistic way to search for such a phenomenon [11].

DOI: 10.1051/epjconf/201714201029

3 Beam experiment at CEBAF

The CEBAF accelerator has ten 180° magnetic arcs with beam energies varying from one to eleven GeV. There are no accelerating cavities in the arcs, and in each arc the beam energy is constant (after correction for a small and stable synchrotron radiation effect). Formulated in the Ref. [9] plan is a measurement of the momentum of the beam at two points at the beginning and the end of each arc by using beam deflection in arc magnets and beam position monitors. The analysis of the beam momenta ratio at these two points as a function of time would be independent of beam energy variation, which is typically of 10^{-4}.

The data for such an experiment would be collected during normal CEBAF operation for hadron physics experiments [12]. Similar measurements could be performed at the KEK high energy electron/positron rings and at LHC for protons, as well as at a number of X-ray facilities.

The main limitation in such an experiment is due to instabilities in the magnetic system and beam line geometry due to a number of effects such as ground movement (a tidal effect), oscillation in power supplies for magnetic elements, daily variation of the tunnel temperature and many others. However, analysis of the data over many days should allow effective selection of the signal with sidereal periodicity.

4 Beam experiments at positron/electron rings

There are storage rings where one can take advantage of closely located orbits of counter-propagating beams and analyze the ratio of the beam momenta in which the instabilities of the magnetic system and geometry drifts are canceled out because they have identical effects on both beams [13].

Two electron/positron storage rings, CESR and VEPP-4, currently operate with a common magnetic system for both beams and have large beam gamma factors. In these storage rings, the electron-to-positron orbit differences, originating mainly from beam energy loss due to synchrotron radiation and coherent effects, are less than 0.5 mm. Such a configuration of the experiment is able to probe the CPT-even terms of the Lorentz violation with precision limited only by the short term stability of the beam position monitors (assuming that the same BPMs and electronics are used from both beams).

Analysis of the first CESR data allows us to put a limit on $\delta c_1/c_1 \leqslant 10^{-14}$. Currently, experiments are active in both accelerators [14, 15].

5 Experiments at the X-ray synchrotron radiation storage rings

There are several storage rings with multi GeV electron beams which are used as X-ray sources to do photon physics [16–18]. Each of them provides dozens of X-ray beam lines with very stable direction (on the order of 0.1 micro radian) and position (on the order of a micro meter) of photons at users' instruments.

The beam orbit in a typical X-ray storage ring is subject to regular corrections which allow the X-ray beam to be kept stable. These corrections complicate the use of the beam orbit data for the speed of light investigation. However, stabilization of the X-ray beam directions leads also to stabilization of the electron beam directions at the location of X-ray emission in the insertion devices.

As a result, the magnetic field integral on the beam orbit between two X-ray beam emission points would provide a perfect measure of the beam momentum and open the possibility of SOLA investigation at many synchrotron radiation facilities.

6 Experiment with a muon storage ring

A high precision measurement of the muon anomalous magnetic moment, the muon $(g-2)$ experiment, was performed at a specialized storage ring with decay in flight of 3.09 GeV/c muons [19]. It was done via analysis of the observed in the detectors muon decay rate with the 5-parameter fit: $N(t) = N_0\, e^{-(t/\gamma\tau)}\,[1 + A\cos(\omega_a t + \phi)]$. The data were also used to evaluate possible Lorentz invariance violations in a muon sector of mSME [20]. In that analysis, the spin precession frequencies ω_a for positive and negative muons were analyzed for potential sidereal time variation.

The muon spin rotation frequency at a given magnetic field value is defined by the muon magnetic moment and its momentum. The relative angle between the momentum and spin directions is proportional to the anomaly magnetic moment and momentum rotation angle. The change in the observed muon decay rate between detectors could be used as a measure of the momentum rotation angle advance. As the speed of light variation changes the particle momentum, the local spin rotation frequency also varies. Assuming that the value of the muon momentum changes along the particle orbit, the effect of interest should be searched for as a function of the orbit phase in contrast to the performed analysis [20], where the frequency ω_a was averaged over all detectors along the orbit before a search for a realtime variation.

We propose that in the fit function above, the ϕ should be replaced by $\phi_0 + B \times \cos(\Omega t + \psi)$, where Ω is the Earth rotation frequency and ψ is a constant defined by the locations of the detector along the beam orbit and the laboratory. The proposed analysis will provide a limit on the parameter B and, as a result, a first estimate for a sidereal time variation of the maximum attainable speed of a muon.

7 Analysis of a resonance production rate

The asymmetric electron-positron collider allows an interesting opportunity to monitor the speed of light by observing the production rate variation when the combined beam energy is set slightly off resonance [9]. In the case of Υ resonance the observed relative width is $\Gamma/M \sim 10^{-3}$. Indeed, due to a large difference in gamma factors, the speed of light variation induces changes in the momentum for a high energy beam (*h.e.b.*) much larger than for a lower energy beam.

As a result, the total energy in the lab system varies with sidereal time and the resonance production rate also varies. The size of the rate variation could be estimated as $\frac{\delta c_1}{c_1} \times \gamma^2_{h.e.b.} \times M/\Gamma$. At the projected luminosity of Belle-2 experiment [21], statistics collected every day would be of 10^8 Υ events, which would correspond to a sensitivity for $\delta c_1/c_1$ on the level of 10^{-15}.

8 Polarized electron-positron beams at a storage ring

The measurement of the electron and positron anomalous magnetic moments ratio, μ'_+/μ'_-, at VEPP-2m [22] found that precession frequencies are equal to each other at the relative level $1.2 \cdot 10^{-7}$ (a record precision for the time of that experiment). The current best limit for $(g_+ - /g_-)/g_{aver.}$ of $2 \cdot 10^{-12}$ was obtained using the particle trap method [23].

As a result, we can consider [9] the data from the storage ring experiment as a comparison of the electron and positron beam momenta when the beams are moving in opposite directions in a transverse magnetic field and make an estimate of the achievable sensitivity for the c_1 variation on the level of $1 \cdot 10^{-13}$ from those 1987 data. Obviously, this could be improved by two to three orders of magnitude using a higher beam energy, e.g. at VEPP-4, and advanced spin resonance detection methods [24].

Acknowledgments

The author would like to thank the organizing committee of the "Advances in Dark Matter and Particle Physics" workshop for support and the invitation to give a talk. I greatly appreciate comments from D. Budker, Yu. Shatunov and V. Zelevinsky, and collaborations with W. Bergan, G. Franklin, V. Lebedev, S. Nikitin, G. Nowak, Y. Roblin, D. Rubin, B. Schmookler, and B. Vlahovic. This work was supported in part by the NCCU NSF CREST Center, and by Department of Energy (DOE) contract number DE-AC05-06OR23177, under which the Jefferson Science Associates operates the Thomas Jefferson National Accelerator Facility.

References

[1] A. Einstein, Annalen der Physik, **17**, 891 (1905).
[2] A.A. Michelson, American Journal of Science, **22**, 120 (1881).
[3] M. Nagel *et al.*, Nature Communications, **6**, 8174 (2015).
[4] A.A. Michelson and E.W. Morley, Am. J. Sci., **34**, 333 (1887).
[5] D. Mattingly, Living Rev. Rel. **8**, 5 (2005).
[6] M.A. Hohensee *et al.*, Phys. Rev. Lett. **102**, 170402 (2009).
[7] J.-P. Bocquet *et al.*, Phys. Rev. Lett. **104**, 241601 (2010).
[8] V.A. Kostelecky and N. Russell, Rev. Mod. Phys. **83**, 11 (2011).
[9] B. Wojtsekhowski, Euro Physics Letters, **108**, 31001 (2014); arXiv:1409.6373.
[10] M. Pospelov *et al.*, Phys. Rev. Lett. **110**, 021803 (2013).
[11] D. Budker and A. Derevianko, Physics Today 68(9), 10 (2015).
[12] B. Wojtsekhowski and Y. Roblin, proposal PR12-15-002, JLab PAC43, 2015.
[13] B. Wojtsekhowski, arXiv:1509.02754.
[14] D. Rubin *et al.*, private communication, Aug. 2016.
[15] S. Nikitin *et al.*, private communication, Oct. 2016.
[16] http://www.spring8.or.jp/en/
[17] https://www1.aps.anl.gov
[18] http://www.lightsources.org/facility/esrf
[19] G.W. Bennett *et al.*, Phys. Rev. D **73**, 072003 (2006).
[20] G.W. Bennett *et al.*, Phys. Rev. Lett. **100**, 091602 (2008).
[21] T.E. Browder *et al.*, arXiv:0802.3201.
[22] I.B. Vasserman *et al.*, Phys. Lett. B, **198** 302 (1987).
[23] C. Patrignani *et al.* (Particle Data Group), Chin. Phys. C, 40, 100001 (2016).
[24] Yu.M. Shatunov, private communication, Sept. 2016.

Excited lepton baryogenesis

Dmitry Zhuridov[1,a]

[1] *Institute of Physics, University of Silesia, Uniwersytecka 4, 40-007, Katowice, Poland*

Abstract. The excited leptons that share the quantum numbers with the Standard Model leptons but have larger masses are widespread in many promising new physics theories. A subclass of excited leptons that at low energies interact with the SM fermions dominantly through the effective coupling to lepton and fermion-antifermion pair can be referred as leptomesons. I introduce possible generation of the baryon asymmetry of the universe using these new particles. The discussed baryogenesis mechanisms do not contradict to the small neutrino masses and the proton stability, and can be interesting for the collider experiments.

1 Introduction

In spite of the present success of the Standard Model (SM) several observations indicate its possible nonfundamentality: large number of the SM fermions, their arbitrary masses and mixings, fractional electric charge of quarks, etc. Among the diversity of new physics models the theories of compositeness [1–8] try to solve these problems by introducing a substructure of the SM particles, which subcomponents are commonly referred as preons [1]. The composite models may include in the particle content radial, orbital, topological, structural, and other excitations of the ground state particles, e.g., an *excited lepton* that shares leptonic quantum number with one of the existing leptons, has larger mass and no color charge.

Besides the outlined issues on the particles and their interactions that come from the laboratory studies, another opened questions (including the dark matter problem) arise from the astrophysical observations of the universe around us. In particular, our universe appears to be populated exclusively with baryonic matter rather than antimatter [9]. However this baryon asymmetry can not be explained within the Big Bang cosmology and the SM. Possible scenarios of dynamical generation of the baryon asymmetry during the evolution of the universe from a hot early matter-antimatter symmetric stage are referred as the baryogenesis (BG) mechanisms [10, 11], and include new physics.

In the next section we discuss one example of the composite models in question. The interactions and the mass bounds for the excited leptons are outlined in section 3, and the new BG scenarios that involve these particles are discussed in section 4. Finally, we conclude in section 5.

[a] e-mail: dmitry.zhuridov@gmail.com

Table 1. The haplon quantum numbers

Haplon	Spin [$\hbar$]	Q [$\lvert e\rvert$]	C1	C2	$SU(2)_h$
α	1/2	+1/2	3	1	2
β	1/2	−1/2	3	1	2
ℓ	0	+1/2	$\bar{3}$	1	2
c_k	0	−1/6	3	3	2

2 Composite model example

Consider the haplon models [7, 12], which are similar to the earlier models with wakems and chroms [4, 13, 14], and are based on the symmetry $SU(3)_c \times U(1)_{em} \times SU(N)_h$, where the new haplon group $SU(N)_h$ has the confinement scale of the order of 0.3 TeV, and denotes, e.g., $SU(2)_L \times SU(2)_R$. These models contain the two categories of preons (haplons): the fermions $\alpha^{+1/2}$ and $\beta^{-1/2}$, and the scalars $\ell^{+1/2}$ and $c_k^{-1/6}$, where $k = r, g, b$ ("red, green, blue"). Their quantum number assignment is given in Table 1, where Q is the electric charge, C1 is the choice for the $SU(3)_c$ representations in Ref. [7][1] and C2 is an alternative choice [12]. Then the haplon pairs can compose the SM particles as $\nu = (\alpha\bar{\ell})$, $e^- = (\beta\bar{\ell})$, $u = (\alpha\bar{c}_k)$, $d = (\beta\bar{c}_k)$, $W^- = (\bar{\alpha}\beta)$, $W^3 = (\bar{\alpha}\alpha - \bar{\beta}\beta)/\sqrt{2}, \ldots$, and the new particles, e.g., a scalar leptoquark $S_\ell^{+2/3} = (\ell\bar{c}_k)$, and the neutral scalars $S_\ell^0 = (\ell\bar{\ell})$ and $S_c^0 = (c_k\bar{c}_k)$. W^3 mixes with the photon γ similarly to the mixing between γ and ρ^0-meson. H scalar can be a p-wave excitation of the Z, and the second and third generations can be dynamical excitations. Notice that $S_\ell^{+2/3}$, S_c^0 and S_ℓ^0 states (if their masses are small) may contribute to the low-energy observables, e.g., so-called, XYZ states [16]. [2]

[1] Notice that C1 assignment does not provide a spin-charge separation [15].

[2] Notice that the SM results can be reproduced in some composite models, at least at the tree level, due to the complementarity between Higgs phase and confining phase [17, 18].

However, new questions arise: Where does this peculiar haplon picture come from? Can it, in turn, result from a substructure of haplons? Consider the two scalar "prepreons"[3] π_k and $\bar{\pi}_k$, which are SU(3) triplets and have the electric charges of $-1/6$ and $+1/6$, respectively. Then the set of haplons with their electric and C2 color charges can be reproduced by the triples of "prepreons" ($\bar{\pi}_{\bar{r}}\bar{\pi}_{\bar{g}}\bar{\pi}_{\bar{b}} \to \{\alpha, \ell\}$, $\pi_r\pi_g\pi_b \to \{\beta, \bar{\ell}\}$, $\bar{\pi}_{\bar{i}}\pi_j\pi_l \to c_k$), while additional mechanism of spin generation is required. One can think of possible relation of spin to a circular color currents similarly to some discussions in the context of the condensed matter [20] and gravity [21] theories, taking into account that the distribution of matter in a composite state can be imagined (in particular, in the SU(3) Yang-Mills theory) in terms of the wave functions or probability distributions for the effective subcomponents of a finite size [22–25]. Then a "spinning" and "nonspinning" states of the same preon (e.g., α and ℓ) may form a supersymmetric multiplet.

Notice that the possibility of multihaplon states such as $(\beta\bar{c}_k\bar{\ell}c_k)$, $(\alpha\bar{\ell}\beta\bar{c}_k\bar{\beta}c_k)$, etc., gets more points from recent discoveries of the multiquark hadrons [26].

3 Excited leptons

The excited lepton states defined in the introduction can be particularly important if their masses are smaller than the leptoquark and leptogluon masses, which can be natural due to the absence of the color charge. The contact interactions among the SM fermions f and the excited fermions f^* can be generically written as [9]

$$\mathcal{L}_{\rm CI} = \frac{g_*^2}{2\Lambda^2} \sum_{\alpha,\beta=L,R} \Big[\eta_{\alpha\beta}(\bar{f}_\alpha\gamma^\mu f_\alpha)(\bar{f}_\beta\gamma_\mu f_\beta) + \eta'_{\alpha\beta}(\bar{f}_\alpha\gamma^\mu f_\alpha)(\bar{f}^*_\beta\gamma_\mu f^*_\beta) + \tilde{\eta}'_{\alpha\beta}(\bar{f}^*_\alpha\gamma^\mu f^*_\alpha)(\bar{f}^*_\beta\gamma_\mu f^*_\beta) + \eta''_{\alpha\beta}(\bar{f}_\alpha\gamma^\mu f_\alpha)(\bar{f}^*_\beta\gamma_\mu f_\beta) + \text{H.c.} + \ldots\Big], \quad (1)$$

where Λ is the contact interaction scale, $g_*^2 = 4\pi$, and the new parameter values are usually taken of $|\eta^j| \leq 1$.

Assuming nearly maximal couplings of $|\eta^j| \simeq 1$ and the excited fermion masses of $M_{f^*} \simeq \Lambda$, the present lower bounds for $\Lambda/\sqrt{|\eta^j|}$ ratios are of the order of few TeV [9]. However, if Eq. (1) expresses a "residue" effective interactions between the composites (with respect to the fundamental interactions among their subcomponents) then $|\eta^j|$ couplings can be small, and even the case of $M_{f^*} \lesssim \Lambda <$ 1 TeV is not excluded for $|\eta^j| \ll 1$ and $\Lambda/\sqrt{|\eta^j|} >> M_{f^*}$.

A particular type of excited leptons that at low energies interact with the SM fermions dominantly through the contact terms we refer as *leptomesons* (LM).[4] The relevant contact terms (with η'' couplings) can be realized, e.g., through the leptoquark exchange.

[3] For supersymmetric models with "prepreons" see, e.g., Ref. [19].

[4] Notice that the same term "leptomeson" was used in the literature for the bound states of colored excitations of e^+ and e^- [27].

4 Baryogenesis

Possible BG and the dark matter generation by a scalar 4-haplon state was considered in Ref. [12]. In this proceedings we discuss if fermionic LM states can provide a successful BG [28]. Similarly to the sterile neutrino ν_R case, depending on the LM properties, deviation from thermal equilibrium can occur at either production or freeze-out and decay (compare to the BG via ν_R oscillations [29] and the usual leptogenesis [30], respectively). In both scenarios one should replace the Yukawa interactions of ν_R by the contact interactions of LMs, which may result in promising effects.

4.1 BG from LM oscillations

Once created in the early universe neutral long-lived LMs oscillate and interact with ordinary matter. These processes do not violate the total lepton number $L^{\rm tot}$ (for Dirac LMs). However the oscillations violate CP and therefore do not conserve individual lepton numbers L_i for LMs. Hence the initial state with all zero lepton numbers evolves into a state with $L^{\rm tot} = L_0 + \sum_i L_i = 0$ but $L_i \neq 0$, where L_0 is the lepton number of other particles.

At temperatures $T \ll \Lambda$ the LMs communicate their lepton asymmetry to neutrinos ν_ℓ and charged leptons ℓ through the effective interactions, e.g., B-conserving (and L-conserving for Dirac LMs) vector couplings

$$\sum_{\psi_\ell,f,f'} \sum_{\alpha,\beta=L,R} \left[\frac{\epsilon^{\alpha\beta}_{ff'\psi_\ell}}{\Lambda^2}(\bar{f}_\alpha\gamma^\mu f'_\alpha)(\bar{\psi}_{\ell\beta}\gamma_\mu N_{\ell\beta}) + \frac{\tilde{\epsilon}^{\alpha\beta}_{ff'\psi_\ell}}{\Lambda^2}(\bar{\psi}_{\ell\alpha}\gamma^\mu f'_\alpha)(\bar{f}_\beta\gamma_\mu N_{\ell\alpha})\right] + \text{H.c.}, \quad (2)$$

where $\psi_\ell = \ell, \nu_\ell$ ($\ell = e, \mu, \tau$), the constants $\overset{(\sim)}{\epsilon} = 4\pi\eta''$ can be real, f and f' denote either quarks or leptons such that $Q_{f_\alpha} + Q_{f'^c_\alpha} + Q_{\psi_{\ell\beta}} = 0$, and N_ℓ is the neutral LM flavor state related to the mass eigenstates N_i as $N_{\ell\alpha} = \sum_{i=1}^n U^\alpha_{\ell i} N_i$.

Suppose that LMs of at least one type N_i remain in thermal equilibrium till the electroweak symmetry breaking time $t_{\rm EW}$ at which sphalerons become ineffective, and those of at least one other type N_j come out-of-equilibrium by $t_{\rm EW}$. Hence the lepton number of former (later) affects (has no effect on) the BG. In result, the final baryon asymmetry after $t_{\rm EW}$ is nonzero. At the time $t \gg t_{\rm EW}$ all LMs decay into the leptons and the quarks (hadrons). For this reason they do not contribute to the dark matter in the universe, and do not destroy the Big Bang nucleosynthesis.

The system of n types of singlet LMs of a given momentum $k(t) \propto T(t)$ that interact with the primordial plasma can be described by the $n \times n$ density matrix $\rho(t)$. In a simplified picture it satisfies the kinetic equation [29]

$$i\frac{d\rho}{dt} = [\hat{H}, \rho] - \frac{i}{2}\{\Gamma, \rho\} + \frac{i}{2}\{\Gamma^p, 1 - \rho\}, \quad (3)$$

where Γ (Γ^p) is the destruction (production) rate, and the effective Hamiltonian can be written as

$$\hat{H} = V(t) + U\frac{\hat{M}^2}{2k(t)}U^\dagger, \quad (4)$$

DOI: 10.1051/epjconf/201714201030

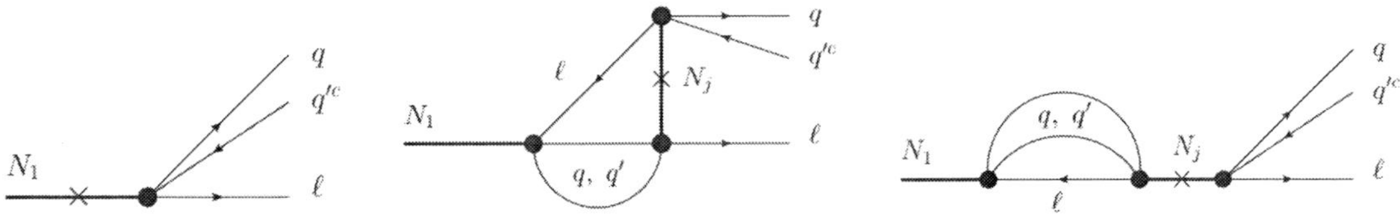

Figure 1. Feynman diagrams for the discussed contributions to the CP asymmetry, where the line direction shows either L or B flow, "×" represents a Majorana mass insertion, and the black bulb represents a subprocess (e.g., a leptoquark exchange).

where $\hat{M}^2 = \mathrm{diag}(M_1^2, \ldots, M_n^2)$ with the masses M_i of N_i, and V is a real potential.

One of the main features of the discussed BG from LMs is that the 4-particle interaction cross section that contributes to the destruction rate is proportional to the total energy of the process s instead of the inverse proportionality that takes place in the BG from ν_R oscillations. Indeed, this cross section can be written as

$$\sigma(a+b\to c+d) = \frac{C}{4\pi}|\epsilon|^2\frac{s}{\Lambda^4} \propto s, \quad (5)$$

where a, b, c and d denote the four interacting particles (f, f', ψ_ℓ and N_ℓ), and $C = O(1)$ is the constant that includes the color factor in the case of the interaction with quarks. In result, the interaction rate that equilibrates LMs,

$$\Gamma \propto |\epsilon|^2\frac{T^5}{\Lambda^4}, \quad (6)$$

is suppressed by $(T/\Lambda)^4$ with respect to the Higgs mediated interaction rate in usual BG via ν_R oscillations.

The conditions that LMs of type N_i remain in thermal equilibrium till t_{EW}, while N_j do not, can be written as

$$\Gamma_i(T_{\mathrm{EW}}) > H(T_{\mathrm{EW}}), \qquad \Gamma_j(T_{\mathrm{EW}}) < H(T_{\mathrm{EW}}), \quad (7)$$

where $H(T)$ is the Hubble expansion rate. Due to the suppression factor of $(T_{\mathrm{EW}}/\Lambda)^4$ the successful BG can be realized with the relatively large couplings $|\epsilon|$ with respect to the sterile neutrino Yukawas of $Y \sim 10^{-7}$ in the BG via ν_R oscillations [29]. In particular, for $\Lambda \gtrsim 10$ and 30 TeV we have $|\epsilon| \gtrsim 10^{-4}$ and 10^{-3}, respectively. Hence the considered BG scenario can be relevant for the LHC and next colliders without unnatural hierarchy of the couplings.

4.2 BG from LM decays

Suppose that the neutral LMs are Majorana particles ($N = N^c$). Consider their out-of-equilibrium, CP- and L-violating decays in the early universe. The relevant interactions can be written as

$$\frac{\epsilon^{\alpha R}_{ff'\psi_\ell}}{\Lambda^2}(\bar{f}_\alpha\gamma^\mu f'_\alpha)(\bar{\psi}_{\ell R}\gamma_\mu N_{\ell R}) + \frac{\epsilon^{S}_{ff'\psi_\ell}}{\Lambda^2}(\bar{f}_R f'_L)(\bar{\psi}_{\ell L}N_{\ell R}) + \frac{\epsilon^{T}_{ff'\psi_\ell}}{\Lambda^2}(\bar{f}\sigma^{\mu\nu}f')(\bar{\psi}_{\ell L}\sigma_{\mu\nu}N_{\ell R}) + \mathrm{H.c.} \quad (8)$$

To be more specific in the following we consider the term

$$\frac{\lambda_{\ell i}}{\Lambda^2}(\bar{q}_\alpha\gamma^\mu q'_\alpha)(\bar{\ell}_R\gamma_\mu N_{iR}), \quad (9)$$

where $\lambda_{\ell i} = \epsilon^{\alpha R}_{qq'\ell}U^R_{\ell i}$ is a complex parameter. Consider the interference of tree and one-loop diagrams[5] shown in Fig. 1. The final CP asymmetry that is produced in decays of the lightest LMs N_1

$$\varepsilon_1 = \frac{1}{\Gamma_1}\sum_\ell[\Gamma(N_1\to\ell_R q_\alpha q'^c_\alpha) - \Gamma(N_1\to\ell^c_R q^c_\alpha q'_\alpha)], \quad (10)$$

can be non-zero if $\mathrm{Im}[(\lambda^\dagger\lambda)^2_{1j}] \neq 0$. Using the width [33],

$$\Gamma_1 = \sum_\ell[\Gamma(N_1\to\ell_R q_\alpha q'^c_\alpha) + \Gamma(N_1\to\ell^c_R q^c_\alpha q'_\alpha)] \simeq \frac{1}{128\pi^3}(\lambda^\dagger\lambda)_{11}\frac{M_1^5}{\Lambda^4}, \quad (11)$$

the condition for the decay parameter $K \equiv \Gamma_1/H(M_1) > 3$ (strong washout regime) translates into the limit of

$$(\lambda^\dagger\lambda)_{11} \gtrsim 4\times10^{-7}\times\left(\frac{\Lambda}{10\,\mathrm{TeV}}\right)^4\times\left(\frac{1\,\mathrm{TeV}}{M_1}\right)^3. \quad (12)$$

The final baryon asymmetry can be written as

$$\frac{n_B - n_{\bar{B}}}{s} = \left(-\frac{28}{79}\right)\times\frac{n_L - n_{\bar{L}}}{s} = \left(-\frac{28}{79}\right)\times\frac{\varepsilon_1\kappa}{g_*}, \quad (13)$$

where n_B (n_L) is the baryon (lepton) number density, s is the entropy density, $-28/79$ is the sphaleron lepton-to-baryon factor, and $\kappa \lesssim 1$ is the washout coefficient that can be determined by solving the set of Boltzmann equations. The observed baryon asymmetry of the universe of [9]

$$\eta_B = \frac{n_B - n_{\bar{B}}}{n_\gamma} = 7.04\times\frac{n_B - n_{\bar{B}}}{s} \simeq 6\times10^{-10}, \quad (14)$$

where n_γ is the photon number density, can be produced, e.g., for $K \sim 100$ with the degeneracy factor of

$$\mu \equiv \frac{M_2 - M_1}{M_1} \lesssim 10^{-6}\left(\frac{M_1}{1\,\mathrm{TeV}}\right), \quad (15)$$

which enters the resonant CP asymmetry of

$$\varepsilon_i \sim \frac{\mathrm{Im}\{[(\lambda^\dagger\lambda)_{ij}]^2\}}{(\lambda^\dagger\lambda)_{ii}(\lambda^\dagger\lambda)_{jj}}\frac{\Gamma_j}{M_j}\frac{M_iM_j}{M_i^2 - M_j^2} \sim \mu^{-1}\frac{\Gamma_1}{M_1}. \quad (16)$$

Notice that the discussed effective LM-quark-quark-lepton vertex can be realized, e.g., through the exchange

[5] Same two-loop self-energy graph as in Fig. 1 (right) was discussed in the resonant BG scenarios of Refs. [31, 32], where the baryon asymmetry is directly produced in the three-body decays of the sterile neutrinos. Although these mechanisms involve B-violating interactions of $qqq\nu_R$ type, they do not lead to fast proton decay due to the large values of ν_R mass and the B-violating interaction scale of $O(1)$ TeV.

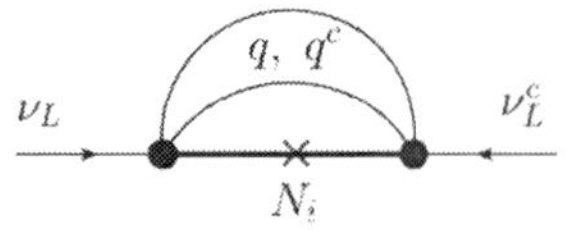

Figure 2. The discussed contribution to the neutrino masses (line direction shows L flow).

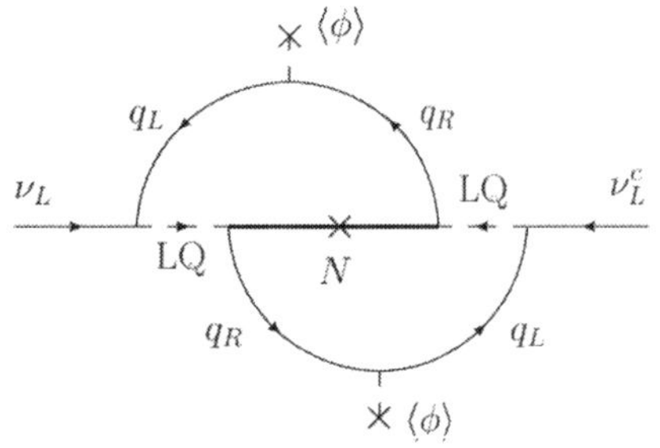

Figure 3. Leptoquark (LQ) contribution to the neutrino masses.

of the SU(2)$_L$ singlet scalar leptoquark S_{0R} with $Y = 1/3$. Relevant interaction terms can be written as

$$-\mathcal{L}_{\rm int} \;=\; (g_{ij}\, \bar{d}^c_R N_{iR} + f_j\, \bar{u}^c_R \ell_R) S^j_{0R} + \text{H.c.} \qquad (17)$$

Then the above expressions are valid with the replacements of $\lambda \to g f^*$ and $\Lambda \to M_{S_{0R}}$. The relevant to the successful BG values of the new couplings of $|g| \sim |f| \sim 0.01 - 0.1$, can be interesting for the collider searches.

4.3 Neutrino masses

For Majorana LMs the effective terms of

$$\frac{\epsilon^S_{ff\nu_\ell}}{\Lambda^2}\bar{f}_R f_L\, \bar{\nu}_{\ell L} N_{\ell R} + \frac{\epsilon^T_{ff\nu_\ell}}{\Lambda^2}\bar{f}\sigma^{\mu\nu} f\, \bar{\nu}_{\ell L}\sigma_{\mu\nu} N_{\ell R} + \text{H.c.} \quad (18)$$

can generate the two-loop contributions to the observable light neutrino masses m_{ν_ℓ} that is illustrated for $f = q$ in Fig. 2, and for a particular model with the leptoquarks in Fig. 3. A simple estimate of this contribution is

$$m_{\nu_\ell} \sim \sum_i \frac{(\epsilon\, U^R_{\ell i})^2}{(16\pi^2)^2}\, \frac{M_i^3 m_q^2}{\Lambda^4}, \qquad (19)$$

where m_q is the quark mass. Hence the present bound of $m_\nu \lesssim 2$ eV [34] can be easily satisfied.

5 Conclusion

The two new testable baryogenesis scenarios in the models with excited leptons are introduced, which do not contradict to the observable neutrino masses. First, *the BG from LM oscillations* may take place for relatively light and long-lived LMs, which do not all decay before $t_{\rm EW}$. Second, *the BG from LM decays* can be realized if all LMs decay before $t_{\rm EW}$. A particular models based on the former (later) BG proposal require detailed study of the neutrino potential (of the Boltzmann equations) to be verified in the future experiments. In both scenarios the baryon number is violated only by the sphaleron processes that does not affect the proton stability. Due to the relatively low temperatures of the discussed BG mechanisms, an analog of the gravitino problem [35, 36] does not exist there.

Acknowledgements

The author thanks Marek Zrałek, Henryk Czyż, Bhupal Dev and Yue Zhang for useful discussions. This work was supported in part by the Polish National Science Centre, grant number DEC-2012/07/B/ST2/03867. The author used JaxoDraw [37] to draw the Feynman diagrams.

References

[1] J. C. Pati and A. Salam, Phys. Rev. D **10**, 275 (1974) [Erratum-ibid. D **11**, 703 (1975)].
[2] H. Terazawa, K. Akama and Y. Chikashige, Phys. Rev. D **15**, 480 (1977).
[3] M. A. Shupe, Phys. Lett. B **86**, 87 (1979).
[4] H. Terazawa, Phys. Rev. D **22**, 184 (1980).
[5] E. J. Squires, Phys. Lett. B **94**, 54 (1980).
[6] H. Harari and N. Seiberg, Phys. Lett. B **98**, 269 (1981).
[7] H. Fritzsch and G. Mandelbaum, Phys. Lett. B **102**, 319 (1981).
[8] R. Barbieri, R. N. Mohapatra and A. Masiero, Phys. Lett. B **105**, 369 (1981) [Erratum-ibid. B **107**, 455 (1981)].
[9] C. Patrignani *et al.* (Particle Data Group), Chin. Phys. C **40**, 100001 (2016).
[10] A. D. Dolgov, Phys. Rept. **222**, 309 (1992).
[11] M. Dine and A. Kusenko, Rev. Mod. Phys. **76**, 1 (2003)
[12] H. Fritzsch and J. Sola, Adv. High Energy Phys. **2014**, 361587 (2014).
[13] H. Terazawa, Conf. Proc. C **840719**, 63 (1984).
[14] H. Terazawa, J. Mod. Phys. **5**, 205 (2014).
[15] C. Xiong, arXiv:1606.01883 [hep-ph].
[16] Q.-p. Ji [BESIII Collaboration], Nucl. Part. Phys. Proc. **270-272**, 127 (2016).
[17] G. 't Hooft, "Topological aspects of quantum chromodynamics," in From the Planck Length to the Hubble Radius, ed. A. Zichichi (World Scientific, Singapore, 2000), 216–236 [arXiv: hep-th/9812204].
[18] X. Calmet and H. Fritzsch, Phys. Lett. B **496**, 161 (2000).
[19] J. C. Pati and A. Salam, Nucl. Phys. B **214**, 109 (1983).
[20] C.-P. Chuu, M.-C. Chang and Q. Niu, Solid State Commun. **150**, 533 (2010).
[21] A. Burinskii, Phys. Lett. B **754**, 99 (2016).
[22] S. D. Glazek, Few Body Syst. **52**, 367 (2012).
[23] M. Gómez-Rocha and S. D. Głazek, Phys. Rev. D **92**, 065005 (2015).

[24] S. D. Glazek and P. Kubiczek, Few Body Syst. **57**, 425 (2016).

[25] S. D. Glazek and A. P. Trawinski, arXiv:1612.06211 [hep-ph].

[26] R. Aaij *et al.* [LHCb Collaboration], Phys. Rev. Lett. **115**, 072001 (2015).

[27] M. Pitkänen and P. Mähönen, Int. J. Theor. Phys. **31**, 229 (1992).

[28] D. Zhuridov, Phys. Rev. D **94**, 035007 (2016).

[29] E. K. Akhmedov, V. A. Rubakov and A. Y. Smirnov, Phys. Rev. Lett. **81**, 1359 (1998).

[30] M. Fukugita and T. Yanagida, Phys. Lett. B. **174**, 45 (1986).

[31] P. S. B. Dev and R. N. Mohapatra, Phys. Rev. D **92**, 016007 (2015).

[32] H. Davoudiasl and Y. Zhang, Phys. Rev. D **92**, 016005 (2015).

[33] O. Cakir, C. Leroy, R. R. Mehdiyev and A. Belyaev, Eur. Phys. J. C **32**, 1 (2004).

[34] V. N. Aseev *et al.* [Troitsk Collaboration], Phys. Rev. D **84**, 112003 (2011).

[35] M. Yu. Khlopov and A. D. Linde, Phys. Lett. B **138**, 265-268 (1984).

[36] F. Balestra, *et al.*, Sov. J. Nucl. Phys. **39**, 626-631 (1984).

[37] D. Binosi and L. Theussl, Comput. Phys. Commun. **161**, 76 (2004).

AUTHOR INDEX